Contents

Plumbing

For Level 2 Technical Certificate and NVQ

2nd Edition

Steve Muscroft

AMSTERDAM • BOSTON • HEIDELBERG • LONDON • NEW YORK • OXFORD
PARIS • SAN DIEGO • SAN FRANCISCO • SINGAPORE • SYDNEY • TOKYO
Newnes is an imprint of Elsevier

Newnes is an imprint of Elsevier
The Boulevard, Langford Lane, Kidlington, Oxford, OX5 1GB
30 Corporate Drive, Suite 400, Burlington, MA 01803, USA

First published 2005
Reprinted 2006
Second edition 2007
Reprinted 2009

British Library Cataloguing in Publication Data
A catalogue record for this book is available from the British Library

Library of Congress Cataloging-in-Publication Data
A catalog record for this book is available from the Library of Congress

ISBN: 978-0-7506-8434-7

Printed and bound in China

09 10 11 12 10 9 8 7 6 5 4 3 2

Preface

About this book

First, welcome to an exciting and rewarding career in plumbing. Plumbing is one of the oldest trades, its first origins dating back to the Roman times, from where the term plumber was derived from the term 'Plumbum' which is Latin for 'worker of lead'.

The industry has come a long way since then and whilst plumbers still need to be very skilful, there is an increasing requirement for in-depth technical knowledge as legislation and product technology becomes more complex.

This book is designed to support the knowledge and understanding required for the 'technology' content of the ***City and Guilds Level 2 Certificate in Basic Plumbing Studies.*** We will refer to this as the 'Technical Certificate' from now on to keep things simple.

In addition to knowledge and understanding, the Technical Certificate has a practical element which you will cover at a City and Guilds accredited training centre or college.

The Technical Certificate content also covers the job knowledge of the NVQ. To achieve an NVQ, you will need to have evidence of work experience and will be assessed on your competence to do plumbing work. Some of you will be doing a plumbing apprenticeship, in which case the Technical Certificate, NVQ and Key Skills (Application of Number and Communication) will all be relevant.

We will explain much more about the qualifications as well as the other progression routes in more detail in Unit 1 'Welcome to Plumbing'.

How to use this book

Research shows that one of the most powerful ways that people learn is 'by doing'. We have made this the basis for this book by making it 'interactive', which means that in addition to providing the information about the various plumbing topics, we will expect you to get involved in the various activities. You will also be expected to find somethings out for yourself, which is another good way of learning.

About the activities

Throughout the text you will come across a number of activities. They will vary in terms of what you will be asked to do, for example the activity could be:

- Finding out information for yourself to answer a question(s) – not all the activities have the answer!
- Carrying out a piece of research
- Questions and answers.

Other key features

In addition to the activities, we have included 'test yourself' questions at the end of a section and a selection of multi-choice questions at the end of a unit. These are designed to prepare you for the City and Guilds Technical Certificate assessments and are referred to as 'checking your learning'.

From time to time, you will find a reference to key points, these are things that we feel are particularly important to a topic area or may be just a helpful tip. For example:

Key point

As a plumber it is really important that you have access to as much technical information as possible. This will give you more confidence, help towards your competence and will make you more attractive to prospective employers.

If eventually you do work for yourself, it will also give you a competitive advantage. Here are a few examples of information resources

- Internet
- Library
- Manufacturers' catalogues
- Manufacturers' websites

- British Standards
- Water Regulations
- Building Regulations
- Trade associations
- Professional bodies
- Sector Skills Council
- Trade magazines
- Your college or centre tutor
- Other plumbers and building workers

This is not an exhaustive list as we hope you will find other resources as you carry out your studies. Remember, this book can only supplement your college or centre studies as you work towards the Level 2 Technical Certificate.

Activity 0.1

Just a bit of fun, but if you have internet access (if not try your library), try typing either 'plumbing training material' or 'plumbing information' into the search engine and see what comes out! Then try some of your own phrases or words for the search; the internet is a rich source of plumbing information.

Finally, to put some of the knowledge and understanding into the context of the industry, we have included a few case studies.

Key point

It is important, that when you complete any work on the activities, the 'test yourself' questions or the 'check your learning' questions are recorded on sheets of paper, or notepad for inclusion in your portfolio.

Remember to cross reference your work to the unit and activity it occurred in. Ask your tutor or assessor if you need further guidance.

Acknowledgements

I would personally like to thank all the organisations referred to in this book that have kindly given not only their permission to use their images, but also their time and effort in supplying them.

I would also like to thank my colleague Kevan Holmes for all his support in the production of the book.

Finally, I would like to dedicate this book to my friend and mentor Laurie Croft. Laurie has been at the heart of most of the things that are good about education and training in the plumbing industry, and has worked tirelessly throughout his career towards the advancement of plumbing and plumbers.

An excellent ambassador for the industry at all levels including his work on the International Skills Competition, Laurie was and always will be someone that I and many of his colleagues greatly respect and admire.

UNIT 1

WELCOME TO PLUMBING

Summary

- An overview of the plumbing industry:
  - What is plumbing?
  - The plumbing industry and its links with the construction industry and the building engineering services sector
  - About the plumbing industry
  - Industry organisations and how to find them.
- The plumbing training scheme:
  - Technical Certificates
  - NVQs and SVQs
  - Modern apprenticeships (MAs)
  - Training and assessment
  - Career opportunities.

About plumbing

Introduction

You have made the decision to learn about plumbing and you are probably looking forward to working in the industry. In this section, we explore what a plumber actually does and give you an insight of what to expect on site.

The section will also provide an overview of the industry structure, and information about the key organisations; this will be a useful reference as you work through the course material because they provide a wealth of information and further contacts.

What do we mean by 'plumbing'?

Anyone new to the industry will probably have a view of what a plumber does. How would you describe plumbing? Think about this for a while and jot down a few notes.

We tend to find when speaking with people that their view of plumbing is of someone clearing a blocked drain, or repairing burst pipes. Whilst a plumber does this work, the job is much more involved than that.

Generally, we think that plumbing meets people's basic requirements by:

- Keeping them healthy and clean by providing cold water for drinking, hot water for washing and sanitation systems for the removal of waste products
- Keeping them warm with the help of hot water heating systems
- Meeting the above requirements together to ensure maximum comfort and convenience for them by providing heating, hot and cold water systems and sanitation 365 days a year, and 24 h a day.

Explore the plumber's role a little further and you will see that most plumbers usually carry out the:

- Installation
- Service
- Maintenance

of a wide range of domestic systems such as:

- Cold water, including underground services to a dwelling
- Hot water
- Heating systems fuelled by gas, oil or solid fuel

- Sanitation (or above ground drainage) including the installation of baths, hand wash basins, water closets (WCs) and sinks
- Rainwater systems, gutters and fall pipes
- Associated electrical systems
- Sheet lead weatherings (at Level 2, this includes things, such as chimney weatherings, and soil vent pipe weatherings).

In fact, you will see that the above list matches the content of the Technical Certificate.

In addition to the systems, the plumber will also have to work on the appliances and components contained within them. Here are a few examples:

- Storage vessels, cylinders, cisterns
- Sanitary appliances, sinks, baths, WCs, washbasins
- Domestic appliances including washing machines and dishwashers
- Heat exchangers, boilers, radiators
- Pumps, accelerators and motorised/isolating valves
- Gas appliances (natural or Liquefied Petroleum Gas (LPG), boilers, water heaters, cookers and fires)
- Pipe materials, fittings, fixings, controls that constitute the above systems.

Plumbers also work on:

- Cabling and electrical components, only if qualified to do so
- Sheet weatherings, aprons, back gutters, step flashings, soakers, lead slates.

The Level 2 Technical Certificate focuses on all aspects of domestic plumbing work. Some plumbing companies, however, specialise in specific work such as central heating, or industrial/commercial, which is basically working on much bigger systems in non-domestic premises.

So, we are now beginning to build up a picture of the extent of a plumber's job. What skills and knowledge do you think a plumber needs in order to be able to carry out the job competently?

Here are our thoughts on what a competent plumber should be able to do:

- Follow health and safety legislation and guidance at all times
- Thoroughly plan the job including making sure that all the tools, materials and equipment are present on the job

- Agree a schedule of work with the customer or client
- Provide a cost estimate for the job
- Prepare the work location, making sure that there is adequate access
- Protect the customer's property
- Mark out, measure and work out the installation requirements
- Fabricate, position and fix system components
- Pre-commission (including testing), commission and de-commission systems
- Service and maintain system components
- Work effectively with customers, workmates and other site visitors
- Work in an environmentally friendly manner
- Promote the products and services of the plumbing business.

Can you think of anything else?

We also think that the plumber should have a working knowledge of:

- Regulations
- Codes of practice
- Principles of plumbing systems including basic design
- Where to find manufacturers' technical data
- Health and safety legislation and guidance
- Commercially agreed standards.

Figure 1.1 Examples of typical plumbing materials. Hot and cold feed in copper to the right, plastic heating pipes to the left, and plastic drain connection below copper pipes

A plumber must also be able to read and interpret details contained within a number of information sources including drawings, specifications and manufacturers' catalogues.

Finally, plumbers have to work with a range of materials and fittings. Figure 1.1 on the previous page shows plastic and copper pipework in a first fix situation. First fix means installing all the pipework runs prior to the wall and floor surfaces being completed.

Try this

Now that you are beginning to pick up a good knowledge of the work carried out by domestic plumbers, it is time for a little bit of practical application. Consider your place of residence, draw up a table detailing all the plumbing connections (cold and hot water connections, appliances, central heating components, etc) within your place of residence. This should help to dispel the notion that all plumbers do is to clear blocked drains!

Next we will take a look at the overall structure and some of the key organisations within the plumbing industry.

The plumbing industry

Plumbing, building engineering services and construction

In Unit 12, Effective Working Relationships, we will look at how the construction industry is structured in more detail.

Primarily, the plumbing industry comes under the Building Engineering Services (BES) industry umbrella, which includes

Activity 1.1

What are the main trades that you think are covered within the construction sector? Jot down your thoughts on a separate piece of paper and check out the answer at the end of this book.

heating, ventilating, air conditioning and refrigeration and electrical sectors. The gas supply from the consumers' meter to the appliances is also seen as being a part of the plumber's job. Similarly BES is further classified as part of the construction sector.

Your list will not cover all the trades, but it should give you an idea of the main ones. The training and development needs of the sector are looked after by an organisation called the Construction Industry Training Board or CITB. You will be doing a bit of research about CITB and other sector-related bodies later.

There is also an organisation that looks after the training and development needs of the building engineering services sector. This is called SummitSkills, and looks after plumbing, heating, ventilating, air conditioning, refrigeration and electrical sectors.

One area that the organisation is responsible for is setting the standards that are used to develop NVQs, SVQs and Technical Certificates. You will find that SummitSkills is an important source of information. If you have access to the internet, you can view their website at www.summitskills.org.uk. Other contact details are mentioned below

SummitSkills Ltd
Vega House
Opal Drive
Fox Milne
Milton Keynes
MK15 0DF
Tel: 01908 303960
E-mail: enquiries@summitskills.org.uk

So far in our description of plumbing, BES and construction sectors have been mentioned as two organisations that look after the industry's training and development needs. Trade Associations have a central role in looking after the interests of plumbing, BES or construction businesses.

In simple terms they are like a 'club', made up of a number of members who pay annual subscriptions in return for membership services. Most operate a committee structure for managing the various association activities; examples would include finance, marketing, etc.

Trade associations provide a number of services to their members including technical support, legal advice, representation to

Government on industry related matters, regular updates on the latest industry developments through their own magazines, raising the profile of the membership to potential customers, etc. Their membership is made up of businesses ranging from sole traders to larger organisations. Most trade associations operate using tough membership selection criteria or licensing schemes to ensure that a high standard of competent membership is maintained.

Industry organisations and how to find them

The key industry organisations are:

- SummitSkills
- Association of Plumbing and Heating Contractors (APHC)
- Amicus
- BPEC Certification Limited
- British Plumbing Employers Council (BPEC) Training Limited
- Construction Industry Training Board (CITB)
- Council for Registered Gas Installers (CORGI)
- Electrical Contractors Association (ECA)
- Institute of Plumbing and Heating Engineering (IPHE)
- JTL
- Joint Industry Board for Plumbing and Mechanical Engineering Services (JIB for PMES)
- National Association of Plumbing Teachers (NAPT)
- Heating and Ventilating Contractors Association (HVCA)
- Plumbing and Heating Industry Alliance (PHIA).

Working in the plumbing industry and the competent persons schemes

What are competent persons schemes?

The idea behind the schemes is that they will authorise members who are adjudged sufficiently competent in their work to self-certify that their work has been carried out in compliance with all relevant requirements of the Building Regulations. To a plumber this means things like Part H Drainage and Waste Disposal and Part L Conservation of Fuel and Power.

The schemes offer benefits to both consumers and industry. Consumers will benefit from lower prices as building control fees are not payable. They also benefit from reduced delays and from the ability to identify competent firms. Firms who join these schemes will avoid the time and expense of submitting a building notice. The schemes will also allow local authority building control departments to concentrate their resources on the areas of highest risk.

Scheme membership

Membership of these schemes is not compulsory, apart from the CORGI gas scheme (see below). Businesses carrying out work covered by the Building Regulations may choose to join the schemes if they judge membership to be beneficial. Alternatively they may choose to continue to use Local Authority Building Control or to employ a private sector Approved Inspector.

If a company or individual chooses to join a competent persons scheme, they are first vetted to ensure they meet the conditions of membership, including appropriate and relevant levels of competence. If they meet these conditions they are classified as 'competent persons'.

The work of organisations or individuals accepted as members of a scheme is not subject to Building Control inspection. Instead, the competent person self-certifies that the work is in compliance with the Building Regulations. They issue a certificate to the consumer to this effect. In some schemes they then report the work to the scheme organisers who in turn inform the Local Authority that work has taken place.

The following competent persons schemes relevant to the scope of work carried out by plumbers are currently in operation:

- Heat Producing Gas Appliances and Associated Heating and Hot Water Systems (CORGI gas scheme)
- Combustion Appliances – Oil
- Combustion Appliances – Solid Fuel
- Electrical Safety in Dwellings
- Plumbing, Heating Systems and Hot Water Service Systems
- Installation of a sanitary convenience or bathroom in a dwelling.

I will make further reference to specific competent persons schemes at appropriate points in the text. Or if you want to find out more visit the website: www.communities.gov.uk.

Try this

It is time for you to do a little research assignment. Using the above list of key organisations, you need to do two things:

- Using either the internet or other research methods, find out the contact details of the key organisations and record them here. If you are not sure of how to go about this, and do not have internet access at home or college/centre, your local library should be a starting point. If you are still struggling, ask your tutor for help.
- Then, from your research, write a short report on what each organisation does.

Write your notes, including the contact details of each of the organisations on a separate sheet for inclusion in your portfolio. You do not have to write long paragraphs, try to summarise your findings in a series of bullet points.

Why not tackle this research activity with a classmate? You could share the work and then discuss the information for your report.

Test yourself 1.1

Time to check your progress. Have a look at the following exercises which relate to the nature of the plumbing industry.

1. Fill in the gaps in the text below:

   Most plumbers usually carry out the:
   - Installation
   - ___________
   - And maintenance of a wide range of ___________ systems such as:
   - Cold water, including underground services to a dwelling
   - ___________
   - ___________ systems fuelled by ___________, oil or solid fuel
   - ___________ (or above ground drainage) including the installation of baths, wash-hand basins, WCs and sinks
   - Rainwater systems, ___________ and fall pipes
   - Associated ___________ systems
   - ___________. At Level 2, this includes things such as chimney weatherings, and soil vent pipe weatherings.

2. The organisation that oversees the training and development needs of the plumbing industry is:
   a. CITB ☐
   b. SummitSkills ☐
   c. JIB for PMES ☐
   d. APHC ☐

3. The acronym IPHE stands for:
   a. Institute of Plumbing and Heating Engineering ☐
   b. Independent Organisation for Plumbing and Heating Engineering ☐
   c. International Organisation of Plumbers and Heating Engineering ☐
   d. Indentured Organisation of Plumbing and Heating Engineering ☐

Once you have completed the 3 questions, check your answers by going back to the text. The model answers for these questions are at the end of this book.

Plumbing qualifications and training

Introduction

The plumbing qualification and training infrastructure is made up of three main elements:

- The Technical Certificate
- The NVQ or SVQ
- The Apprenticeship and Advanced Apprenticeship (Key Skills are a requirement of Apprenticeship schemes).

The Technical Certificate covers the off-the-job knowledge (what is taught in a college or centre) aspect of an NVQ (Technical Certificates are not available in Scotland). It also covers practical off-the-job training. The Technical Certificate is awarded by

Key point

Your tutor should deliver a detailed session about plumbing qualifications and training, the following text is included to provide you with an insight into what is happening in the industry.

City and Guilds, who also produce knowledge and practical assessments which must be undertaken in a City and Guilds accredited centre.

The material contained in this course is designed to cover all the knowledge requirements of the Technical Certificate, and to help to prepare you for the knowledge assessments. As the knowledge content of a Technical Certificate is consistent with the NVQ, it also covers the knowledge requirement of the NVQ.

The NVQ (SVQ in Scotland) is available at Levels 2 and 3 in England, Wales and Northern Ireland. In Scotland, the plumbing SVQ is available at Level 3 only.

A Level 2, NVQ in plumbing is the minimum level of qualification recognised by the plumbing industry.

This section will deal with the difference between training and assessment and give you an insight into how the training for the Technical Certificate will be delivered.

Although you may not be directly involved in apprenticeships, it will be helpful if you know a little bit about this, as, once in employment, you may get involved in training apprentices. Currently, apprenticeships are aimed at those entering the industry straight from school.

Finally, we will look at the career opportunities and progression routes that are available to you, once you have qualified.

The MES certificate in Plumbing Level 2: plumbing studies

An SVQ/NVQ is about proving competence in the workplace. This is achieved by a City and Guilds assessor making judgements about your ability to do the job competently. Underpinning knowledge is also assessed using multi-choice assessment questions.

The **Technical Certificate** is a qualification based on what you are taught **off-the-job**, and covers practical and theory courses. The practical courses are assessed by a number of workshop assignments, and the theory by a number of written assessments that are the same as those used in the SVQ/NVQ knowledge assessments.

What the Technical Certificate contains

You will be aiming to complete the Level 2 certificate, which contains:

Level 2

Unit	Title
1	Safety
2	Key Plumbing Principles
3	Common Plumbing Processes
4	Cold Water Systems
5	Hot Water Systems
6	Sanitation Systems
7	Central Heating Systems (pipework only)
8	Electrical Supply and Earth Continuity
9	Sheet Lead Weathering
10	Environmental Awareness
11	Effective Working Relationships
12	Practical

There is a knowledge assessment for each subject area, ranging from 20 to 60 questions depending on the subject. Your assessment centre (normally your place of study) will provide more details on this at that time. The mock assessments in this course will have fewer questions, but will deal with all the key areas.

Practical training covers the full range of domestic plumbing installations, starting with basic jointing, bending and fixing exercises. You will get more of a picture of what to expect as you work through the technical units.

Key point

NVQs/SVQs in plumbing are available to trainees of all ages, but to complete an NVQ/SVQ you must be employed in the plumbing industry. For adults wishing to retrain as plumbers we'd recommend contacting local employers and your local college/training centre to help facilitate progression onto a plumbing NVQ/SVQ scheme.

Levels 2 and 3 SVQ/NVQ in Mechanical Engineering Services (MES) Plumbing (Domestic)

MES Plumbing SVQ/NVQs are referred to as vocational qualifications which means they are work based. The qualifications are recognised by the plumbing industry and are used to grade plumbers by employer/employee organisations such as the JIB for PMES, one of the organisations you researched earlier. Level 2 SVQs are not available in Scotland.

Qualification development involves detailed discussions with a wide range of people involved with plumbing. This is done to make sure that the content of the qualifications meets the requirements of all the interested parties which includes:

- Employers
- Union representatives
- Trainers
- Professional bodies
- City and Guilds (C&G) England and Wales
- SQA (Accreditation) Scotland.

What is included in the plumbing SVQ/NVQ?

The aim of the SVQ/NVQ is to cover all the work activities that the majority of plumbers are likely to do. Since a Level 2 NVQ is quite a high qualification, it is broken down into a number of units to make the assessment process more manageable and user friendly. Some of the larger units are broken down further into elements for the same reason.

The unit titles describe *what a candidate must be able to do*. Let us take a look at the units at Level 2.

Level 2 Content
All candidates have to do all of these units:

- Maintain the Safe Working Environment when Undertaking Plumbing Work Activities
- Install Non-Complex Plumbing Systems and Components
- Decommission Non-Complex Plumbing Systems
- Maintain Non-Complex Plumbing Systems and Components
- Maintain Effective Plumbing Working Relationships
- Contribute to the Improvement of the Plumbing Work Environment.

Obtaining the qualifications

NVQ qualifications are accredited by the Qualification and Curriculum Authority (QCA) for England and Wales and SVQs by the Scottish Qualification Authority in Scotland. Submission for accreditation has to be made by an approved awarding body.

Awarding Bodies

City and Guilds (C&G) and the Scottish Qualification Authority Accreditation (SQA) are the Awarding Bodies for the MES Plumbing NVQs (C&G) or SVQs (SQA). These organisations are responsible for certifying candidates who have successfully completed their NVQ or SVQ. City and Guilds is also the Awarding Body for the Technical Certificate.

The Awarding Body is responsible for the quality assurance of the scheme and has set up an infrastructure of assessment centres, assessors, internal verifiers and external verifiers. Assessment centres are usually independent colleges or private centres approved by City and Guilds or SQA.

Internal verifiers are appointed by the assessment centres and are responsible for making sure the assessments are consistent and to the required standard. The external verifier works directly for the Awarding Body, and will look after several centres. Their job is effectively to 'police' the centres by sampling the quality of assessments, and making sure the scheme criteria are being met.

Remember, SVQs/NVQs are designed to show if someone can do a job competently in the workplace, so most of the assessments will be based on what a candidate actually does whilst at work.

Some of the work will be assessed whilst a candidate is actually doing their job. This will be done by a qualified SVQ/NVQ assessor, and is known as observed assessment. Because it is not possible to observe all work activities, due to time and cost, all other work carried out on-site will be recorded in a workplace activity record. This will be used by an assessor as a basis for completing the site assessments.

Simulations

There is also some practical work that is assessed in the centre and not in the workplace, which is known as *simulated assessment*. Simulated assessment is utilised for activities where it is difficult to get on-site evidence, and includes things like health and safety emergencies. It also includes sheet lead work, which has become a fairly specialised area of plumbing work, but is still required by some employers.

The SVQ/NVQ also requires the job knowledge, or theory aspect of the job, to be assessed. This is done by using multi-choice questions, where you have to select the correct answer from four or possibly five options. These are produced by the Awarding Body and are the same multi-choice questions used to assess the Technical Certificate that we mentioned earlier. The content of this text is designed to underpin this knowledge requirement.

Training and assessment

Training and assessment forms an integral part of both the Technical Certificate and SVQ/NVQ. The training is the 'how to get there' part, and assessment is the 'have you achieved' part.

Training

Training is usually referred to as 'on-the-job' (what you receive at work) and 'off-the-job' (what you receive in a college or centre). The training is designed to cover practical and theory aspects of the job, usually with the theory training taking place exclusively in the college or training centre.

You will be given instruction on how to do something correctly, as well as the theory that underpins the job. Training also gives you the opportunity to practise things until you can do the job competently (see for example Figure 1.2).

Figure 1.2 Student experiencing off-the-job training in a college

Assessment

Assessment in SVQ/NVQ terms is about finding out if someone can do a job competently, and if they understand the theory that underpins the job. There are various forms of assessment.

SVQs/NVQs include assessments that are done at the workplace, either watching a person doing the job, or assessing what they have included in their workplace recorder/portfolio. It also includes the knowledge assessment, which in the main is done by multi-choice questioning. The SVQs/NVQs also use assessments which are linked to the simulations such as the sheet weathering assessments.

For the Technical Certificate, you will be assessed on your job knowledge and practical ability off-the-job. The end assessments in both cases will be in a City and Guilds centre.

The Apprenticeship scheme

Apprenticeship schemes have been in existence in the plumbing industry for well over a hundred years. In recent times, the Government has seen the need to revamp the apprenticeship system across all industries, and have attached funding to apprenticeship scheme delivery in order to boost uptake and thus raise the skills level of the nation. These were termed Modern Apprenticeship Schemes.

The Government introduced a standard apprenticeship framework template that all industry sectors were required to follow, but to tailor to meet their specific requirements.

The Modern Apprenticeship schemes have evolved into the Apprenticeship and Advanced Apprenticeship schemes which exist today. The name given to apprenticeship schemes varies between England, Wales and Scotland, as follows:

- England – Apprenticeship (delivers an NVQ Level 2) – Advanced Apprenticeship (delivers an NVQ Level 3)
- Wales – National Traineeship (delivers an NVQ Level 2) – Advanced Apprenticeship (delivers an NVQ Level 3)
- Scotland – Modern Apprenticeship (delivers an SVQ Level 2 and 3).

Whilst the name of the apprenticeship schemes in each country varies, the content varies very little in principle. To save confusion, this next section will concentrate on how the scheme works in England and Wales.

The MES Plumbing Apprenticeship or Advanced Apprenticeship is made up of three main parts which the apprentice must achieve:

- An NVQ in MES Plumbing (Level 2 for apprenticeships and Level 3 for advanced)

- An MES Certificate in Plumbing: Plumbing Studies (Level 2 for apprenticeships and Level 3 for advanced)
- Key skills at Level 2 in Communication and Application of Number, this applies for both schemes.

You can keep up to date with the latest developments in apprenticeship schemes by logging on to www.apprenticehips.org.uk or by phoning 08000 150 400.

Career opportunities available to plumbers

We appreciate that you have only just started on this course, so you are focusing your thoughts on completing your technical certificate or SVQ/NVQ.

However, it is worth taking a few moments just to think about what your future in plumbing could hold. Have a look at the chart (Figure 1.3). It should be pretty self-explanatory. If there is anything you are not sure about, the Institute of Plumbing and Heating Engineering (IPHE) should be able to help. The IPHE is the professional body for the plumbing industry which you should have encountered already by completing *Activity 1.4*.

That completes the section on plumbing training. Complete the following exercises as a quick review of your progress so far.

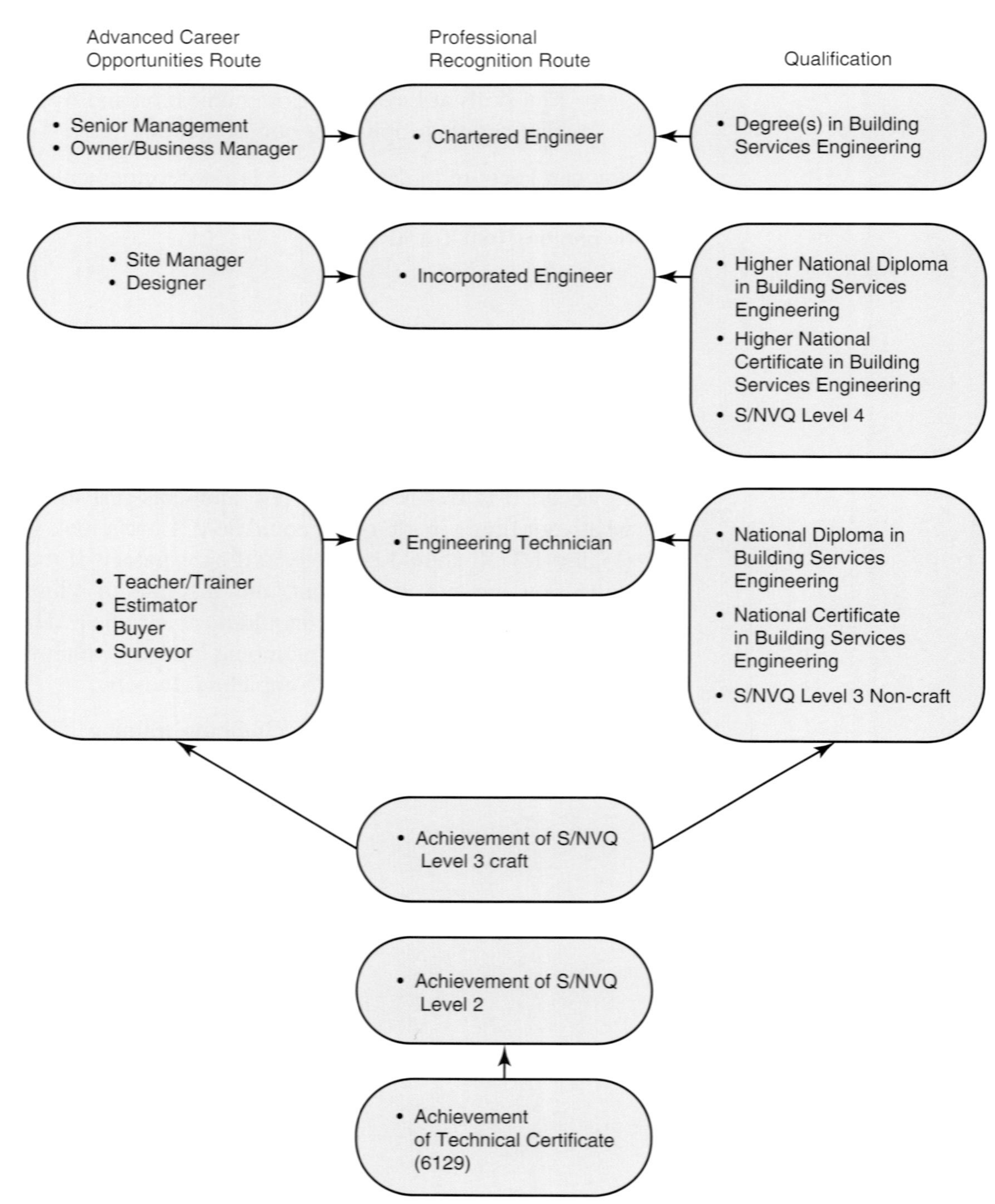

Figure 1.3 Career paths

Test yourself 1.2

1. Complete the missing words in the paragraph from the words given below:

   Whilst the ___________ is about proving competence in the workplace, the ___________ Level 2, Plumbing Studies, is a qualification based on what you are taught ___________. It covers ________ and theory. The _________ is assessed by a number of workshop assignments, and the theory is the same assessments as used in the NVQ ________ assessments.

   SVQ/NVQ, Certificate in Plumbing, workplace, workshop, on-the-job, off-the-job, practical, knowledge

2. Tick which one of the following statements you think is correct:
   a. The main method of assessing the job knowledge aspect of the Technical Certificate is by using multi-choice questions ☐
   b. The main method of assessing the job knowledge aspect of the Technical Certificate is by using simulations ☐
   c. The main method of assessing the job knowledge aspect of the Technical Certificate is by conducting practical tests ☐
   d. The main method of assessing the job knowledge aspect of the Technical Certificate is by a written examination ☐

3. What does the acronym NVQ stand for?

4. Write a short paragraph on what you understand about the term assessment.

5. What qualification would you need to become an Incorporated Engineer?

Once you have completed the 5 questions, check your answers by going back to the text. The model answers for these questions are at the end of this book.

Check your learning Unit 1

Aim at completing the questions in 15 minutes.

Complete your answers on a separate sheet for inclusion in your portfolio.

Example: Once you have completed the questions, check out the answers at the end of this book.

1. Which of the following components is a plumber most likely to install?
   a. Hot water storage vessel ☐
   b. Air conditioning unit ☐
   c. Refrigerated display cabinet ☐
   d. Underground gas service ☐
2. Apart from domestic systems, a plumber will also be required to work on:
   a. Sheet lead weatherings ☐
   b. Underground drainage pipework ☐
   c. Steam pipework installations ☐
   d. Replacement of roof slates and tiles ☐
3. In the context of the plumbing industry, the acronym BES stands for:
   a. British Employment Services ☐
   b. Building Engineering Specialist ☐
   c. Building Engineering Services ☐
   d. British Engineering Services ☐
4. The Sector Skills Council responsible for the plumbing industry is known as:
   a. The Institute of Plumbing ☐
   b. CITB ☐
   c. SkillsSmart ☐
   d. SummitSkills ☐
5. Which organisation is the official trade union for the plumbing industry?
   a. APHC ☐
   b. Amicus ☐
   c. BPEC ☐
   d. City and Guilds ☐
6. The plumbing qualification and training infrastructure is made up of three main elements, the apprenticeship scheme, NVQ or SVQ, and:
   a. Training Certificate ☐
   b. Technical Certificate ☐
   c. Technical Diploma ☐
   d. Key Skills Certificate ☐
7. NVQs and SVQs are designed to:
   a. Measure someone's ability to pass a knowledge exam ☐
   b. Demonstrate someone's competence to do the job ☐
   c. Measure someone's ability to pass a practical exam ☐
   d. Enable an employer to register with a trade association ☐
8. The organisations responsible for certifying SVQ/NVQ candidates are known as:
   a. Certificating Bodies ☐
   b. Accreditation Bodies ☐
   c. Awarding Bodies ☐
   d. Regulatory Bodies ☐
9. To enrol on a plumbing NVQ or SVQ, candidates must be:
   a. under 16 years old ☐
   b. over 25 years old ☐
   c. employed in the plumbing industry ☐
   d. unemployed ☐
10. What is the minimum level of qualification recognised by the plumbing industry?
    a. Plumbing NVQ Level 2 ☐
    b. Plumbing NVQ Level 3 ☐
    c. Plumbing Technical Certificate Level 2 ☐
    d. Plumbing Technical Certificate Level 3 ☐

UNIT 2

HEALTH AND SAFETY

Summary

Health and Safety forms a very important part of your everyday working life. This unit will include sessions on the following topics:

- Personal safety
- Access
- Risk assessment
- COSHH
- Fire safety and emergencies
- Basic first aid procedures and other plumbing-related safety issues
- Regulations and the safety of others.

The induction unit showed you how important Health and Safety considerations are in the Domestic Plumbing Industry. This unit covers the requirements that need to be met by you and your employer in respect to health and safety at work.

Personal safety

The British construction industry employs 2.2 million people making it the country's biggest industry.

Every year in the construction and plumbing industry, there are tens of thousands of accidents (some of which are fatal; 2800 people have died from injuries they received as a result of construction work). These figures are not intended to put you off your new career, but to make sure you understand that accidents do happen if health and safety guidance procedures are not followed.

In this unit, you will learn more about these and other regulations designed to keep you safe at work.

You will also learn about the various hazards that you may face, implementing safe methods of working, including the correct actions to be taken in the event of accidents.

There is a great deal of information available about health and safety. Here are some useful contacts:

- HSE Books
  Tel: 01787 881 165
  Website: hsebooks.co.uk

HSE Publications:

- Essentials of health and safety at work
- Introduction to health and safety for small firms
- Five steps to risk assessment
- Basic advice on first aid at work
- RIDDOR explained
- COSHH: a brief guide to the regulations
- Getting to grips with manual handling
- Safe working with flammable substances.

- The Health and Safety Executive
  Tel: 0870 154 5500
- Royal Society for the Prevention of Accidents (ROSPA)
  Tel: 0121 248 2000
  Website: www.rospa.co.uk
  E-mail: help@ospa.co.uk
- Institution of Occupational Safety and Health (IOSH)
  Tel: 0116 257 3100
  www.iosh.co.uk

Hot tips

There are two main bodies responsible for the establishment and implementation of Health and Safety laws:

- The Health and Safety Commission (HSC)
- The Health and Safety Executive (HSE).

Hot tips

It would be a good idea to obtain a copy of the HSE publications. These should be available in your college or centre or better still, get your own.

- Incident Contact Centre
  Tel: 0845 300 9923
  www.riddor.gov.uk
- British Safety Council
  Tel: 0208 741 1231
  www.britishsafetycouncil.org.uk
- The British Red Cross
  Tel: 0207 235 5454
  www.redcross.org.uk
- The Institution of Electrical Engineers
  Tel: 0207 240 1871
  www.iee.org.

Safety signs

During your domestic plumbing career, you will spend time on various domestic building/construction sites. All sites are potential hazard areas. Any site you enter will display safety signs. Safety signs are designed to keep visitors and workers safe, pass on useful information (regarding exits, fire extinguishers, etc) and to warn personnel about possible dangers they may be exposed to during their time on site.

There are four groups of safety signs:

- Mandatory signs
- Prohibition signs
- Warning signs
- Information signs.

Here is a description of each group.

Figure 2.1 Mandatory sign

Mandatory signs

These signs are circular and show white symbols on a blue background. Figure 2.1 shows what must be done by all site personnel, e.g. wear head protection.

Figure 2.2 Prohibition sign

Prohibition signs

These signs are circular and show black symbols on a white background, the signs also have a red border with a red line passing through the centre of the circle. Figure 2.2 shows what is not allowed, e.g. no smoking.

Figure 2.3 Warning sign

Figure 2.4 Information sign

Warning signs

These signs are triangular, have a black border and show black symbols on a yellow background. Figure 2.3 shows potential dangers, e.g. flammable material.

Information signs

These signs are square or rectangular and show white symbols on a green background. Figure 2.4 shows information usually relating to access or health and safety, e.g. first aid.

The following section contains more examples of signs you will see at times on work sites. Some signs are just symbols, while other signs may have words or other additional information, such as heights and distances (as shown in Figure 2.5).

Some of these may be familiar to you, while you might not have seen others before. It is up to you to learn and understand what

Figure 2.5 Safety sign gallery

they mean and to take notice of them. Again, when looking at the signs, try to make a mental note of which group they fit into – it is not a test, but will be good practice for future!

Protective clothing

Refer to the free HSE publication 'A Short Guide to the PPE at Work Regulations 1992'.

General introduction to 'PPE'

The guidelines that govern protective clothing are set out in the Personal Protective Equipment at Work Regulations (1992). The Regulations came into force on 1 January 1993, and you will possibly hear workmates referring to personal protective equipment as 'PPE'. The equipment is defined as all equipment that is designed to be worn or held to protect against risks to health and/or safety. 'PPE' includes most types of protective clothing and equipment, such as eye, hand, foot and head protection. All PPE clothing and equipment must carry the CE marking.

Your employer has a duty to assess any work-related risks that you may face, and where possible, minimise the risks you may face at work although some work tasks that you will be asked to complete will carry a certain degree of risk. In these situations, PPE will be used as a last resort to keep you as safe as possible. Your employer should ensure:

- All PPE is provided free of charge
- PPE is suitable for its purpose
- Information on procedures to maintain, clean and replace damaged PPE is freely available
- Storage for PPE is provided
- PPE is maintained in an efficient state and is in good repair
- That PPE is properly used.

(a)
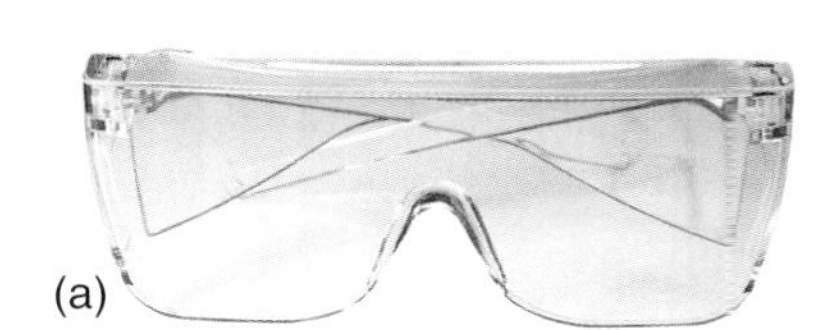

(b)
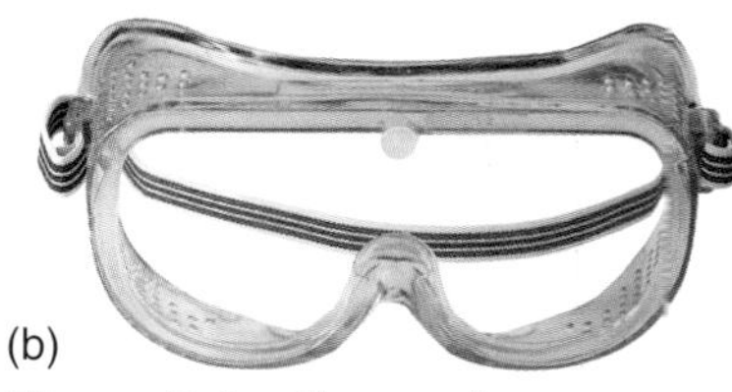

Figure 2.6 Types of eye protection (Reproduced with permission of Draper Tools Limited)

Eye protection

Your eyes are two of your most precious possessions. You need them for learning and for earning, because without your eyesight you probably cannot perform your present job! Your eyes are extremely vulnerable to injury at work. Figure 2.6 shows the types of eye protection.

Every year thousands of workers suffer eye injuries, resulting in pain, discomfort, loss of income and even blindness. Following the safety procedures correctly and wearing eye protection, as shown in Figure 2.6, can prevent these injuries. There are many types of eye protection equipment available, e.g.:

- Safety glasses
- Safety goggles, these can also be used for eye protection while carrying out lead welding
- Full face masks/shields.

Foot protection

Figure 2.7 Safety footwear (Reproduced with permission of Draper Tools Limited)

As you probably guessed, Figure 2.7 refers to protecting your toes, ankles and feet from injury. However, wearing the correct footwear can also protect your whole body from injury, e.g. from electric shock. Toe and foot injuries account for 15% of reported accidents in the workplace. More than 30,000 injuries to the feet are reported each year and many more accidents go unrecorded. Accidents to the feet have serious consequences and can result in pain and suffering, disability, loss of work and income.

Protective footwear can help prevent injury and reduce the severity of any injuries that do occur (Figure 2.7). The basic universal form of foot protection is the safety shoe which will usually include some kind of metallic toe protection, rubber soles and sturdy leather uppers. The illustration shows typical safety boot/shoe protection.

Hand protection

Figure 2.7 shows the protection of two irreplaceable tools – your hands, which you use for almost everything including working, playing, driving, eating, etc. Unfortunately, hands are often injured. Almost one in four work-related injuries happens to the hands and fingers. One of the most common problems other than cutting, crushing or puncture wounds to the hand is dermatitis. Dermatitis is an inflammation of the skin normally caused by contact with irritant substances.

The irritation of the skin may be indicated in several ways, sores, blisters, redness or dry cracked skin, which easily become infected. Over 700,000 workdays per year are lost due to dermatitis. To protect your hands from irritating substances, you need to keep them clean by regular washing using approved cleaners. Make good use

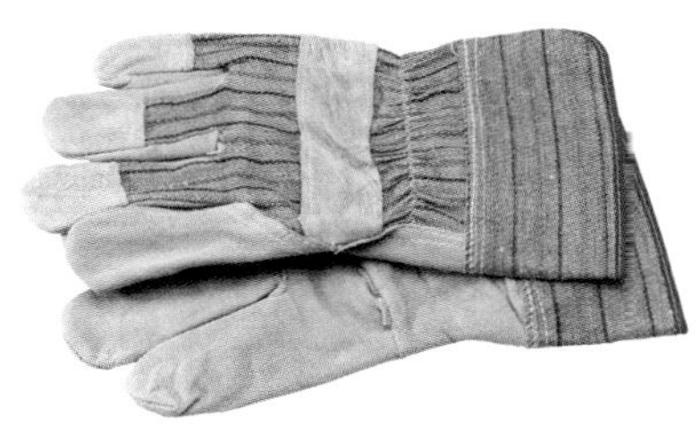

Figure 2.8 Safety gloves (Reproduced with permission of Draper Tools Limited)

of barrier creams where provided, and wear appropriate personal protection (usually a strong pair of gloves (see Figure 2.8)) when required.

Head protection

This is another vitally important form of PPE and is especially important to plumbers who need to spend time on rooftops and up ladders. A hammer accidentally dropped from a roof location could cause serious injury or death if it landed on someone's unprotected head (see Figure 2.9).

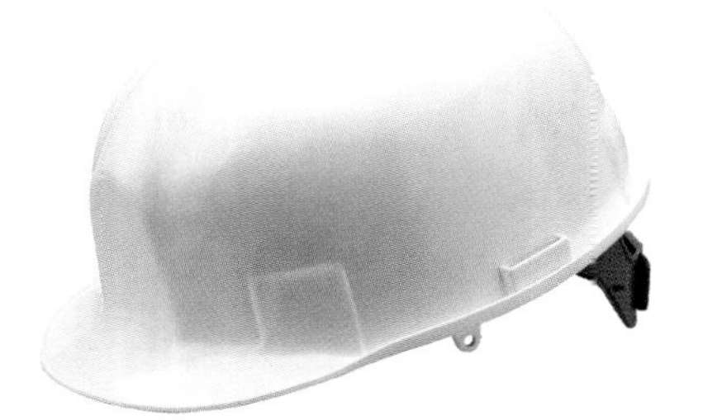

Figure 2.9 Safety helmet (Reproduced with permission of Draper Tools Limited)

There are also other potential hazards such as objects (e.g. scaffold tubes), which often jut out at head height – just waiting for you to walk into them. Hard hats, constructed from high-density polystyrene, like the one in Figure 2.9, will provide some protection from falling objects and accidental head knocks.

Head protection is extremely important because it guards your most vital organ – your brain. As we have previously said, a head injury can physically disable a person for life and can even result in death. Head protection can help prevent such injuries. Approximately 80% of industrial head injuries are a result of people not wearing 'hard hat' protection.

You need to wear your safety helmet whenever it is required and should ensure that it is worn correctly, i.e. always wear it the correct way round despite what your workmates might be doing with theirs.

To keep safe, use your head. Know the potential hazards of your job and what protective gear to use. Follow safe working procedures. Take care of your protective headgear. Notify your employer of unsafe conditions and equipment. Seek medical help/advice promptly in the event of head injury.

Activity 2.1

What safety guidelines can you think of to make your helmet as comfortable and safe as possible? Write your answer down, before checking the suggested one at the end of this book.

UNIT 2

Tools and equipment

You will use a wide variety of tools and equipment as a plumber, all of which can be potentially dangerous if misused or neglected. Typical hand and power tools used by plumbers are covered in Unit 3. Hand tools and manually operated equipment are often misused. You should always use the right tool for the job; never just make do with whatever tool you may have at hand.

Example: Screwdrivers are not intended to be used as an alternative to wood chisels or bolsters for lifting floorboards!

- Cutting tools, saws, drills, etc must be kept sharp and in good condition. In the course of domestic plumbing work, you will frequently need to use cutting tools, such as hacksaws and wood saws. You should ensure that blades used are always fitted properly and are sharp. Hacksaw teeth should be pointing in the forward direction of cut. After use, guards should be fitted wherever possible.
- Handles should be properly fitted to tools, such as hammers and files, and should be free from splinters. Hammer heads should be secured correctly using metal or wooden wedges. 'Mushroom heading' of chisels is also a dangerous condition which can lead to serious eye injury. Unprotected file tangs present a serious danger of cuts and puncture wounds.
- It is particularly important to check that the plugs and cables of hand-held electrically operated powered tools are in good condition. Frayed cables and broken plugs should be replaced. Electrically powered tools of 110 or 230 V must be PAT tested in accordance with recommended procedures. It is a good practice to check all electrical equipment for test labels to ensure that they are in safe working order.
- Other common items of equipment, e.g. barrows, trucks, buckets, ropes and tackle, etc are all likely to deteriorate with use. If they are damaged or broken, sooner or later they will fail in use and may cause an accident; non-serviceable tools and equipment should not be used, faulty tools or equipment should be repaired or replaced, and the unsafe equipment must be removed from the site.

Key point

PAT (Portable Appliance Testing) tests are maintenance records of all portable equipment to ensure it is in safe working order.

You may have to use cartridge-operated tools during your career. If you do, you will be given the necessary instruction on the safe and correct methods in which to use them. They can be dangerous, especially if they are operated by accident, or used as toys. This could cause ricochets, which could lead to serious injury. People under the age of 18 are not permitted to use cartridge-operated tools.

Trips and fire hazards

Trip hazards are responsible for numerous accidents every year and almost everything in the workplace is a potential trip hazard:

- Carelessly discarded tools, equipment and materials
- Spilled material (oil, lubricants)
- Trailing cables and welding hoses
- Uneven terrain.

Clutter and debris, oily rags, paper, etc should be cleared away to prevent fire hazards. As an individual, you may have no control over the general state of the workplace but you should ensure that your own work area is kept clear and tidy, as it is the mark of a skilled and conscientious tradesman.

Electricity on site

The safe use of electricity on site is covered by the Electricity at Work Regulations 1989, which came into force on 1 April 1990, and in turn are covered by the Health and Safety at Work Act 1974. These Regulations impose specific duties on employers to put into place measures to protect their employees against death or personal injury from the use of electricity at work.

Employers are required to have specific codes of practice for their employees. These practices must include the maintenance records of all portable equipment (PAT tests). The HSE suggest that portable electrical equipment should be tested every three months. These records must show that the equipment is tested regularly by a competent person using suitable test equipment.

The scope of what is classified as portable equipment is very wide and includes everything from kettles to 110 V equipment for industrial use. Typically, the supply of electricity to a work site may be from the local public supply. If the supply is taken from a generator, the generator should be sited to ensure that noise and fumes are reduced to a minimum. However, whatever the source of electrical supply, it must be routed to where it is required on site. This could involve electric cables being buried underground, or more often by suspended overhead cables on poles.

Figure 2.10 Transformer (Reproduced with permission of Draper Tools Limited)

The installation of this system must conform to the requirements of the current edition of BS 7671. You should never interfere with or alter any installation.

Figure 2.10 shows that, when working on site, you should use a transformer to reduce the voltage from 240 to 110 V, this is much safer as shocks from 240 V supplies can often prove fatal.

Electricity on site safety check

- Use equipment with a voltage of 110 V whenever possible
- Do not use lighting circuits for power tools
- Power tools should be double insulated
- Never carry a portable electric tool by its cable
- Check to ensure that equipment you are about to use is not damaged before you plug it in
- Always have adequate lighting for the job
- Keep lights clean
- Check if all cables are correctly insulated and not damaged or frayed
- Check if plugs and sockets are clean and in sound condition
- Check for current PAT labels
- Check that RCD protection is provided wherever necessary
- Remember to visually inspect all electrical equipment before use even if it carries a valid test label.

Lifting and carrying (manual handling)

The manual handling or lifting of objects is the cause of more injuries on work sites than any other factor. Back strains and associated injuries are the main cause of lost hours in the services sector. Within the context of the domestic plumbing industry, manual handling can involve pushing, pulling, lifting and lowering of loads (baths, boilers, cylinders, radiators, tools, etc).

Hot tips

The free HSE publication 'Getting to grips with manual handling' gives more information on manual handling – www.hse.gov.uk.

The movement of large or heavy loads requires careful planning in order to identify potential hazards before they cause injury. You must follow safety precautions and codes of practice at all times.

Figure 2.11 shows how extreme care must be taken when lifting or moving heavy or awkward objects manually. The general average acceptable maximum lifting load for a fit male is 20 kg,

Figure 2.11 Steps to safe manual handling

and 15 kg for a female. The rules for correctly lifting a load are as follows:

- Ensure that the path where you need to move the load to is clear from obstructions, that any doors you have to pass through are opened and that you have a clear area for placing the load
- Test the load by gently applying force with your foot, this will tell you if the load feels heavy and difficult. If this is the case you may need to seek help for a double lift.
- If it feels comfortable to move, start from a good base and stand with the feet hip width apart
- Maintain the back straight and upright; bend the knees and let the strong muscles of the legs and thighs do the work
- Keep the arms straight and close to the body
- Balance the load using both hands if possible
- Avoid sudden movements and twisting of the spine
- Take into account the position of the centre of gravity of the load when lifting
- Use gloves to avoid injuries from sharp or rough edges

Note: The person carrying the load must always be able to see over or around it

- Never obstruct your vision with the load that you are carrying
- All obstacles should be removed from the vicinity
- In cases where team lifting is required, the following points must be remembered:
  - Team members should ideally be of similar height and build
  - All team members must know the lifting sequence
  - One member must be nominated to act as a co-ordinator
  - Good communication when lifting should reduce the risk of an accident occurring.

Movement of loads and methods of transport

Figures 2.12–2.15 show use of various methods of transport that can make the movement of loads much simpler and safer. Examples of transport are:

Fork lift trucks which are used on site for moving heavy loads, but can only be operated by a qualified person.

Figure 2.12 Wheel barrow

Figure 2.13 Sack barrow

Figure 2.14 Flatbed trailer

Figure 2.15 Forklift truck

Test yourself 2.1

1. What type of safety sign shows a black symbol on a yellow background bordered by a black triangle?
   a. Warning ☐
   b. Information ☐
   c. Prohibition ☐
   d. Mandatory ☐

2. Which of the following in Figures 2.16–2.18 is a mandatory sign?
   a. Figure 2.16 ☐
   b. Figure 2.17 ☐
   c. Figure 2.18 ☐

3. Electricity on site is normally reduced to a lower voltage, which is the normal lower voltage found on site?
   a. 230 V ☐
   b. 24 V ☐
   c. 110 V ☐
   d. 400 V ☐

4. All the following are important considerations before lifting a load except one
   a. Making trial lifts on similar loads ☐
   b. Getting help if the load is too heavy ☐
   c. Checking that doors are opened if you have to carry the load through them ☐
   d. Checking the load for any sharp edges ☐

Figure 2.16

Figure 2.17

Figure 2.18

Risk assessment

> **Hot tips**
>
> The HSE provides much information on risk assessment, a good general guidance document is '5 steps to risk assessment' – available at www.hse.gov.uk.

Introduction

Risk assessments are a kind of checking system, and their purpose is to help keep you safe at work. The aim of risk assessments is to make you think about the possible hazards you may encounter while performing a specific work task. At this stage of your development, you will not be required to write risk assessments but you must have a working knowledge of them and understand their principles.

What is risk assessment?

A risk assessment is nothing more than a careful look at aspects of work that could cause harm to yourself or others. It is a system which should help enable you to sum up whether you have taken enough precautions, or if you could do more to prevent harm. The main aim of a risk assessment is to ensure that no one gets hurt or becomes ill.

An employer must decide whether a hazard is significant and whether satisfactory precautions have been taken so as to minimise any potential risks. An employer is also legally bound to assess the risks in the workplace. Legislation which governs the requirements of risk assessment is contained within the Management of Health and Safety at Work Regulations (1999).

In order to assess potential risks, processes are usually undertaken to categorise hazardous operations. The format risk assessments vary from organisation to organisation, but they will all be based on the basic risk assessment premise that:

$$\text{Likelihood} \times \text{Consequence} = \text{Risk}$$

The list below shows some of the most common risks you will face throughout your plumbing career.

List of possible hazardous operations

- working with non-electrical-powered plant
  - hand tools
  - specialist tools
  - pneumatic tools
  - hydraulic tools.
- working with non-powered tools
- working with non-powered tools
  - specialist equipment.
- manual handling of loads
  - specialist equipment.
- manual handling of loads
  - general lifting.
- working with hazardous substances
- working in areas below the ground
  - physical conditions
  - interruption services
  - general.
- working with powered industrial trucks
  - fork lift trucks
  - dump trucks

  - jcb/tractors
  - tail lift vehicles and driving road vehicles.
- working with highly flammable liquids
- working with liquefied petroleum gases
- working with lead
  - manual handling.
- working at heights
  - ladders
  - scaffolds
  - ropes/harnesses.
- working with demolitions
- working with electrical installations
- controlling work with fumes
- controlling work with noise
- controlling work with dust
- working with asbestos
- working within vessels
- working in confined spaces
- clearing of hazardous waste
- cartridge-fixing devices
- working on suspended timber floors.

If you are personally carrying out risk assessments, there are five important factors to consider, these are:

1. Look for hazards
2. Decide who might be harmed and how
3. Evaluate the risks and decide whether the existing precautions are adequate, or should more be taken
4. Document your own findings
5. Review your assessment and revise it if necessary.

Try this

Write about any additional possible hazardous operations.

Test yourself 2.2

1. The purpose of a risk assessment is to:
   a. Assess the levels of risk to safety of operatives that may be present in work operations ☐
   b. Assess the extent of the risk that a particular job might result in a loss of profit to the company ☐
   c. Assess the value of a customer's property for insurance purposes when damage has been caused to property ☐
   d. Assess the possible costs of an insurance claim by operatives who have been injured at work ☐

2. Which is the correct risk assessment formula?
   a. Time × Cost = Risk ☐
   b. Cost × Likelihood = Risk ☐
   c. Consequence × Cost = Risk ☐
   d. Likelihood × Consequence = Risk ☐

3. Who is responsible for carrying out written risk assessments?
   a. The HSE ☐
   b. You ☐
   c. The client ☐
   d. Your employer ☐

Access

Introduction

Access equipment refers to items, such as ladders, roof ladders, trestle scaffolds, independent scaffolds, putlog scaffolds and tower scaffolds. Access also includes working in excavations. Access equipment forms a very important part of site work, so it is vital that it is kept in good order. For this reason all access equipment should be regularly checked and its condition recorded.

Hot tips

For a good introduction to basic access safety, there is also the HSE construction information sheet 49 'General access scaffolds and ladders' – www.hse.gov.uk.

Ladders

As a plumber you will need to use ladders frequently, either working from the ladder directly, or using it to gain access to the place of work or a scaffold. Mainly, the ladder should only be used as a means of access, or for working for a short period of time (up to 30 min). As the ladders are frequently used, their condition tends to be neglected which can lead to defects. It is advisable to closely inspect any ladder before use. Ladders on construction sites tend to be made either of wood or aluminium. In domestic situations, aluminium ladders and steps are probably more common than wooden ladders, but the chances are that during your plumbing career you will spend a lot of time upon both!

You should always check wooden ladders for the following to ensure that they are safe for use, as shown in Figure 2.19.

Figure 2.19 Typical wooden site ladder

- The stiles/strings are not cracked or have any warping
- The rungs are not split or dirty
- Tie-rods are not missing and are not damaged
- There is no wood rot or temporary repairs
- The ladder should not be painted as the paint may be hiding defects.

For metallic (aluminium/stainless steel) ladders, the following should be checked:

- The stiles and rungs should be checked for signs of corrosion
- The ladder 'feet' must be level and intact
- The ladder must be free from dents
- Rungs and stiles must be free from cracks and other structural defects.

Short ladders can be carried by one person on the shoulder, in either the horizontal or vertical position. Longer ladders should be carried by two people horizontally on the shoulders, one at either end holding the upper stile. When carrying ladders you should take care when rounding corners and when passing between or under obstacles.

There are certain rules to be followed when erecting ladders, which must be followed to ensure safe working. These are:

- The ladder must be placed on firm level ground. Bricks or blocks should not be used to 'pack up' under the stiles to compensate for uneven ground.
- If using extension ladders, they must be erected in the closed position and extended one section at a time. There must be at least a three rung overlap on each extension for ladders up to 6 m and a four rung overlap for ladders over 4 m.
- If the ladder is placed in an exposed position it must be guarded by barriers.
- The angle of the ladder to the building must be in the proportion 1 out 4 up 75°.
- The ladder must be secured at the top (where possible) and also at the bottom by lashing to a stake (where possible) to prevent it from slipping or falling sideways. Alternatively, the ladder may be 'steadied' by someone holding the stiles, and placing one foot on the bottom rung, this is commonly known as 'footing' the ladder.
- When the ladder provides access to a roof or working platform the ladder must extend at least 1.05 m or five rungs above the access point.

Key point

This person must not under any circumstances move away while there is someone up the ladder.

Figure 2.20 Safely lashed ladder

- Ladders must not rest against any fragile surface (e.g. glass window) or against fittings, such as gutters or drainpipes – these could easily give way and result in an accident.
- When climbing up a ladder you must use both hands to grip the rungs. This will give you better protection if you slip.
- All ladders, stepladders and mobile tower scaffolds must be inspected before and after use, and must be tested and examined by a competent person on a regular basis.

The results of the tests and inspection must be recorded, any ladder found to be defective must be suitably marked and removed from service. Figure 2.20 shows a correctly positioned ladder, securely lashed to the scaffold, with adequate extension past the access point (1.05 m or 5 rungs).

Step ladders

Step ladders are often used by plumbers. The first essential check before using step ladders is to make sure the ground is level and firm. All four legs of the step ladder should rest firmly and squarely on the ground (Refer Figure 2.21). They will do this provided that the floor or ground on which they stand is level and the floor or ground or steps themselves are not worn or damaged. The top of the steps should not be used unless it is constructed as a working platform.

On wooden step ladders, check that the hinges are in good condition and that the rope is of equal length and not frayed.

Key point

When accessing a scaffold on a construction site where heavy materials and equipment are moved or there is regular foot access, only a Class 1 ladder should be used. There are three classes of ladder, 1, 2 and 3. A Class 1 ladder is intended for industrial use and should always be used on construction sites.

Trestles

Some jobs cannot be carried out safely by using step ladders. In these cases, a working platform known as a 'trestle scaffold', as shown in Figure 2.22, should be used. This consists of two pairs of trestles or 'A' frames spanned by scaffolding boards which provides a simple working platform. Alternatively, purpose made frames can be used like the ones in Figure 2.22.

When setting out and erecting trestle scaffolds, the following rules must be observed:

- Trestle scaffold should be erected on a firm level base (as with ladders) with the trestles fully opened
- The platform must be at least two boards or 450 mm wide. The platform should be no higher than two-thirds of the way up the trestle, this ensures that there is at least one-third of the trestle above the working surface.

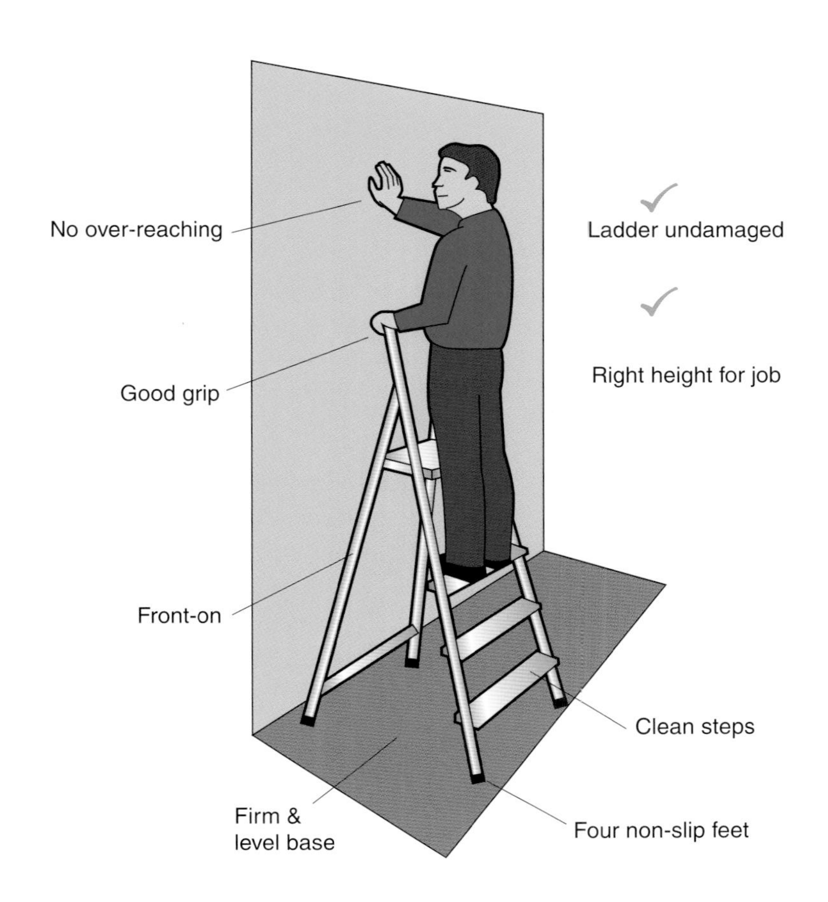

Figure 2.21 Step ladder safety

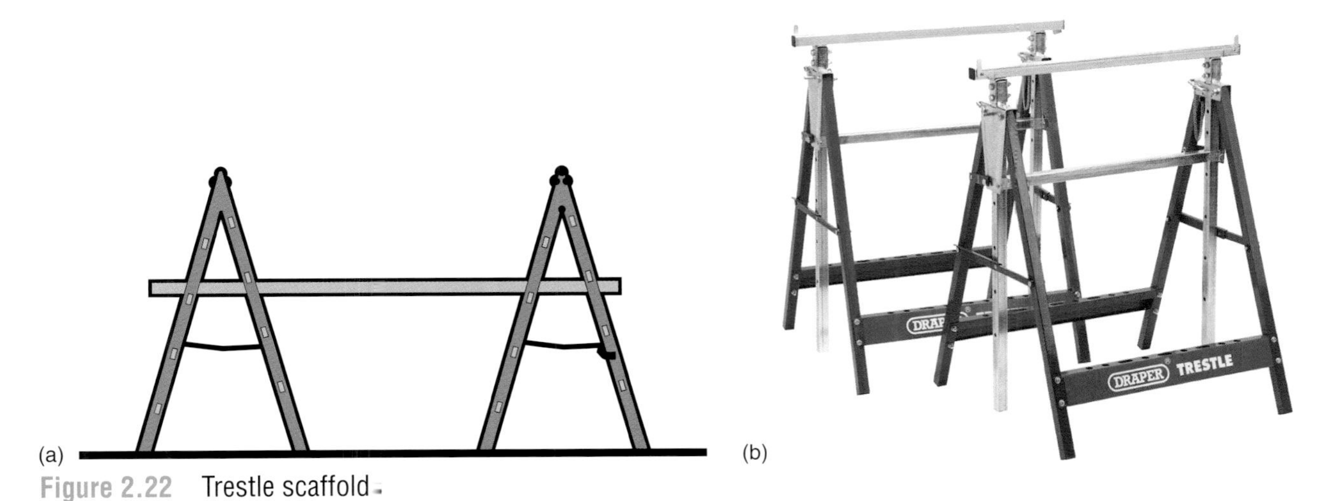

Figure 2.22 Trestle scaffold

- The scaffold boards must be of equal length and should not overhang the trestle by more than four times their own thickness, e.g. a 40 mm board must not overhang by more than 160 mm

- The maximum span for boards is 1.3 m for 40 mm thickness and 2.5 m for 50 mm thickness
- If the platform is more than 2 m above the ground, toe boards and guardrails must be fitted and a separate ladder provided for access
- Trestles must not be used where anyone can fall from a height of more than 4.5 m.

Trestle scaffold boards

Scaffold boards are manufactured to satisfy the requirements of BS 2482/70 and these are the only types of boards that should be used. The maximum length is usually no greater than 4 m. If a length greater than this is required, then special staging is used. Scaffold boards should be:

- Clean, free from grease, oil and dirt and straight in length
- Free from decay, damage or any splits
- Unpainted, as paint could hide defects.

Independent and putlog scaffolds

As a plumber, probably, you would not be expected to erect an independent or putlog scaffold. You will almost certainly work from it at some stage, may be to install or renew guttering, or to inspect or renew chimney or roof weatherings.

It is therefore very important that you are happy that the scaffold has been erected correctly (Figure 2.23), and it is safe for you to work from: as a rule of thumb, stand back, look at the scaffold and ask yourself the following questions:

- Does it look safe?
- Are the scaffold tubes plumb and level?
- Is there a sufficient number of braces and scaffold boards?
- Are adequate toe boards and guardrails fitted?
- Is it free from excessive loads such as bricks?
- Is there proper access from a ladder?

Make sure that there are ledger to ledger braces on each lift for independent scaffolds, and that putlog scaffolds are tied to the building.

Figure 2.24 shows that you also need to be careful to check the boarded area of the scaffold as this is the area from where you will

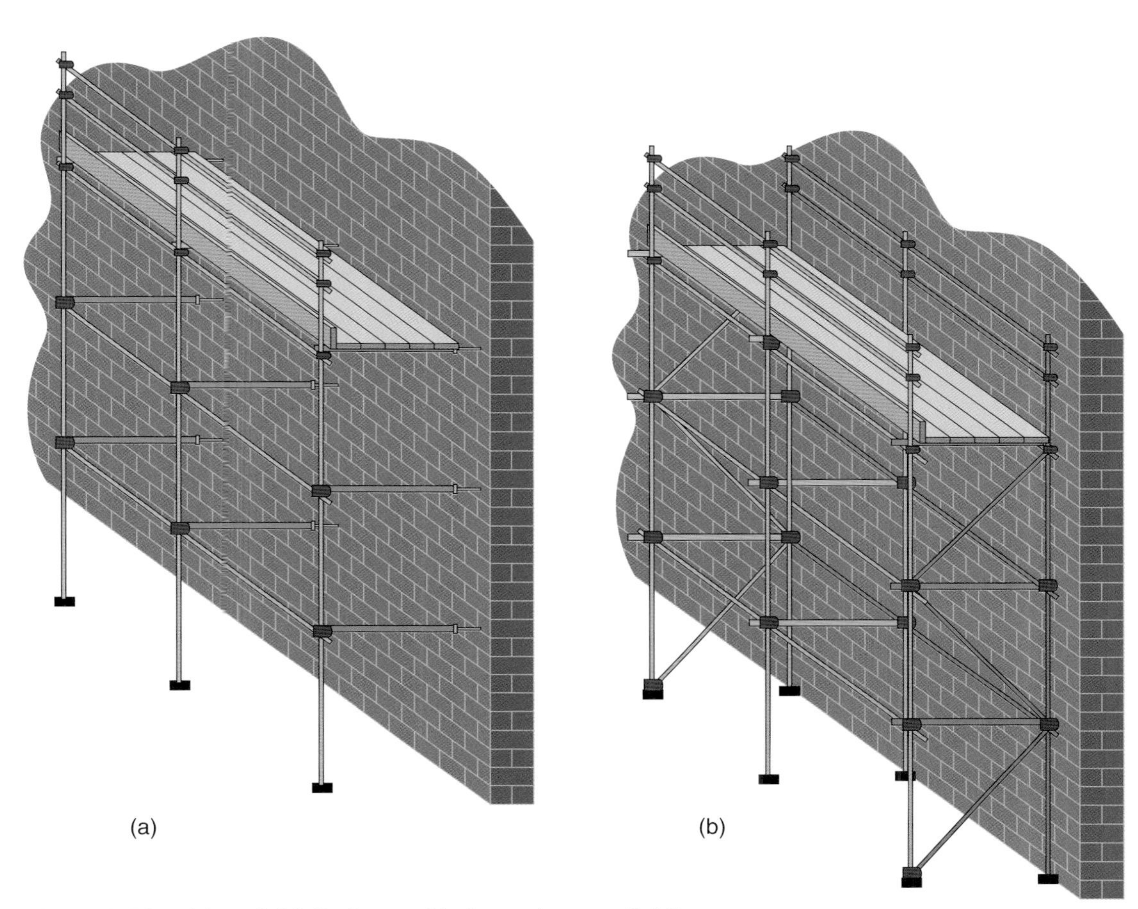

Figure 2.23 (a) and (b) Putlog and independent scaffolding

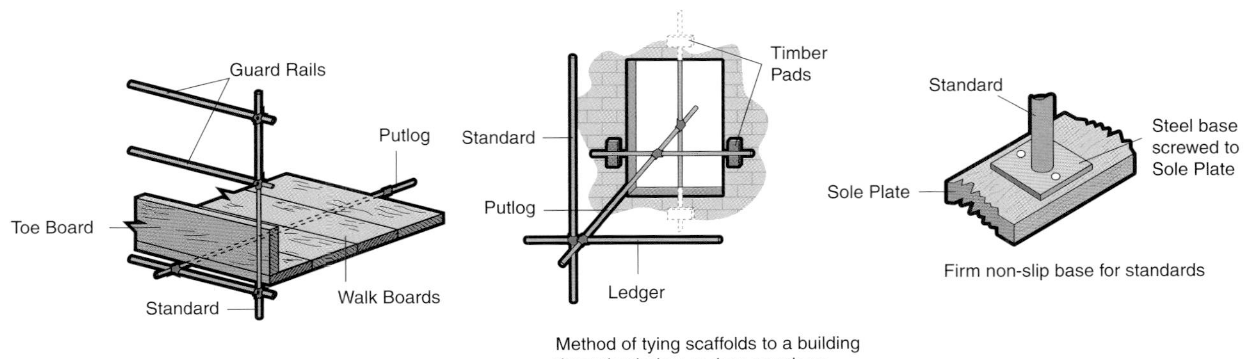

Figure 2.24 Scaffolding features

be working. Look for any gaps between boards, and make sure a toe board and guardrail are fitted.

Mobile scaffold towers

Mobile scaffold towers (Figure 2.25) may be constructed of basic scaffold components or may be specially designed 'proprietary' towers made from light alloy tube. The tower is built by slotting the sections together until the required height is reached. Mobile towers can be provided with wheels, or as static towers, fitted with base plates.

Before working with mobile tower scaffolds, it is very important to take into consideration the height of the tower. The taller the tower, the more likely it is to become unstable, where towers are to be used in exposed or external conditions, the height of the working platform must be no greater than 3 times the minimum base dimension.

(a)

(c)

(b)

Figure 2.25 Mobile scaffold tower in a framing centre. (a) Detail at head of tower; (b) Side elevation; (c) Detail at base of tower (wheels locked).

If the tower is to be used internally on firm level ground, the height ratio can be increased to 3.5 that of the base dimension. A typical example would be a tower which has a base dimension of 2 m × 3 m, so the maximum would be 6 m for external use and 7 m for internal use.

When working with mobile tower scaffolds, the following points must be followed:

- If the working platform of a tower scaffold is above 2 m from the ground it must be fitted with guardrails and toe boards
- The guardrail, also known as a 'knee rail', must be between 0.4 and 0.7 m from the working platform and must consist of two horizontal bars
- There must also be a handrail which must be no more than 910 mm from the working platform
- When the platform is being used, all four wheels must be locked
- The platform must *never* be moved unless it is clear of tools, equipment and people. It should be pushed at the bottom of the base.
- The stability of a tower depends upon the ratio of the base width to height. A ratio of base to height of 1:3 gives good stability.

Key point

Never attempt to work off a scaffold or tower scaffold if you are uncertain about its safety. If in doubt contact your supervisor.

Outriggers can increase stability by effectively increasing the area of the base but if used, must be fitted diagonally across all four corners of the tower and not on one side only. When outriggers are used they should be clearly marked (e.g. with hazard marking tape) to indicate a trip hazard is present

- Towers taller than 9 m should be secured to the building
- Towers must not be built taller than 12 m unless they have been especially designed for that purpose
- Access to the working platform of the tower should be by a ladder securely fastened inside the tower. Ladders must never be leaned against a tower as this may push the tower over.
- Before you use a tower scaffold on a pavement, check whether you will need a 'pavement licence' from the local authority.
- Always inspect the tower scaffold before first use, and after any substantial alteration and also after any event that is likely to have affected its stability.

The height of a mobile tower must not be more than 12 m (unless it is specially designed) and any tower over 9 m in height must be tied to the building.

When using a roof ladder, you should check the following:

- Stiles must be straight and in sound condition
- Rungs must be in good sound condition
- The roof ladder hook must be firmly fixed over the ridge
- Wheels of the roof ladder must be firmly fixed and run freely
- Pressure plates must be sound (they are the bits that rest on the roof surface)

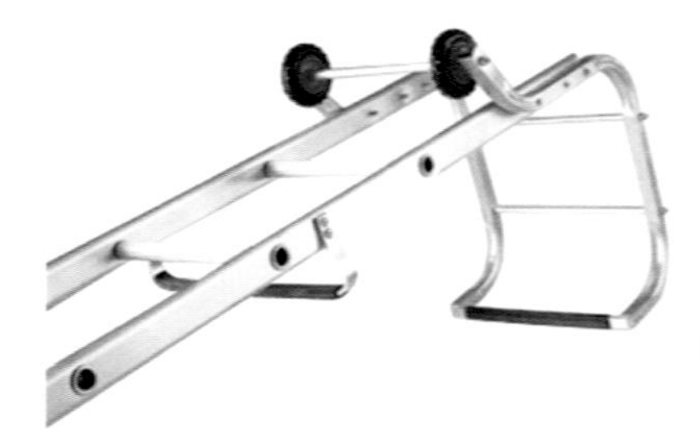

Figure 2.26 Roof ladder

Activity 2.2

Figure 2.26 is what a roof ladder looks like. When would a plumber have to use a roof ladder? Write down some typical situations, then check your answer against the suggested one at the end of this book.

Working in excavations

As a plumber you may occasionally be required to work in excavations. When working in or around the excavations, it is necessary to work in accordance with the specific Health and Safety guidelines that apply. The Health and Safety legislation that exists is mainly designed to prevent the following eventualities:

- Collapse of the excavation walls
- Persons falling into the excavation.

Excavations are often classed as confined spaces and control measures may include a permit to work system.

Hot tips

For further guidance on basic safety when working in excavations, check out the HSE publication Construction Information Sheet No 8 – www.hse.gov.uk.

Wall collapse prevention

Material from an excavation must be piled a safe distance away to ensure that it does not begin to spill back in, risking the safety of those who may be working there. When an excavation is deeper than 1.2 m the sides should be either sloped or shored. Where shoring is required strong wooden timbers or steel plates should be used (steel plates are usually found in deeper excavations). The amount of these materials required will normally depend upon the

Key point

Note on Liquefied Petroleum Gas (LPG):

LPG is heavier than air; this is a particular danger when working in excavations. If LPG is being used in an excavation, any leakage could result in the formation of pockets of the highly explosive gas – potentially a very dangerous situation. You will learn more about LPG later in this unit.

soil/ground type in a particular area or the depth of the excavation. It is recommended that excavations should be inspected by a competent person:

- At the start of each shift before work begins
- After any event likely to have affected the strength or stability of the excavation
- After any accidental fall of rock, earth or other material.

Fall prevention

Fencing must be erected around any excavation into which someone could fall more than 2 m, but it stands to reason that precautions should also be taken for holes/trenches less than this depth in order to protect the safety of workers and the public. If you are working in flats under construction, there should be fencing around stairwells and lift shafts if there is a danger of people falling. You may not be directly responsible for erecting the barriers, but you should be aware of situations in which they should be present.

Test yourself 2.3

1. Which of the following would not be a necessary safety check on a ladder?
   a. No damage to toe boards ☐
   b. No dirt on rungs ☐
   c. No damage to tie-rods ☐
   d. No splits in rungs ☐
2. What is the correct angle that a ladder should be in relation to the ground?
   a. 65° ☐
   b. 45° ☐
   c. 55° ☐
   d. 75° ☐
3. When a ladder provides access to a working platform how much of the ladder must extend above the access point?
   a. 4 rungs ☐
   b. 3 rungs ☐
   c. 5 rungs ☐
   d. 6 rungs ☐
4. The two long sections of a ladder, which hold the rungs together, are known as
   a. Tie-rods ☐
   b. Stiles ☐
   c. Toe boards ☐
   d. Scaffold boards ☐
5. Trestle platforms should be at least two boards wide or how many mm?
   a. 550 ☐
   b. 350 ☐
   c. 250 ☐
   d. 450 ☐
6. What is the usual maximum length of a scaffold board?
   a. 4 m ☐
   b. 5 m ☐
   c. 6 m ☐
   d. 7 m ☐

(*Continued*)

UNIT 2

Test yourself 2.3 (Continued)

7. Is this statement true or false?
   An independent scaffold uses putlogs to fix it to the building.

8. What should the end of the standard be placed on to help the stability of the scaffold?
   a. Kicker plate ☐
   b. Flat plate ☐
   c. Back plate ☐
   d. Base plate ☐

9. Name the component parts of this scaffold, as shown in Figure 2.27.

   a.
   b.
   c.
   d.
   e.

10. Which of the following should have toe boards fitted?
    a. Mobile scaffold ☐
    b. Trestle scaffold ☐
    c. Roof ladder ☐
    d. Step ladder ☐

11. What is the maximum height a tower scaffold that can be erected before it has to be secured to a building?
    a. 8 m ☐
    b. 7 m ☐
    c. 9 m ☐
    d. 6 m ☐

Figure 2.27 Typical scaffolding erection

12. State three safety checks when using roof ladders.

13. On tower scaffolds, which of the following ratios of base to height would ensure it to be most stable for safe working?
    a. 1 to 4 ☐
    b. 1 to 5 ☐
    c. 1 to 3 ☐
    d. 1 to 8 ☐

14. Excavations must have sloped or shored sides if they exceed what depth?
    a. 1 m ☐
    b. 1.2 m ☐
    c. 2 m ☐
    d. 2.2 m ☐

COSHH 2002

COSHH stands for the Control Of Substances Hazardous to Health. The Regulations were first introduced in 1988 and provided a new legal framework for the Control of Substances Hazardous to Health in all types of business, including factories, farms, offices, shops and plumbing installations.

The Regulations require that the employer makes an assessment of all work, which is liable to expose any employee to hazardous substances, solids, liquids, dusts, fumes, vapours, gases or micro-organisms, to prevent ill health. Any risks to health must be evaluated, and a decision must be taken on what action to take to remove or reduce the risks. For example, if you were replacing a cistern and associated pipework in an insulated roof space, you will expect that protective clothing and respiratory protective equipment would be provided to prevent injury to yourself whilst carrying out your work.

What is a substance hazardous to health?

Substances that are 'hazardous to health' include those labelled as dangerous (i.e. very toxic, toxic, harmful, irritant or corrosive). They also include micro-organisms and substantial quantities of dust and indeed any material, mixture or compound used at work, or arising from work activities, which can harm a person's health.

What are the requirements of COSHH?

- The risk to health arising from work related activities involving potentially hazardous substances must be assessed
- There must be appropriate measures introduced to prevent exposure to the hazardous material and to ensure precautions are taken to control the risk.

Control measures must be taken to ensure that equipment is properly maintained and safety procedures are observed.

- Employees must be informed and instructed (and where necessary receive specialist training) about the risks, associated with hazardous substances following all the necessary precautions as required.

Usually prevention will involve the use of protective clothing and equipment which could involve the use of respiratory equipment masks, dust masks, goggles, gloves or other equipment.

Hot tips

More information on COSHH regulations is available from the HSE document 'COSHH: a brief guide to the regulations' www.hse.gov.uk or contact HSE books on Tel: 01787 881165.

Chemical safety

Figure 2.28 shows chemical warning signs. Some chemicals can be harmful to the people working with them, if they accidentally come into contact with the body. Chemicals when not contained or handled properly can be:

- Inhaled as dust or gas
- Swallowed in small doses over a period of time (airborne vapour)

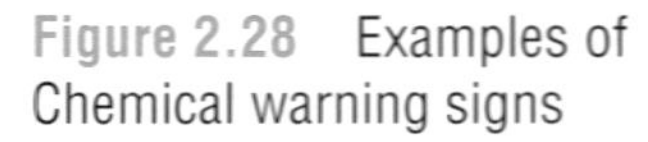
Figure 2.28 Examples of Chemical warning signs

UNIT 2

- Absorbed through skin or clothing
- Touched or spilled onto unprotected skin.

Some chemicals can also cause:

- Injury to eyes, skin, organs from fires, burns, etc
- Illness which can sometimes leave you feeling all right after the exposure but cause health problems after several months or years of exposure (e.g. cancer)
- Allergy to the skin, rash, respiratory problems
- Instant death – some poisonous chemicals can kill outright.

These are also risks from the following biological dangers when working on above/below ground waste systems:

- Tetanus
- Hepatitis (passed through human fluids)
- Leptospirosis (passed from rats)
- Psittacosis (from pigeon droppings)
- Tubercolosis

Asbestos dust safety

Asbestos is a mineral found in many rock formations. When separated from rock, it becomes a fluffy, fibrous material that has many uses. About two-thirds of all the asbestos produced was used in the construction industry. It was formerly used in cement production, flue pipes, gas appliance seals (ropes, etc), roofing, cold water tanks, gutters, rain water pipes and decorative plaster finishes. You may still discover asbestos materials in buildings built or refurbished before 1985. Although asbestos is a hazardous material it can only pose a risk to health if the asbestos fibres become airborne so that they can be inhaled into the lungs.

There is no cure for asbestos-related diseases. Breathing in asbestos fibres can lead to you developing one of three fatal diseases:

- Asbestosis – which is a scarring of the lung leading to shortness of breath
- Lung cancer
- Mesothelioma – a cancer of the lining around the lungs and stomach.

There are three main types of asbestos:

- Chrysotile – White (accounts for about 90% of asbestos in use)
- Amosite – Brown
- Crocidolite – Blue.

Asbestos cannot be identified by colour alone, a laboratory analysis is required to establish its type. Blue and brown are the two most dangerous forms of asbestos and have been banned from use since 1985. White asbestos has been banned from use since 1999.

To protect yourself

If you have been given the all clear signal to work on the job involving contact with asbestos, then you will need to work to the following guidelines:

Key point

Extensive work with, or the removal of asbestos must only be carried out by specialist licensed contractors.

- Keep asbestos materials damp when working on them
- Do not use power tools on asbestos materials as they create dust. Use hand tools instead.
- Always wear and maintain any personal protective equipment provided
- Make sure that you know how to correctly wear and use any personal protective equipment provided correctly
- Always practise good housekeeping. Only use special 'H-type' vacuums and dust collecting equipment.
- Asbestos waste must be suitably sealed in a container or double bagged in heavy duty polythene bags. These must be clearly labelled to show that they contain asbestos.
- Report any hazardous conditions, i.e. unusually high-dust levels to your supervisor.

Test yourself 2.4

1. The abbreviation COSHH means?
   a. Control of oxidising substances harmful to health ☐
   b. Control of substances harmful to health ☐
   c. Control of substances hazardous to health ☐
   d. Control of oxidising substances hazardous to health ☐
2. Which of the following labels *would not* be expected if the substance were hazardous to health?
   a. Very toxic ☐
   b. Irritant ☐
   c. Corrosive ☐
   d. Smells ☐
3. Asbestos dust can become a health hazard if it is . . .
   a. Washed from clothing ☐
   b. Inhaled into the lungs ☐
   c. Vacuumed away ☐
   d. Used in its solid form ☐
4. Asbestos waste must be . . .
   a. Disposed off with other general waste. ☐
   b. Taken back to the depot for disposal. ☐
   c. Swept down the drain. ☐
   d. Sealed in suitable containers and correctly labelled. ☐

Fire safety and emergencies

Introduction

All the topics covered within this module area are designed to keep you safe, but this is a very important session. Fire is a constant risk in any industry, but especially so for plumbers who will come into contact with electricity, gas and heating equipment.

Fire safety

Figure 2.29 The fire triangle

Fire or burning is the rapid combination of a fuel with oxygen (air) at high temperature. A fire can reach temperatures of up to 1000°C within a few minutes of starting. Figure 2.29 shows that for a fire to start, there are three requirements:

1. Combustible substances (called the fuel)
2. Oxygen (usually air)
3. Source of heat (spark, friction or match).

The fire triangle

Fires can spread quite rapidly and once established, even a small fire can generate sufficient heat energy to spread and accelerate

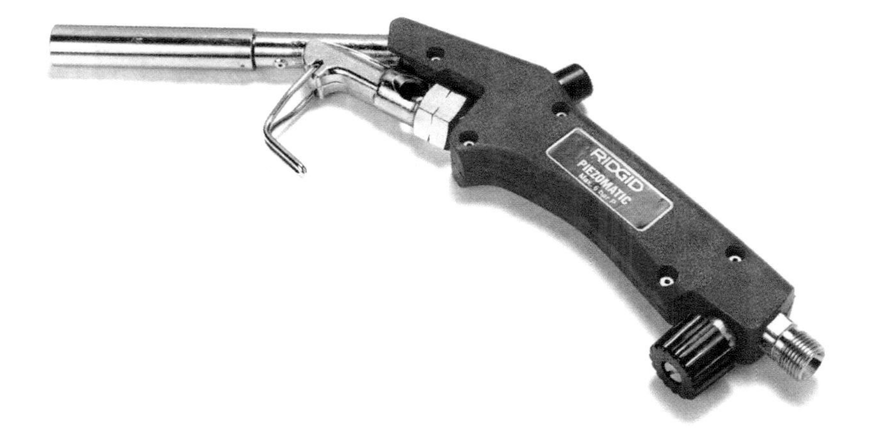

Figure 2.30 Blow torch (Reproduced with permission of Ridgid Tools)

the fire to surrounding combustible materials. Fire prevention is largely 'good housekeeping' and common sense.

Plumbing work is a particularly risky job in terms of potential fire hazards. Blow torches (see Figure 2.30) and welding equipment are often used near combustible materials, and in tight positions or areas with difficult access, which means that you are particularly vulnerable when using this equipment. As it is such a risky job, an employer is required to ensure strict working methods when hot working; this is not only common sense but also a requirement for insurance purposes. This must include:

- Fire extinguisher in the immediate work area
- Hot working must be completed a minimum of an hour before leaving the site.

The following points must be remembered when undertaking 'hot work':

- If you are working close to a combustible material (e.g. timber skirting, joists, etc) always make sure you protect the area around the fitting you are soldering with a heat-resistant mat. You can get these from your local merchant. If the pipe is insulated, remove the insulation from about 300 mm each side of the fitting from the area where you are using the torch.
- If you think you may have caught any material with the flame, wet the area, and check again after a few minutes to make sure it is OK. If you are working in someone's home, make sure you pull back carpets, or place curtains away from the area you are going to work in.

- Another major cause of fire is due to electrical faults. All alterations and repairs in electrical installations must only be carried out by a qualified person, and must be to the standards laid down in the IEE Regulations.
- Finally, construction sites can be dirty places. Timber shavings and other combustible material may find their way under floors, etc. Make sure that the area where you are using the blow torch is clean before you start.
- On occasions, you may be working in occupied buildings, such as office blocks or flats. You must be aware of their fire safety procedures, and be familiar with normal and alternate escape routes. You must also know where your assembly point is located and report your presence to your supervisor in case of evacuation due to fire.

Classes of fire

Fires are commonly classified into five groupings according to fuel type as listed below:

- Class A – Fires involving solid fuels – can be extinguished by water
- Class B – Fires involving flammable liquids – should be extinguished by foam or carbon dioxide
- Class C – Fires involving flammable gases – should be extinguished by dry powder
- Class D – Fires involving flammable metals – should be extinguished by dry powder
- Class E – Fires involving cooking oils and fats – should be extinguished by wet chemical, dry chemical or foam based with special additives.

Key point

Personal safety must always come before your efforts to contain a fire.

Fire fighting equipment

If it is a small fire, it could be put out quickly and safely. However, your efforts to contain a fire that is getting out of control could prove fruitless, so when you finally decide to escape, you could have difficulty in finding your way through the smoke and the fumes. Remember, smoke and fumes from a fire are just as lethal as the fire itself. Fire fighting equipment, including extinguishers (see Figure 2.31), fire buckets of sand, or water and fire-resistant blankets should be readily available in buildings. In larger premises you will find automatic sprinklers, hose reels and hydrant systems.

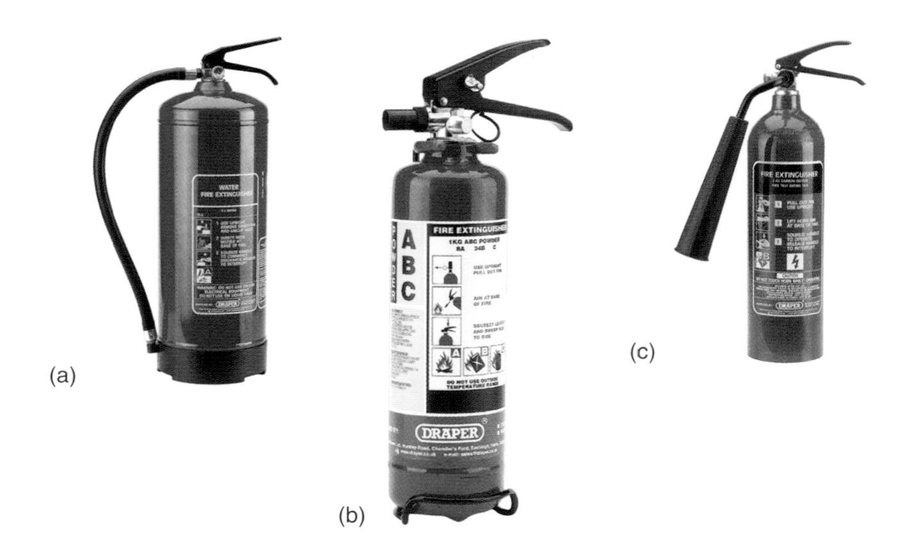

Figure 2.21 Fire extinguishers (Reproduced with permission of Draper Tools Limited)

Table 2.1

Type of extinguisher	Colour	Main use
Water	Red	Wood, paper or fabrics.
Foam	White or cream	Petrol, oil, fats and paints.
Carbon dioxide	Black	Electrical equipment.
Dry powder	Blue	Liquids, gases, electrical equipment.
BCF halon	Green	Motor vehicles and electrical equipment.

Table 2.1 shows different types of fire extinguishers and the use to which they should be put.

Note: The colour coding given may not be immediately obvious. Older extinguishers may have their whole body painted in the appropriate colour (e.g. black ones are filled with carbon dioxide). Recent European legislation dictates that new extinguishers must be coloured red whatever substance they contain, but will carry 5% of the colour if the extinguisher would have been in the original system, e.g. a carbon dioxide extinguisher will be red with a black stripe, triangle or lettering. Care should, therefore, be taken when choosing an extinguisher that the correct type is being used prior to attempting to extinguish a fire.

Key point

Always read the operating instructions on the extinguisher before use.

Here are a few safety points to remember:

- Never use a fire extinguisher unless you have been trained to do so
- Do not use water or water-type extinguishers on electrical fires due to the risk of electric shock and explosion
- Do not use water or water-type extinguishers on oils or fats, as this too can cause an explosion
- Do not handle the nozzle when using carbon dioxide extinguishers as this can cause freeze burns to the hands
- Do not use the carbon dioxide extinguisher or halon types in a confined space or small room, as this could cause suffocation.

Accidents

You will receive some basic first aid training when in the centre, but here are some important pointers when dealing with accidents.

First actions

No matter how many safety precautions are taken or observed, sooner or later you are more than likely to witness an accident. Whatever the accident be, i.e. a cut finger or a fall from a ladder, you must know what to do. Consider a situation where you are working with another plumber who has just cut his hand, and the victim is showing signs of distress. What are you going to do? More importantly, what are you going to do first?

There are many things that need to be done following an accident:

- Seek or administer immediate first aid
- Get help if necessary, i.e. phone for an ambulance
- Report the accident to the site supervisor
- Write down the details in an accident report book
- Complete a company accident report form.

Which one of these things will you carry out first will depend on several things:

- Are you a qualified first aider?
- How severe is the injury?
- Is your qualified first aider or supervisor immediately available?

Figure 2.32 The recovery position

There are several important things to remember. In this case, we are dealing with a cut hand; the patient is bleeding quite heavily so it would be necessary to administer first aid as soon as possible.

However, if we were dealing with, for example, an electric shock casualty you must first make sure that the area is safe, that is to ensure that whatever caused the accident is not going to injure you. In this case, you must first isolate the supply of electricity, or break the electrical contact, then immediately attend to the casualty. Call an ambulance if necessary, and inform your supervisor.

As we have said already, people falling from heights is one of the major causes of injury in the building industry. If you witness this and the person is unconscious, put them in the recovery position. If they are conscious, make them as comfortable as possible until help arrives, this may include treating minor wounds. Do not move the patient. If they have seriously injured themselves your actions could make things worse.

Figure 2.32 shows what the recovery position looks like – there is more about this in the next section.

Accident reporting

If you are injured in an accident you will have to enter the details in the accident book. You may also have to enter details on a casualty's behalf if they are unable to do so themselves. If the accident involved a piece of faulty equipment, do not tamper with it, it may be subject to an inspection from the Health and Safety Inspectorate.

Full name of injured person	
Home address	Sex Male/Female Age
Status Employee Contractor Visitor	
Date of accident	
Time of accident	
Precise location	
What was the accident and its cause? (You may have to give a detailed written description.) Name and address of witness if any	
Details of apparent injuries	
Summary	

Figure 2.33 Accident report form

Accidents that result in three or more days of absence from work must be reported to the Health and Safety Executive. Figure 2.33 shows a typical example of an accident report form.

Remember, after an accident – seek first aid, inform your supervisor, complete the accident report form.

Reporting of Injury, Diseases and Dangerous Occurrences Regulations (RIDDOR)

The RIDDOR regulations are commonly referred to as RIDDOR 95 and stands for the Reporting of Injuries, Diseases and Dangerous Occurrences Regulations 1995, which came into force on 1 April 1996.

HSE Guide to Riddor Booklet L73
(HSE Books) P.O. Box 1999
Sudbury, Suffolk
CO10 2WA
Tel: 01787 881165
Fax: 01787 313995

RIDDOR 95 requires the reporting of work-related accidents, diseases and dangerous occurrences. The regulations apply to all the work activities, but not to all incidents. The reporting of accidents and ill health at work is a legal requirement, and the information enables the enforcing authorities to identify and investigate how risks and serious accidents arise.

Key point

An act of aggression by a colleague whilst at the centre or at work, which causes a situation of conflict, would also be considered a dangerous occurrence.

What to report

As a plumber, relevant examples that should be reported are:

- Injury lasting more than three days: In a work-related injury that lasts for more than three days (including non-working days), a written accident report (F2508) form must be submitted to the enforcing authority within ten days of occurrence
- Disease: If a doctor informs an employee of a work-related disease that is reportable, a written disease report (F2508A) form must be submitted to the enforcing authority
- Dangerous occurrence: If an occurrence happens, which does not result in a reportable injury, but clearly could have been done, this can be categorised as a dangerous occurrence which must be reported immediately. Also, a written report form must be submitted to the enforcing authority within ten days of occurrence.

Here are a few examples of reportable major injuries:

- Fracture other than to fingers, thumbs or toes
- Amputation
- Loss of sight (temporary or permanent)
- Dislocation of the shoulder, hip, knee or spine
- Hot metal or chemical burn to the eye or any penetrating injury to the eye.

Here are a few examples of reportable dangerous occurrences:

- Explosion, collapse or bursting of any closed vessel or associated pipework
- Collapse, overturning or failure of load-bearing parts of lifts and lifting equipment
- Collapse or partial collapse of a scaffold over five metres high, or erected near water where there could be a risk of drowning after a fall
- Plant or equipment coming into contact with overhead power lines
- Accidental release of a biological agent that is likely to cause severe illness to humans.

Here are a few examples of reportable diseases:

- Skin diseases, such as occupational dermatitis, skin cancer, oil folliculitis – acne, chrome ulcer
- Certain types of poisonings
- Lung diseases including occupational asthma, asbestosis, mesothelioma, pneumoconiosis
- Infections, such as legionellosis, tetanus, leptospirosis, hepatitis and tuberculosis
- Other conditions, such as occupational cancer, types of musculoskeletal disorders and hand-arm vibration syndrome.

Test yourself 2.5

1. When dealing with a fire in an open oil storage tank, which extinguisher would be most suitable?
   a. Foam ☐
   b. Water ☐
   c. BCF halon ☐
   d. Carbon dioxide ☐
2. Under the RIDDOR Regulations which of the following is not a reportable occurrence?
   a. A major injury to a person ☐
   b. A fatal injury to a person ☐
   c. An accident at home ☐
   d. An explosion of a storage vessel ☐
3. The three elements that all fires require are?
   a. Fuel, oxygen and heat ☐
   b. Fuel, heat and paper ☐
   c. Heat, air and oxygen ☐
   d. Fuel, air and oxygen ☐
4. European legislation requires fire extinguishers to be of a certain colour no matter what their intended uses. What is this colour?
   a. Cream ☐
   b. Black ☐
   c. Red ☐
   d. Silver ☐
5. On an accident report form which one of the following is not required?
   a. Name of the parent or guardian ☐
   b. Name of the injured person ☐
   c. Location of the accident ☐
   d. Date of the accident ☐
6. Which of the following types of fire extinguisher must not be used on live electrical equipment if there is a fire?
   a. Water-type ☐
   b. Carbon dioxide ☐
   c. Dry powder ☐
   d. BCF halon ☐

Basic first aid procedures and other plumbing-related safety issues

Introduction

In this section, you will be looking first at some of the steps you should be capable of taking to treat injured persons at work. You will be looking at resuscitation procedures and basic first aid principles for the treatment of burns, shock and breaks/fractures. Also covered in this section are safety considerations that should be observed when working with LPG and sheet lead.

The aim of the first part of this section is not to turn you into a fully proficient first aider, but to give you the basic knowledge you will need if you ever find yourself in a situation where you have to administer emergency first aid, which could ultimately save someone's life. The following steps are laid out to cover some of the situations in which you may encounter a work colleague who requires first aid. We will begin with what to do if you come across a colleague who appears to be unconscious:

Surveying casualties and resuscitation procedures

1. *Check for danger*
   Are you or the casualty in any danger? If you have not already done so, make the situation safe and then assess the casualty, as shown in Figure 2.34.
2. *Check the casualty for consciousness*
   If the casualty appears unconscious check this out by shouting: 'Can you hear me?', 'Open your eyes' and gently shaking their shoulders. If there is no response, shout for help then follow the Airway, Breathing and Circulation (ABC) procedure outlined below:
3. *Airway*
   Open the airway by placing one hand on the casualty's forehead and gently tilting the head back. Check the mouth for obstructions and then lift the chin using two fingers only, as shown in Figure 2.35.
4. *Breathing*
   Spend 10 s to see if the casualty is breathing:

   - Look to see if the chest is rising and falling. Listen for breathing.
   - Feel for breath against your cheek
   - If the casualty is breathing, place them in the recovery position.
   - Check for other life-threatening conditions
   - For casualty's who are not breathing, follow the guidelines given at 6.

Figure 2.34 Check casualty for consciousness

Figure 2.35 Open the airway

5. *Circulation*
   Spend 10 s checking for signs of circulation: look, listen and feel for breathing, coughing, movement or any other signs of life.
6. *Decision making*
   Depending on the first five steps your next actions should be based on the following criteria. If the casualty is:

Conscious and breathing

- Check circulation (including a check for severe bleeding)
- Treat any injuries
- Get help if necessary.

Unconscious but breathing

- Place the casualty in the Recovery Position
- Check circulation (including a check for severe bleeding)
- Treat any life-threatening conditions
- Call for an ambulance.

Unconscious, not breathing but has circulation

- If the condition is due to injury, drowning or choking:
  – Give 10 rescue breaths
  – Call for an ambulance
  – On return to casualty follow the resuscitation sequence again, acting on your findings.
- If the condition is not due to injury, drowning or choking:
  – Call for an ambulance
  – On return to casualty follow the resuscitation sequence again, acting on your findings.
- Check for circulation every 10 breaths.

Unconscious, not breathing and has no circulation

- If the condition is due to injury, drowning or choking:
  – Give chest compressions together with Rescue Breaths (CPR) for 1 min
  – Call an ambulance, then return to casualty and follow resuscitation sequence again, acting on your findings.
- If circulation is absent, and the condition is not due to injury, drowning or choking:
  – Call an ambulance, then return to casualty and follow resuscitation sequence again, acting on your findings.
- Continue to give chest compressions together with CPR until help arrives.

The checking procedure makes reference to three important first aid procedures that you should be familiar with.

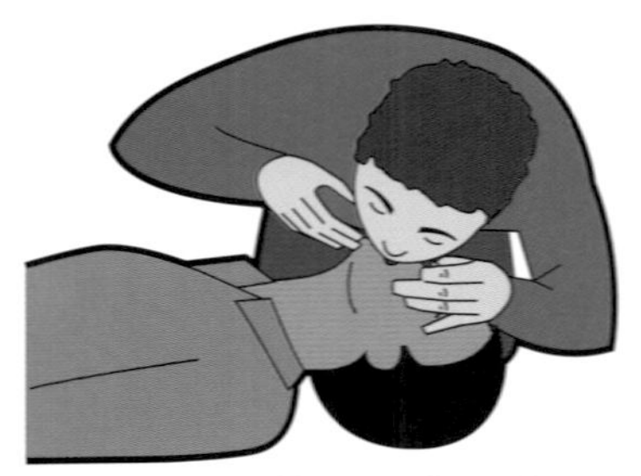

Figure 2.36 Rescue breaths

Rescue breaths

Rescue breaths should be administered in the following way if a casualty is not breathing (Figure 2.36):

- Ensure the airway is open
- Pinch nose firmly closed
- Take a deep breath and seal your lips around the casualty's mouth
- Blow into the mouth until the chest rises
- Remove your mouth and allow the chest to fall
- Repeat once more then check for circulation
  - If circulation is absent commence Chest Compressions (CPR).
- Check for circulation after every 10 breaths
- If breathing starts, place in Recovery Position.

Chest compressions

Chest compressions should be administered to the casualty with no circulation, as shown in Figure 2.37.

Note: Chest Compressions must always be combined with Rescue Breaths.

- Place the heel of your hand two-fingers width above the junction of the casualty's rib margin and breastbone
- Place the other hand on top and interlock fingers. Keeping your arms straight and your fingers off the chest, press down by 4–5 cm; then release the pressure, keeping your hands in place.
- Repeat the compressions 15 times, aiming at a rate of 100/min
- Give two Rescue Breaths

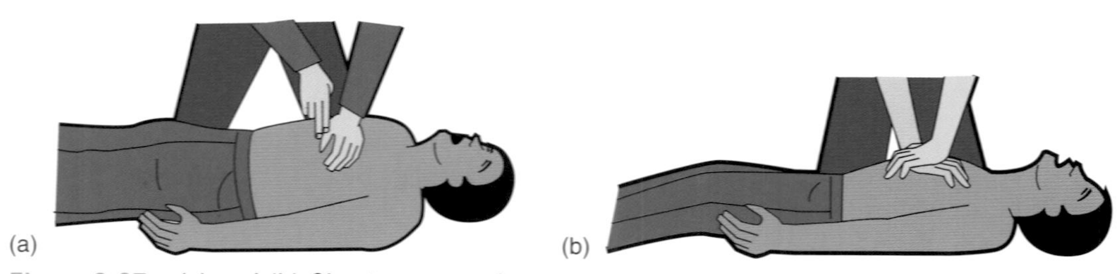

Figure 2.37 (a) and (b) Chest compressions

- Continue resuscitation, 15 compressions to two Rescue Breaths
- Only check for circulation if the casualty's colour improves.

If circulation is present, stop the Chest Compressions but continue Rescue Breaths if necessary.

Recovery position

An unconscious casualty who is breathing but has no other life-threatening conditions should be placed in the Recovery Position, as shown in Figure 2.38.

- Turn casualty onto their side
- Lift chin forward in an open airway position and adjust the hand under the cheek as necessary
- Check if casualty cannot roll forwards or backwards
- Monitor breathing and pulse continuously
- If injuries allow, turn the casualty to the other side after 30 min.

Note: If you suspect spinal injury, use the jaw-thrust technique. Place your hands on either side of the casualty's face. With your fingertips gently lift the jaw to open the airway. Take care not to tilt the casualty's neck.

Other basic first aid procedures

It is likely that the majority of first aid incidents you will encounter will be to do with cuts and scrapes. Much of the guidance given below is common sense, but there is also useful advice for the treatment for cuts, burns and breaks.

Figure 2.38 The recovery position

Bleeding

Minor cuts, scratches and grazes

- Wash and dry your own hands
- Cover any cuts on your own hands and put on disposable gloves
- Clean the cut, if dirty, under running water. Pat dry with a sterile dressing or clean lint-free material. If possible, raise the affected area above the heart.

Cover the cut temporarily while you clean the surrounding skin with soap and water and pat the surrounding skin dry. Cover the cut completely with a sterile dressing or plaster.

> **Key point**
>
> **Remember:** Protect yourself from infection by wearing disposable gloves and covering any wounds on your hands. If blood comes through the dressing **DO NOT** remove it – bandage another over the original. If blood seeps through **BOTH** the dressings, remove them both and replace with a fresh dressing, applying pressure over the site of bleeding.

Severe bleeding

- Put on disposable gloves
- Apply direct pressure to the wound with a pad (e.g. a clean cloth) or fingers until a sterile dressing is available
- Raise and support the injured limb. Take particular care if you suspect a bone has been broken.
- Lay the casualty down to treat for shock
- Bandage the pad or dressing firmly to control bleeding, but not so tightly that it stops the circulation to fingers or toes. If bleeding seeps through first bandage, cover with a second bandage. If bleeding continues to seep through bandage, remove it and reapply.
- Treat for shock
- Dial 999 for an ambulance.

Burns and scalds

Severe burns

- Start cooling the burn immediately under running water for at least 10 min
- Dial 999 for an ambulance
- Make the casualty as comfortable as possible, lie them down
- Continue to pour copious amounts of cold water over the burn for at least 10 min or until the pain is relieved
- Whilst wearing disposable gloves, remove jewellery, watch or clothing from the affected area – unless it is sticking to the skin.

Cover the burn with clean, non-fluffy material to protect from infection. Cloth, a clean plastic bag or kitchen film all make good dressings.

Minor burns

For minor burns, hold the affected area under cold water for at least 10 min or until the pain subsides. Remove jewellery, etc and cover the burn as detailed above.

If a minor burn is larger than a postage stamp it requires medical attention. All deep burns of any size require urgent hospital treatment.

On ALL burns DO NOT

- Use lotions, ointments and creams
- Use adhesive dressings
- Break blisters.

Suspected breaks/fractures

- Give lots of comfort and reassurance and persuade the casualty to stay still
- Do not move the casualty unless you have to
- Steady and support the injured limb with your hands to stop any movement
- If there is bleeding, press a clean pad over the wound to control the flow of blood. Then bandage on and around the wound.
- If you suspect a broken leg, put padding between the knees and ankles. Form a splint (to immobilise the leg further) by gently but firmly bandaging the good leg to the bad one at the knees and ankles, then above and below the injury. If it is an arm that is broken, improvise a sling to support the arm close to the body.
- Dial 999 for an ambulance
- If it does not distress the casualty too much, raise and support the injured limb
- Do not give the casualty anything to eat or drink in case surgery is necessary.

First aid kits

Employers should provide first aid kits (Figure 2.39) based on a risk assessment of its activities. The HSE suggest that a basic first aid kit should contain:

- Twenty adhesive dressings (assorted sizes); two sterile eye pads; six triangular bandages; six un-medicated wound dressings 10 cm × 8 cm; two un-medicated wound dressings 13 cm × 9 cm; six safety pins; three extra sterile un-medicated wound dressings 28 cm × 17.5 cm.

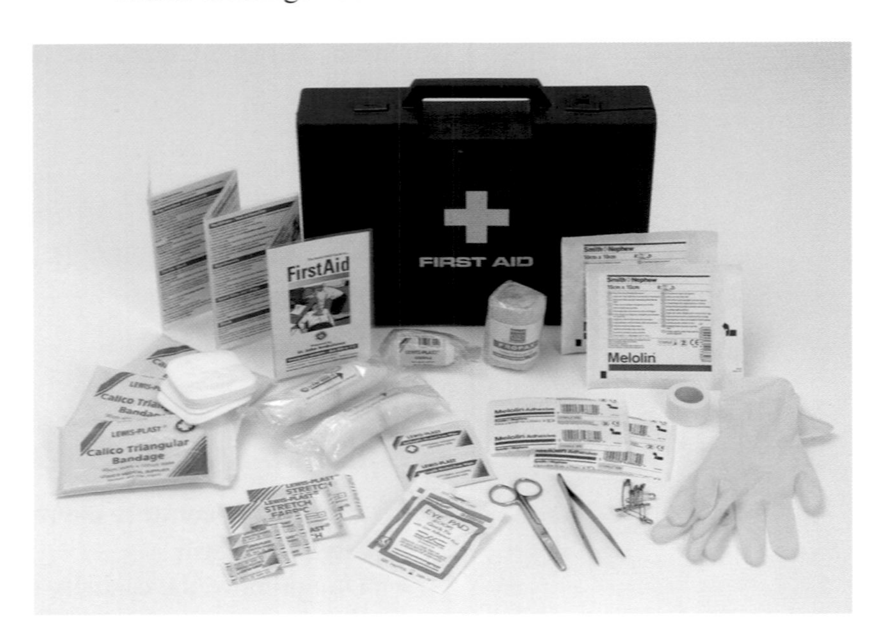

Figure 2.39 Typical first aid kit

> **Key point**
>
> When mixed with air the gas can burn or explode when it meets a source of ignition, even small quantities of LPG gas mixed with air create an explosive mixture. If 1 l of the gas is boiled or evaporated it becomes 250 l of gas (LPG is a gas above minus 42°C). This is enough to make an explosive mixture in a shed, room, store or office.

LPG

Gases that are in any concentration of commercial propane or butane and used in heating processes are generally referred to as LPG. LPG is a colourless, odourless liquid which readily evaporates into gas, and has an odour added to help detect leaks.

LPG is heavier than air having a specific gravity between 1.5 and 2.0 (air being 1) and if released will sink towards the ground. It is able to flow for long distances along the ground, and can collect in cellars, drains and gullies.

These gases are widely used in construction and building work as a fuel for burners, cookers, heaters and gas torches. The liquid, which comes in cylinders and containers, is non-toxic, but is highly flammable and needs careful handling and storage. If leaked onto the skin, liquid LPG can cause cold burns.

Transportation of LPG

Transporting LPG has to meet the requirements of Health and Safety legislation. This covers making sure that:

- The vehicle carrying LPG cylinders is adequately ventilated (in the case of enclosed or box van-type vehicles)
- Information/warning signs advising what the vehicle is carrying is clearly displayed on the outside of the vehicle
- The bottles are secured to prevent them from moving round, preventing risk damage
- Two fire extinguishers are carried in the vehicle (foam or carbon dioxide) and a first aid pack.

Safety precautions for sheet lead work

The safety precautions that must be observed when working with sheet lead will be covered in more detail during the chapter on lead working. However, it is necessary for you to be aware of the hazards involved when working with lead from the outset. The following lists will give you a basic appreciation of some of the safety considerations that must be observed when carrying out sheet lead work.

Safe sheet lead working is covered under the Control of Lead at Work Regulations, these regulations split sheet lead work into two categories:

- Handling tasks – controlled by simple protective methods designed to prevent lead entering the body through the skin, mouth and eyes
- Heating process tasks – require more stringent protective methods as poisonous fumes are given off when lead is heated (as it is during lead-welding activities).

Handling sheet lead safely

- Wear gloves and barrier cream
- Wear eye protection
- Wear overalls

- Do not smoke while working with sheet lead
- Take rest periods well away from worksite.

Key point

It is also imperative that you remember to wash hands, arms and face after a lead-working session. Ingestion or inhalation of lead particles can lead to chronic illness.

Heating sheet lead safely

In addition to the above protective methods

- Use either:
  - Respiratory protective equipment
  - Local extraction ventilation systems.

Test yourself 2.6

1. When 1 l of LPG is boiled or evaporated, it becomes how many litres of gas?
   a. 150 l ☐
   b. 250 l ☐
   c. 350 l ☐
   d. 450 l ☐
2. When working with LPG in cellars, basements, cellars or trenches extra care should be taken because:
   a. The gas could make the trench walls unstable ☐
   b. LPG is stronger in trenches ☐
   c. LPG is heavier than air and a leak can sink to high-concentration levels ☐
   d. The pressure below ground is greater on the tank ☐
3. It is a requirement that where LPG cylinders are being transported in an enclosed van:
   a. Two fire extinguishers must be carried ☐
   b. The driver must have a current HGV licence ☐
   c. The vehicle must carry at least one fire extinguisher ☐
   d. The driver must be above the age of eighteen years ☐
4. What acronym is used to describe the resuscitation procedure?
   a. XYZ ☐
   b. ABC ☐
   c. EZE ☐
   d. CAT ☐
5. What does the acronym stand for?
6. If the casualty was unconscious and lying on the floor what should you check for next?
   a. Colour of skin ☐
   b. Pulse ☐
   c. Bleeding ☐
   d. Breathing ☐
7. What is the correct immediate treatment for a burn?
   a. Remove clothing ☐
   b. Check for pulse ☐
   c. Flood the burn with cold water ☐
   d. Treat for shock ☐

Regulations and the safety of others

Although 'Regulations' is not the most exciting of subject titles, knowing about regulations will form an important part of your job. The following definitions are often used in regulatory documents. Before we get into the regulations in detail, it is important for you to have an understanding of what the terms mean and how they are relevant to persons who have the responsibilities under these regulations.

Key point

The Water Regulations are covered in detail in Unit 5 Cold Water Supply.

1. *Absolute/reasonably practicable*
   Duties in some regulations have a qualifying term called 'reasonably practicable'. Where qualifying terms are absent the requirement in the regulation is said to be absolute. The meaning of 'reasonably practicable' has been well established in law.
2. *Absolute*
   If the requirement in a regulation is 'absolute', for example if the requirement is not qualified by the words 'so far as is reasonably practicable', the requirement must be met regardless of cost or any other consideration.
3. *Reasonably practicable*
   A person who is required to do something so far as is 'reasonably practicable' must consider the amount of risk involved for the particular work activity or task, the costs in terms of the physical difficulty, time, trouble and expense which would be involved in taking steps to reduce the risks to Health and Safety of a particular work process.

For example, in your own home you would expect to find a fireguard in front of a fire to prevent young children from touching the fire and being injured. This is a cheap and effective way of preventing accidents and would be a reasonably practicable situation.

UNIT 2

The Workplace Regulations (1992)

Hot tips

Don't forget that the HSE is a rich resource of information, including information on regulations.

These Regulations apply to all workplaces. The requirements of these Regulations are imposed upon every employer or any person who has, to any extent, control of a workplace. They do not, however, cover domestic premises. The Workplace Regulations include requirements that affect the:

- Working environment
- Safety
- Welfare facilities.

These Regulations may be cited as the Workplace (Health, Safety and Welfare) Regulations 1992.

Working environment

Regulations 5 to 11 from the Health, Safety and Welfare Regulations 1992 require that adequate, effective and suitable provision is maintained in the following areas: (5) maintenance of workplace, e.g. equipment, devices and systems, (6) ventilation, (7) temperatures (internal), (8) lighting, (9) cleanliness, (10) room dimensions, (11) workstations and seating:

- The workplace and the equipment, devices and systems shall be maintained; this includes cleaning as appropriate, keeping in efficient state and working order
- The temperature of indoor workplaces shall be reasonable (approximately 16°C for offices/13°C for manual work) during all working hours, even though this may require the provision of a suitable method of heating and/or cooling the workplace
- The ventilation shall be effective and suitable to ensure that every enclosed workplace is ventilated by sufficient quantities of fresh or purified air.

Safety

Regulations 12 to 19 : (12) floor conditions – traffic routes, (13) falls or falling objects, (14 to 16) windows, (17) organisation of traffic routes, (18) doors and gates, (19) escalators, require that:

- Every floor and surface area be suitable for the purpose for which it is used
- No floor should be slippery, uneven, have holes or a slope, which may expose any person to a Health and Safety risk
- Where necessary every floor must have a means of drainage
- Measures are taken to ensure persons are protected from falling objects and from falling from height
- Workplaces must be organised in such a way that pedestrians and vehicles are separated and can circulate in a safe manner
- Escalators and moving walkways are to be equipped with any necessary safety devices and should be fitted with one or more emergency stop controls, which are easily identifiable and readily accessible.

Facilities

Regulations 20 to 25, (20) sanitary conveniences, (21) washing facilities, (22) drinking water, (23) accommodation for clothing, (24) facilities for changing clothing of which (25) facilities for rest and to eat meals, the general requirements of which are that suitable toilets are provided in easily accessible places. A supply of hot and cold water, soap and towels should be provided and also a supply of clean drinking water should be accessible to all persons within the workplace.

Activity 2.3

What do you already know about HSWA? Jot down here to whom it applies and the key points (you should already be familiar with some aspects of the HSWA from Unit 1). Then check your answer against the one suggested at the end of this book.

The Health and Safety at Work Act (1974)

The HSWA is a crucial piece of legislation which will be referred to throughout this section.

The rules that govern your own and your employer's Health and Safety responsibilities come directly from the Health and Safety at Work Act (HSWA) of 1974. Although we have talked about HSWA already, we are going to concentrate on the requirements and duties that the Act places on employers and employees in greater detail.

Key point

It is an offence under the Act to misuse or interfere with equipment provided for your Health and Safety or the Health and Safety of others. Employers provide you with safety equipment and should instruct you about its use. Your employer also has a duty to ensure that the safety equipment provided is kept in good and safe condition, but they cannot cater for human error or negligence.

Under the HSWA, you, as the employee, can be prosecuted for breaking safety laws. You are legally bound to co-operate with your employer to ensure your company complies with the requirements of the Act. It is the legal responsibility of all workers to take reasonable care of their own health and safety and to ensure that they act in a responsible manner so as not to endanger other workers or members of the public. Some more of the most important aspects of the HSWA are:

- In domestic premises the occupier is normally covered by insurance (Occupier's Liability) for visitors (postman, milkman, etc). When you are working on domestic premises a change occurs whereby you as the plumber become responsible for the health, safety and welfare of the occupant, because it is your place of work.

You should always use Health and Safety equipment correctly and for the purpose it is intended.

- Substances, such as oil, grease, cutting compounds, paints and solvents, are hazardous if spilled on the floor. Items such as off-cuts of pipe, cables, tools and even food are also dangerous if left underfoot. You have a responsibility both to yourself and others to keep the workplace hazard free.

What should your employer do?

Your employer has important Health and Safety responsibilities. You should know what your employer's legal responsibilities are for keeping you safe at work. The HSWA requires your employer to ensure, so far as is reasonably practicable, your health, safety and welfare at work is not put at risk.

The matters to which this duty extends include:

- The provision and maintenance of plant and systems of work are safe and without risk to health
- Safety in the use, handling, storage and transport of articles and substances
- The provision of information, instruction, training and supervision as necessary to ensure the Health and Safety at the workplace of employees
- The provision of access to and exits from the workplace are safe and without risk
- The provision of adequate facilities and welfare arrangements for employees at work.

Additional HSWA employer requirements

A company must produce a Health and Safety policy statement (if more than five people are employed by the company) which must be brought to the attention of all members of staff. Arrangements for ensuring this must be in place.

Under the HSWA employers also have a duty to:

- Carry out an assessment of risks associated with all the company's work activities
- Identify and implement control measures

- Inform employees of the risks and control measures
- Periodically review the assessments
- Record the assessment if over five persons are employed.

Construction Design and Management (CDM) Regulations (1994)

The CDM Regulations aim to improve the overall management and co-ordination of health, safety and welfare throughout all stages of a construction project. They were designed to reduce the large number of serious and fatal accidents and cases of ill health, which happen every year as a result of work related to the construction industry, which includes domestic plumbing work.

Construction (Health, Safety and Welfare) Regulations (1996)

These Regulations are aimed at protecting the health, safety and welfare of everyone who carries out construction work. They also give protection to others, who may be affected by the construction work.

The main dutyholders under these Regulations are employers, the self-employed and those who control the way in which construction work is carried out. Employees too have duties to carry out their own work in a safe way. Also, any person involved in construction work has a duty to co-operate with others on Health and Safety issues, reporting any defects to those in control.

The Regulations are closely related to the workplace regulations but they also contain guidance on working at heights including the use of scaffolds and means of accessing them. The guidance given for this Regulation will be relevant to some of the work you undertake during your career as a plumber. Summary details of the most important aspects of this regulation follow.

Welfare facilities

As far as is reasonably practicable, the person in control of a construction site must provide the following welfare facilities:

- Suitable and sufficient sanitary conveniences (toilets)
- Suitable and sufficient washing facilities (including showers if required due to the nature of the work or for health reasons)

- An adequate supply of wholesome drinking water
- Suitable and sufficient accommodation shall be provided or made available:
  - for the clothing of any person at work on a construction site and which is not worn during working hours
  - for special clothing which is worn by any person at work on a construction site but which is not taken home.
- Suitable and sufficient facilities shall be provided or made available for workers to change clothing in all cases where:
  - a person has to wear special clothing for the purpose of his work
  - that person cannot, for reasons of health or propriety, be expected to change elsewhere.
- Suitable and sufficient facilities for rest shall be provided or made available at readily accessible places.

Duty to take precautions against falls

- Prevent falls from height by physical precautions or, where this is not possible, provide equipment that will check falls
- Ensure that there are physical precautions to prevent falls through fragile materials, e.g. roofs
- Erect scaffolding, access equipment, harnesses and nets, this should be done under the supervision of a competent person
- Ensure that the criteria for using ladders are provided.

Manual Handling Operations Regulations (1992)

Manual Handling Operations Regulations refer to the human effort involved in handling loads. This includes effort applied directly or through straining on a rope or lever. You will already be aware that your job as a plumber involves manual handling of tools, equipment and materials. (The principles of safe manual handling have been covered earlier in this unit.)

The Manual Handling Operations Regulations provide the following requirements to employers and employees:

The employer

- Should endeavour to avoid the need for employees to undertake manual handling operations at work which could involve a risk of them being injured

- Where it is not reasonably practicable to avoid the need for employees to undertake any manual handling operations at work which involves a risk of there being an injury, the employer should:
  - assess all such manual handling operations to be undertaken by the employees
  - reduce the risk of injury to those employees arising out of them undertaking any such manual handling operations
  - provide any of those employees who are undertaking any such manual handling operations with certain information about the loads to be carried by them – to include supply of relevant training and safety equipment.

The employee

- While at work, will make full and proper use of systems of work provided for their use by their employer, in compliance with the employer's duty under the Manual Handling Operation Regulations.

Fire Precautions Act (1971)

Some larger or high-risk premises may require a fire certificate. This includes factory premises employing more than 20 people, or buildings that store/use highly flammable materials. This could affect you if you were working in this type of building. Employers in all premises are required to carry out a fire risk assessment to identify the hazards from fire.

Once identified, precautions must be implemented to minimise the risks, these include:

- Reduction/elimination of ignition sources
- Elimination/isolation of materials likely to assist in the spread of fire
- Means of giving warning of fire
- Provision of means of escape in case of fire
- Provision of fire fighting equipment
- Appropriate signing of fire exits and fire fighting equipment
- Training in what to do in case of fire for employees
- Periodic revision and maintenance of the above.

The Electricity at Work Regulations (1989)

The Electricity at Work Regulations, which came into force on 1 April 1990, lay down general Health and Safety requirements regarding electricity at work for employers, self-employed persons and employees. The Regulations are made under the HSWA 1974 and broadly speaking, impose a duty upon every employer and self-employed person to comply with the provisions of the Regulations. They also impose a duty on every employee to co-operate with their employer so far, as is necessary, to enable the Regulations to be complied with.

As a plumber you will need to be especially careful when dealing with electricity. The Electricity at Work Regulations provide the full safety standards requirement. The Regulations refer to a person as a 'dutyholder' in respect of systems, equipment and conductors. These Regulations are therefore statutory and consequently penalties can be imposed on those people found guilty of malpractice or misconduct.

The Gas Safety Installation and Use Regulations (GSIUR) (1998)

The GSIUR oversees all the work carried out in domestic dwellings and underpin the safe installation/operation of domestic gas systems and appliances. You will cover the precise requirements of the GSIUR if you chose to progress to the Level 3 Technical Certificate or if you go onto complete ACS training and assessment.

Test yourself 2.7

1. The HSWA requires employers to complete which of the following courses of action?
   a. Train every employee in basic first aid ☐
   b. Discipline workers who leave site early ☐
   c. Pay JIB recommended wage rates ☐
   d. Carry out an assessment of risks associated with all the company's work activities ☐
2. Which of the following statements correctly describes the 'CDM Regulations'?
   a. Construction (Design and Management) Regulations (1992) ☐
   b. Construction (Design and Maintenance) Regulations (1994) ☐
   c. Construction (Design and Management) Regulations (1994) ☐
   d. Civil (Design and Maintenance) Regulations (1992) ☐
3. The Construction (Health, Safety and Welfare) Regulations (1996) cover in detail which of the following workplace activities?
   a. Safe use and storage of LPG ☐
   b. Safe use of electrical power tools ☐
   c. Safe storage of COSHH-registered materials ☐
   d. Safe work at height, including use of scaffold ☐
4. Imagine a co-worker and yourself have been asked to move a heavy object into position, which regulations should you consider when thinking about how to tackle the task?
5. When working in domestic premises, who is responsible for the Health and Safety of the occupant?
   a. The customer ☐
   b. Your employer in his office ☐
   c. The HSE inspector ☐
   d. Yourself as the plumber on site ☐
6. If your company employs five or more members of staff, the HSWA states that they should have what?
   a. A Health and Safety policy statement ☐
   b. A company pension scheme ☐
   c. A full-time company safety officer ☐
   d. At least two apprentices ☐

Check your learning Unit 2

Time available to complete answering all questions: 30 minutes

Please tick the answer that you think is correct.

1. The Health and Safety at Work Act 1974 requires that employers must prepare a written safety policy where there are:
   a. More than 5 employees ☐
   b. More than 7 employees ☐
   c. More than 9 employees ☐
   d. More than 12 employees ☐
2. Which one of the following would provide information on the general safety requirements for scaffolds on building sites?
   a. Construction (Lifting Operations) Regulations 1961 ☐
   b. Health and Safety at Work Act 1974 ☐
   c. Construction (Health, Safety and Welfare) Regulations 1996 ☐
   d. The specification for the construction work ☐
3. Which of the following types of ladder should always be used to gain access to a scaffold on a construction site where heavy equipment or materials are to moved, or where there is regular foot traffic?
   a. Class 1 ladder ☐
   b. Class 2 ladder ☐
   c. Class 3 ladder ☐
   d. Triple extension ladder ☐
4. A circular safety sign with a red band enclosing a crossed out symbol, with a white background is a:
   a. Warning sign ☐
   b. Mandatory sign ☐
   c. Prohibition sign ☐
   d. Safe condition sign ☐
5. Which of the following is true of LPG heating gases
   a. Butane is lighter than air and any leaking gas will accumulate near a ceiling ☐
   b. Propane is heavier than air and any leaking gas will accumulate at ground level ☐
   c. Propane is lighter than air and any leaking gas will accumulate near a ceiling ☐
   d. Both propane and butane are lighter than air and any leaking gas will accumulate near a ceiling ☐
6. Only one of the following is correct, identify the correct colour of panel on a fire extinguisher?
   a. Water – red ☐
   b. Foam – black ☐
   c. Powder – green ☐
   d. Carbon dioxide – blue ☐
7. Which of the following should be included in a company safety policy?
   a. Address of the Health and Safety Executive ☐
   b. Maximum hours to be worked by operatives ☐
   c. First aid treatment to be given for minor injuries ☐
   d. When and where personal protective equipment must be worn ☐

Check your learning Unit 2 (Continued)

8. Which one of the following types of accident must be reported to the Health and Safety Executive?
   a. All accidents causing injury to persons while at work ☐
   b. All accidents causing damage to customers' property ☐
   c. Accidents resulting in more than three days absence from work ☐
   d. Accidents resulting in more than one week absence from work ☐
9. Which of the following actions should be taken first in the case of a minor burn?
   a. Wash with soap and water ☐
   b. Cool the area of the burn with clean cold water ☐
   c. Apply antiseptic cream to the burn ☐
   d. Cover the burn with a clean cloth ☐
10. What is the best way to eliminate trip hazards and ensure 'good housekeeping'
   a. Store materials outside ☐
   b. Keep the workspace tidy ☐
   c. Only use cordless tools and equipment ☐
   d. Keep tools and equipment off the floor (lean them up) ☐
11. Hacksaw and wood-saw blades should be
   a. Replaced if rusty ☐
   b. Changed by a supervisor ☐
   c. Replaced after each cut ☐
   d. Only used if sharp ☐
12. In terms of COSHH regulation, a chemical substance is identified by:
   a. Warning sign ☐
   b. The smell ☐
   c. Its name ☐
   d. The colour ☐
13. How should asbestos found on site be prepared for disposal?
   a. Burnt, ensuring the ashes are safely sealed in containers ☐
   b. Labelled and sealed in containers ☐
   c. Placed in a designated building fabric skip ☐
   d. Kept damp ☐
14. A panicked customer has telephoned to report a water leak. What should a plumber's first action be?
   a. Inform the customer that they will be on site as soon as possible ☐
   b. Advise the customer to try to somehow plug the leak ☐
   c. Tell them to phone their water company ☐
   d. Advise the customer to turn off the water ☐
15. What should be the first action of someone who witnesses an accident on a worksite?
   a. Attend to the casualty ☐
   b. Make sure they are not also in danger ☐
   c. Telephone the emergency services ☐
   d. Raise the alarm ☐
16. Which of the following types of extinguisher would it be safe to use on a small electrical fire in an enclosed atmosphere?
   a. Water ☐
   b. Carbon dioxide ☐
   c. Dry powder ☐
   d. Foam ☐

(Continued)

Check your learning Unit 2 (Continued)

17. What do the letters HSE stand for?
    a. Health and Safety Executive ☐
    b. Health and Safety Enforcement ☐
    c. Health and Safety Enquiry ☐
    d. Health and Safety Endorsement ☐

18. Who must supply a plumber's Personal Protective Equipment when working on site, the?
    a. Health and Safety Officer ☐
    b. Plumbing employer ☐
    c. Individual plumber ☐
    d. Site foreman ☐

19. Which area of HSE legislation covers the general health and safety responsibilities of the employer and employee in the work place?
    a. The Health and Safety Workplace Act of 1974 (HSW Act) ☐
    b. The Health and Safety in the Workplace Act 1974 (HSIWA) ☐
    c. The Health and Safety at Work Act of 1974 (HSW Act) ☐

20. What regulations cover the management of construction projects from planning and construction through to demolition?
    a. The Workplace Regulations 1992 ☐
    b. The Construction (Health, Safety and Welfare) Regulations 1996 ☐
    c. The Construction (Design and Management) Regulations 1996 ☐
    d. The Management of Health and Safety at Work Regulations 1999 ☐

UNIT 3

COMMON PLUMBING PROCESSES

Summary

In this unit, we cover the following:

- Using hand tools
- Using power tools
- Measuring
- Marking out
- Cutting
- Bending
- Jointing and
- Fixing

on a range of pipework materials, copper, low carbon steel (LCS) and plastic, used in domestic plumbing systems.

In addition to the items mentioned above, we also look at:

- Generic systems knowledge
- Associated trade skills, such as:
  - Processes for lifting flooring surfaces
  - Requirements for cutting holes and notching timber joists
  - Procedures for cutting holes through a range of materials
  - Making good.

One key feature about this unit is that we ask you to carry out some of your own research. As a plumber, finding information out for yourself is a useful skill to learn because you need to keep up-to-date with changes to regulations and new technology. It is also necessary in this unit because it would have been impossible for us to include every type of fitting or fixing that is available.

UNIT 3

Generic systems knowledge

Introduction

The purpose of this unit is to cover what we refer to as generic systems knowledge, these are areas that are 'general' or common to all the systems used in the domestic plumbing and heating industry and it will save repeating the same information during each systems unit: cold water, hot water, etc.

Generic systems knowledge covers topics such as taking basic site measurements, how to prepare work locations, use of specifications, paperwork for ordering materials, and how to deal with customers and co-workers.

We think it is important that you understand the preparation work that is required before starting a job – that is why we have put this unit first.

Use of documentation

General documentation can include such things as:

- Health and Safety Regulations, covered earlier in the Health and Safety Unit
- Water Regulations
- Requirements of British Standard Specifications and in particular BS 6700 and BS 8000
- Building Regulations, affecting a plumber's work such as the energy efficiency of central heating boilers.

The above documents underpin much of the work that is carried out by the plumber. Other documentation includes manufacturers' instructions, site drawings, job specifications and work programmes.

Key point

Domestic plumbing can vary from carrying out a small maintenance job like re-washering a tap, to carrying out complete system installations on large scale new housing developments. In this book, we make reference to 'larger contracts' and we take this to mean the large scale new housing developments or refurbishments where site drawings, specifications and work programmes are likely to be used.

Manufacturers' instructions

Most appliances and plumbing components are supplied with manufacturers' instructions. These provide information on the installation service and maintenance requirements and once an appliance is installed, these instructions should be left with the customer for future reference. Depending on the type of appliance or component, user instructions are also included.

Site drawings

Site drawings are covered in greater detail at Level 3. Building drawings provide details of how a building is going to be designed, e.g. size and shape of rooms, location of doors and windows, etc. They also show specific design details, such as floor, wall and roof construction. Building services drawings show the layout of pipework systems and the location of appliances and components, such as baths, sinks, water closets (WCs), radiators, boilers and pumps.

Hot tips

Throughout this book, we encourage the reader to research the internet for manufacturers' information as well as provide contact details for manufacturers' catalogues. Try to get into the habit of obtaining and reading the information as this will help to make sure you keep yourself up-to-date with new products and technology.

Job specifications

These usually accompany site and services drawings and details, and are mostly used on larger contracts.

Activity 3.1

What sort of details do you think a job specification might cover? Jot down your thoughts for inclusion in your portfolio and check out your answers at the end of this book.

Specifications form part of the contract documentation, so any alterations required to the specification should not be done by you.

Key point

If for any reason, a part of the job cannot be done to the specification, you should inform your supervisor, foreman or employer immediately. This applies equally to a smaller job, e.g. boiler replacement, where a detailed specification has not been provided but a quote or estimate has been provided, but the job, for whatever reason, can no longer be done as originally planned.

Work programmes

If a plumber was working on a replacement bathroom suite, it is unlikely that a written work programme would have been prepared, but the plumber would have the work programme inside their head, based on an agreed start and finish date and a series of activities required in between to get the job done.

This principle is not very different on a larger contract but in this case, a contract programme will have been written out. This could consist of an overall programme for all site trades, as well as a separate programme for each trade (Figure 3.1).

Other documentation

Other documentation used by plumbers relates to carrying out a plumbing job from an initial customer enquiry to completing the work. Let us take a look at the procedures for a typical job, adding three extra radiators to an existing heating system, replacing the boiler and upgrading the controls.

- Customer asks a plumber for a quote
- Plumber visits the customer, measures up and estimates the materials required

Activity	Time														
		1	2	3	4	5	1	2	3	4	5	1	2	3	4
Plot 1	TD														
First Fix	AD														
Plot 1	TD														
Second Fix	AD														
Plot 1	TD														
Gas	AD														
Plot 1	TD														
Test	AD														
Plot 1	TD														
Snagging	AD														
Plot 1	TD														
Hand over	AD														
Plot 2	TD														
First Fix	AD														
Plot 2	TD														
Second Fix	AD														

TD = Target Date
AD = Achieved Date

Figure 3.1 Example of a programme for a domestic plumbing installation

- Plumber contacts the merchants and gets a quotation for the materials
- This enables the plumber to finalise a quote to the customer
- In our case, the customer accepts and the quotation is confirmed in writing
- The plumber orders the material from the merchants; on delivery of materials, the plumber receives a delivery note
- The work is carried out to meet the customer requirements
- The plumber receives the invoice requesting payment for the materials used, and accordingly invoices the customer for the work that has been completed
- A remittance advice (record of payment) is sometimes issued with payment
- On any job, large or small, if you are advised that the delivery of materials will be delayed or a particular item is not the one ordered, either the site foreman (so he/she is aware of potential delays) or the immediate supervisor/employer is notified.

Preparing for plumbing installations

Estimating the material requirement

As a plumber qualified at S/NVQ Level 2, you are likely to do most of your estimation from site and this is done in the absence of drawings or specifications.

Again, the estimation process will depend on what job you are doing – could be working out what pipe and fittings you will need for a simple repair job to the requirement for a complete plumbing installation from scratch. On large multi-dwelling developments, the material, requirements will probably have already been worked out for you and supplied in 'packs' for each dwelling.

Where you are required to estimate the materials for a job, you will need to have a thorough understanding of what you are going to do so you can 'visualise' the installation, where you are going to install appliances, components and fittings, how you are going to run the pipework and what sizes you are going to use.

Once you have determined all this, it might be a good idea to produce a sketch of the installation with dimensions on it by taking measurements from site. This will help in working out the pipework lengths.

Preparation checklist

See Activity 3.2.

Activity 3.2

How much preparation is required will depend on the size of the job, but there are a number of things that a plumber should prepare before starting any job. Can you think what these might be? Write your answers in your portfolio and then take a look at our list at the end of this book.

Test yourself 3.1

1. Which of the following relates to the requirements for the energy efficiency of central heating boilers?
   a. Building Regulations ☐
   b. Water Regulations ☐
   c. British Standards Specifications ☐
   d. Health and Safety Regulations ☐
2. Give two examples of what a job specification might cover.
3. What action should you take if the job cannot be done to a specification?
4. Fill in the gaps, selecting from the words below:
   _________ can be used to give a price for a _________ by a _________ or from a _________. An _________ is confirmation that a quotation has been accepted. Once a job has been completed an _________ is sent requesting _________. The _________ confirms that _________ has been made.

   remittance advice, invoice, job, plumber, quotations, payment, order, merchant, materials, payment
5. Which are the two main requirements of an installation programme?
   a. Activity against time ☐
   b. Activity against speed ☐
   c. Time against labour ☐
   d. Time against materials ☐
6. State three checks that you would carry out in preparation for an installation. Now check your answers at the end of the book.

UNIT 3

Plumbing tools

Introduction

As a plumber, you will be required to:

- Measure
- Mark out
- Cut
- Fabricate
- Make joint and
- Fix a range of materials.

In most of the cases, this will involve the use of tools to enable you to do this.

Once you have got your tool kit, keep the tools clean and well maintained, this should ensure a long life and keep down the cost of having to regularly buy replacements.

Key point

Good quality tools are expensive, but worth the initial outlay. Our advice is to buy good quality tools of a well-known brand. What might seem a bargain could end up damaged or failing when most needed, simply because they are not up to the job. Cheap tools that are suitable more for the DIY market, will soon wear out under prolonged professional use.

The range of power tools used in domestic plumbing work includes:

- Cordless power drills and screwdrivers
- Power drills
- Combined cordless drills and screwdrivers
- Power saws.

The power tools illustrated in this unit have been reproduced with the kind permission of Draper Tools Ltd., whose contact details are given in the next section.

The hand tool kit

The tools illustrated here have been reproduced with the kind permission of:

Ridgid Tools

Arden Press House
Pixmore Avenue
Letchworth
Herts
SG6 1LH
Tel: 01462 485335
E-mail: sales.uk@rigid.com

Draper Tools Ltd.

Hursley Road
Chandler's Ford
Eastleigh
Hants
SO53 1YF
Tel: 023 8026 6355
E-mail: sales@draper.co.uk
Website: www.draper.co.uk

Key point

Hand tools can be subdivided into a number of categories, such as tools for:

- Measuring and marking out
- Cutting and preparation
- Fabrication
- Jointing
- Fixing and making good
- Specialist tools.

Measuring and marking out

The standard tools include:

- Spirit Level
- Folding Steel Rule
- Tape Measure.

And do not forget, you will need a pencil for marking out!

Cutting and preparation

This will require:

- Hacksaw Frame Straight Tin Snips
- Compact Pipe Cutter
- Trimming Knife
- File
- Wood Chisels

- Floorboard Saw
- Abrasive Cleaning Pads
- Cold Chisel with Guard
- Floorboard/Brick
- Bolster with Guard
- Padsaw
- Junior Hacksaw.

Fabrication

Bending and threading equipment is also available for use with low carbon steel (LCS) pipe. These are covered in more detail in Unit 4, tube bending, measuring and marking out. Fabrication tools include:

- Claw Hammers
- Floorboard/Nail Bar
- Club/Lump Hammer
- Copper Pipe Hand Bender
- Internal Bending
- Spring
- External Bending
- Spring.

Jointing

There are three considerations here (for copper) – jointing by soldering, compression fittings, and jointing using the latest push fit methods. There is also jointing of LCS using threaded fittings. Let us see what they look like in Figure 3.2.

(a) Adjustable Basin Wrench
(b) Adjustable Wrench
(c) Basin Wrench
(d) Adjustable spanner
(e) One Hand Speed Wrench
(f) 250 mm Waterpump Pliers
(g) Blow Lamp-Propane Torch
(h) Combination Pliers
(i) Curved Jaw Locking Pliers

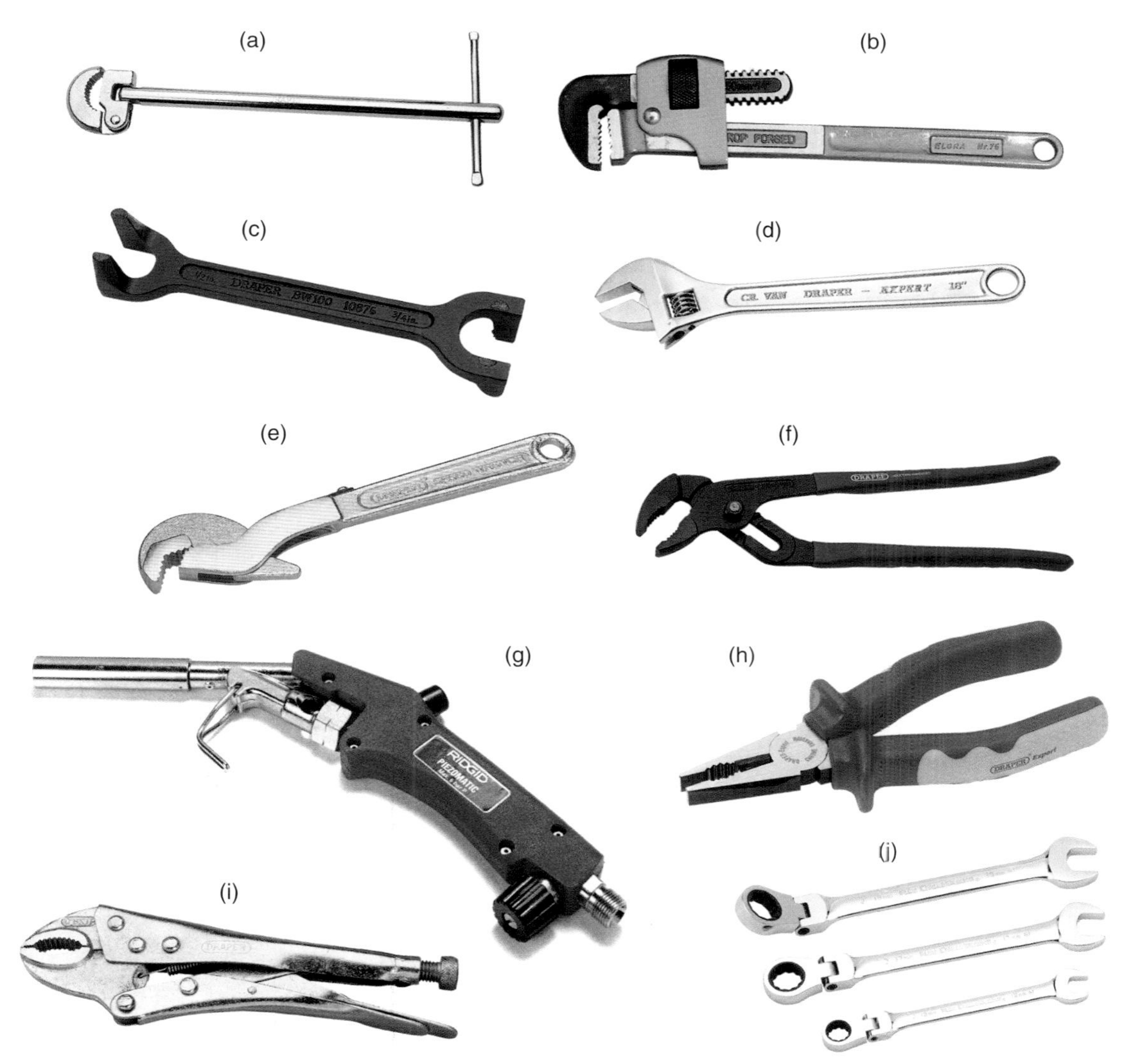

Figure 3.2 Jointing tools (Reproduced with permission of Draper Tools Limited (a–f, h–j) and Ridgid Tool (g))

As you build up your tool kit over time, you may consider investing in a set of spanners.

(j) Spanners.

Fixing and making good

- 8 piece screwdriver set
- Insulated screwdriver
- Flat piece wood bits set

- 150 mm pointing trowel
- Masonry drill bits
- High speed steel drill bits.

Other tools

- Allen keys
- Immersion heater key
- Sink plunger
- Stop valve key
- Tool box.

Specialist tools

Plumbers also use tools and equipment of a more specialist nature and these should be used in accordance with manufacturers' instructions and by personnel who have been properly instructed in their use (Figure 3.3).

Figure 3.3 Pipe freezing kit (Reproduced with permission of Ridgid Tools)

In addition to the tools and equipment, you will need materials such as:

- Solder wire
- Fluxes approved for plumbing work (including those suitable for wholesome water installations)
- Jointing tape approved for plumbing work
- Jointing compound approved for plumbing work (including those suitable for wholesome water installations)
- A range of screws and nails.

Key point

Do not forget to make sure that solder wire used on water supply systems is 'lead free'.

Tool safety maintenance checklist

Activity 3.3

You covered tool safety briefly during the Health and Safety Unit. Using the headings that follow, make a list of bullet points outlining the key safety maintenance points, write them in your portfolio and then check it out with our list at the end of this book.

- General
- Hacksaws
- Pipe cutters
- Wood chisels
- Cold chisels
- Hammers
- Pipe grips and wrenches
- Screwdrivers.

Power tools

Cordless power drills and screwdrivers

The cordless drills/screwdrivers are available in a range of voltages; 12, 14, 14.4 and 24 V being a few examples. Most drills are combined so they can be used as a drill and a screwdriver. They are powered by batteries (usually supplied with two) and a charger, so you can have one working and one on charge.

Figure 3.4 shows an example from a very wide range of what is available.

Key point

These tools are popular because you do not need to carry a transformer around, and if there is no electricity on site you do not have to revert to hand tools. The added advantage is that they can also be used as a screwdriver.

Figure 3.4 Cordless Combi hammer drill (Reproduced with permission of Draper Tools Limited)

Power drills

110 V

There is a wide range of makes and models of power drills; this section is designed to show a cross section of what is available.

Typical power ratings for 110 V drills range from 620 to 1400 W. Most drills are of variable speed and some have a reversible action. Drills vary in power depending on the size of motor; this in turn has a bearing on what the drill can do.

The one shown in Figure 3.5 has a 'hammer action' which when engaged makes drilling through masonry easier. Not only does the drill rotate but it also moves fractionally backwards and forwards at high speed, giving the effect of hammering the drill into the building fabric.

Figure 3.5 110 V hammer drill (Reproduced with permission of Draper Tools Limited)

A range of bits is available for powered and cordless drills for use on metal, wood, brickwork, blockwork and concrete. Core drills are also available for drilling large diameter holes. Screwdriver bits are available in the following designs (Figure 3.6):

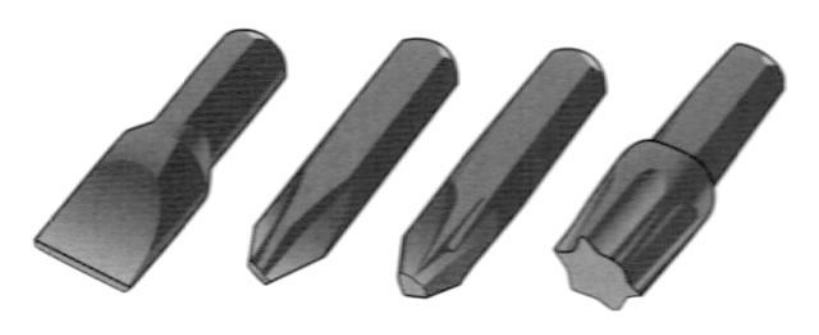

Figure 3.6 Screwdriver drill bits

- These are precision made using hardened steel and are hard wearing
- The bits can be purchased individually or in sets. The type shown slot into a purpose made bit for the drill.

Power saws

Power saws such as circular saws are used by plumbers for taking up floorboards or sheets in order to install pipework under floors (Figure 3.7).

Figure 3.7 Example of a power saw; make sure a guard is fitted (Reproduced with permission of Draper Tools Limited)

Here is a power saw checklist:

- Power saws should run off 110 V
- Never use one without a guard
- Make sure that nails or screws are avoided when cutting with the saw. If possible, take up a board by hand to check what is beneath the area where you are going to use the saw.
- Always make the cutting depth the same depth as the floor thickness.

Key point

Why is making the cutting depth the same depth as the floor thickness a good idea? Think about what might happen if you make it deeper than the depth of the floor thickness.

Jig saws can be used to take up boards, but are mostly used to cut holes in worktops or countertops to fit sinks or washbasins. Battery operated circular saws are also available, usually with an 18 V motor.

Power tools in general

Here are a few points to remember about power tools:

- All electric tools should be double insulated
- Always use 110 V supply
- Check that electrical cables are not damaged or worn out
- Check that plugs are not damaged
- Check for test labels to show the equipment is safe to use
- Remember to wear safety goggles when using drills and saws. These will protect your eyes from dust and any splinters of material that might fly off whilst working.

Key point

All electrical equipment should be PAT tested in accordance with your employer's procedures. PAT tests are maintenance records of all portable electrical equipment to ensure that it is in safe working order.

Try this

Next time you are in a plumbing merchant's or a tool shop, make it a point to look at the range of hand and power tools that are available for plumbing work. Take time to obtain hand and power tool catalogues, and look on the internet to see what is available based on the tools listed in this unit. Prepare a list of tools you will need to include in your plumbing tool kit. [You may have some of the tools already, compare the quality and prices of the various manufacturers].

Cartridge-operated tools

We covered cartridge-operated tools under Health and Safety. Remember, people under the age of 18 are not allowed to use them. If over the age of 18, you must receive proper instructions on their use.

Test yourself 3.2

1. List a typical tool for each of the following:
   - Measuring and marking out
   - Cutting and preparation
   - Fabrication
   - Jointing
   - Fixing and making good
2. State the safety and maintenance requirements for:
   - Hacksaws
   - Wood chisels
   - Pipe grips and wrenches
3. What tools would you use for:
   - The cutting of sheet metal
   - De-burring a pipe
   - Removing a mild steel fitting from a pipe
   - Notching a floor joist
   - Removing an immersion heater
4. Which voltage must power drills run on?
   a. 110 V ☐
   b. 240 V ☐
   c. 225 V ☐
   d. 415 V ☐
5. What is a PAT test used for?
6. What is the minimum age that you can use cartridge-operated tools?
   a. 18 ☐
   b. 20 ☐
   c. 21 ☐
   d. There is no minimum age ☐
7. State three specific safety precautions when using a power saw
8. State three safety precautions when using any power tools

Tube bending, measuring and marking out

Introduction

Bending, measuring and marking out is a basic essential skill for any plumber. In this unit, we will concentrate on the bending of copper pipe, which can be done by hand (with a spring) or machine. We will also look at LCS which, in the main, is bent by hydraulic machines.

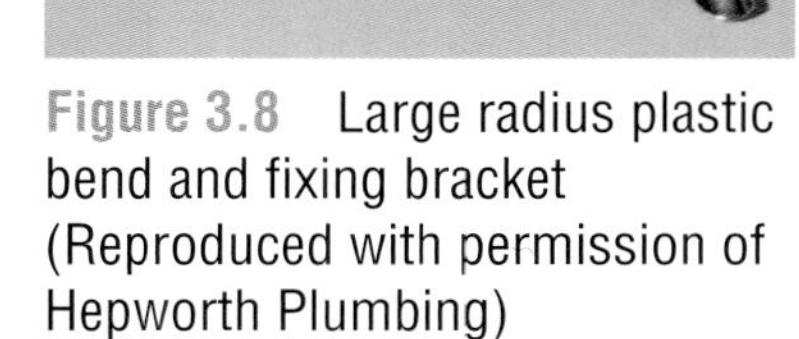

Figure 3.8 Large radius plastic bend and fixing bracket (Reproduced with permission of Hepworth Plumbing)

Bending methods

We will concentrate here on copper and LCS pipes. Plastic pipes used in domestic plumbing systems can be bent, but the main application is restricted to small-bore polythene pipes which can be positioned into large radius 90° bends or offset by hand, and then clipped into position. There are also steel preformed 90° brackets for tighter bends, into which the pipe can be clipped (see Figure 3.8).

This image was reproduced with the kind permission of Hepworth who can be contacted on the numbers below, and you can visit their website www.hepworthplumbing.co.uk

Technical Support Tel: 01709 856 406
Fax: 01709 856 407
Literature Service Tel: 01709 856 408
Fax: 01709 856 409

Key point

What do you think are the advantages of bending pipe rather than using fittings?

Bending pipe rather than using fittings has the following advantages:

- They produce larger radius bends than elbow fittings and larger radius bends have less frictional resistance than fittings
- Using bends costs less than using fittings
- Long sections of pipework can be prefabricated before installation, saving time.

Copper tube bending

The type of copper tube suitable for bending is BS EN 1057–R250 (previously designated as BS EN 1057; Part 1 – Table X). It is termed as half hard, and is available in straight lengths.

Other types are BS EN 1057–R220 (previously designated as BS EN 2871; Part 1 – Table Y), this type has thick walls, and is supplied in coils. It is not usually used internally in dwellings; it is mostly used for underground services and is available with a plastic coating to protect it from corrosion.

There is also BS EN 1057–R220 (previously designated as BS EN 2871; Part 1 – Table W), supplied in coils and generally used for micro-bore heating installations.

We will explore copper tube in a bit more detail when we look at pipe jointing in the next session. The latter two are not suitable for bending by machine. R250 (Table X) copper pipe can be bent by using either:

- Hand or
- Machine.

> **Hot tips**
>
> Yorkshire Copper Tube produce an excellent 'technical guide' covering all aspects of working with copper tube. It can be downloaded as a Pdf from their website on: http://www.yorkshirecopper.com/technical/yc_te_index.php

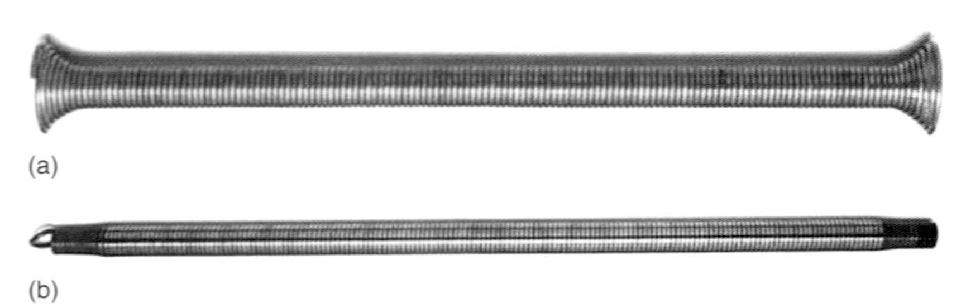

Figure 3.9 (a) Internal bending spring, (b) External bending spring

By hand

This is a popular method of bending pipe when carrying out maintenance and repair work and the copper pipe can be bent using either an internal or external spring. In either case you pull the bend against your (padded) knee to get the desired angle.

There is some excellent technical material available on this topic, for bending both by hand and machine. Here are some contacts where you can find further information to support your learning. We strongly recommend that you obtain as much information as you can.

> **Key point**
>
> Spring bending methods can quickly enable high quality, accurate bends to be formed on copper tubes without wrinkling or flattening as the springs support the wall of the tube as it is being bent.
> Once the necessary skills have been developed and practised, there is little difficulty bending light gauge copper tube (R250) by hand, up to a diameter of 15 mm. A 22 mm pipe can also be bent by spring methods but it is usually advisable to anneal the area forming the radius of the bend first.

Yorkshire Copper Tube

This company produce a publication called 'Yorkshire Tube Systems technical Guide' which can be viewed on: www.yorkshirecopper-tube.com

UK Copper Board

As well as producing a range of technical material, the UK Copper Board runs a Copper Club.

The Copper Club – a loyalty scheme set up by the UK Copper Board in May 2000 to reward supporters of Copper – is proving a great success. The club currently has over 2300 members, including a substantial number of students, and the number of members is continuing to rise!

When signing up to the Copper Club via the application form on their website new members will receive a welcome pack which includes the latest copy of their reference book 'Installation Tips', a 'Make the Right Start CD' and details of our Plumbing Advice Telephone Hotline.

In addition to the pack, members receive regular updates about the industry and the activities of the UK Copper Board.

The UK Copper Board site is at: www.ukcopperboard.co.uk and their address is:

UK Copper Board
c/o Copper Development Association
5 Grovelands Business Centre
Boundary Way
Hemel Hempstead
Hertfordshire HP2 7TE

Copper Development Association

The Copper Development Association provides excellent technical advice in the form of various publications, and also supports further education. Their website is at: www.cda.org.uk

Setting out for hand bends using bending springs

It's pretty obvious that if you tried to bend a piece of copper tube without supporting the wall of the pipe, the pipe would simply collapse, leaving a totally unacceptable result. One way of preventing this is to use a bending spring.

Half-hard copper tubing R250 (formally table X) is the recommended grade for pipe bending, using either an internal or external bending spring. Hand-made bends have to be 'set out' in order to form the radius of the bend, and provide accurate measurements to a fixed point or fitting.

Hand bends should be limited to up to 22 mm diameter; some plumbers may use an internal spring for larger sizes, but this isn't recommended or indeed allowed for in BS 5431(4) (the British Standard for bending springs).

The first thing to bear in mind when setting out for a 90° bend is that there is an apparent gain of the material when the bend is formed. Take a look at Figure 3.10.

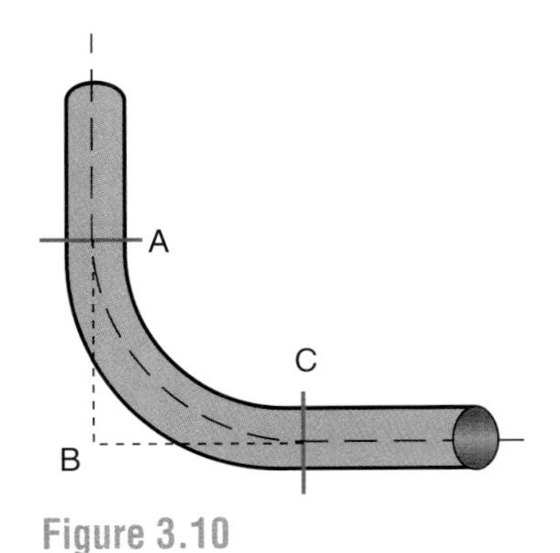

Figure 3.10

The distance from 'A' to 'C' through 'B' along the broken line is in effect the measured length of the bend, but when the bend is actually formed, its path follows the arc A to C which is a shorter distance than the measured length. In summary the gain in length is due to the measured length A-B-C being longer than the actual length A-C.

UNIT 3

When setting out then:

- Allowances have to be made for the 'gain in material'
- The bend must be pulled in the right position in relation to the fixed point.

Step by step to setting out

- Decide on the centre line radius of the bend, which (unless given on a drawing or specification) most practitioners usually determine as four times the diameter of the pipe (4D), although Yorkshire Copper Tube recommend five times
- The length of the pipe occupied by a 90° bend can be calculated using the formula:

$$\frac{Radius \times 2 \times 3.142}{4}$$

- Next we'll assume that a 15 mm pipe is to be bent to a radius of 4D and we need to find out how much pipe will be taken up by the bend, so:

  Radius of bend is 4D, which is $4 \times 15 = 60$ mm.

  Now use the formula:

$$\frac{Radius \times 2 \times 3.142}{4}$$

  So:

$$\frac{60 \times 2 \times 3.142}{4}$$

  Length of bend = 94.26 mm, say 95 mm.

- The next step when making the bend is to measure and mark off the length required from a fixed point (which could be where the pipe is going to enter a fitting for example) to the centre line of the bend (see procedure indicated at Figure 3.11a)
- Then divide the calculated length of pipe by three, which in our case gives three equal measurements of approximately 32 mm
- From the original centre line, mark 32 mm forward and 64 mm back (see procedure indicated at Figure 3.11b)

- The bend is then pulled making sure that it is kept within the confines of the three 32 mm measurements, this will make sure that the centre will be the correct distance from the fixed point
- This setting out technique can also be used for offsets, but a bend of 45°/135° will only require half the length of pipe as that of a 90° bend
- You're now ready to make the bend

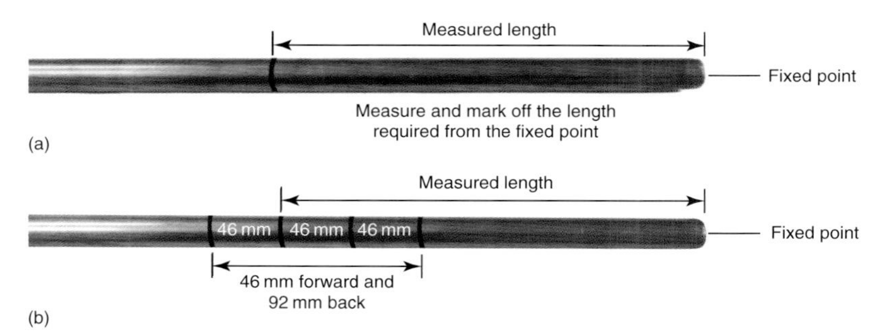

Figure 3.11

- It's advisable to use a template, most plumber's will use a 90° set square or similar
- Insert the spring, it may be an idea to lubricate it first using oil or grease
- Pull the bend gently around the knee to an angle slightly over 90° and then pull it back to 90° and check its accuracy against the template before removing the spring
- Forming offsets, in order to route pipe work around obstacles, is best done by making a template out of strong wire such as welding rod or similar, and then bending the pipe to match the template.

Why do you think it's necessary to overpull the bend? You don't have to write anything down here, but have a think before moving on.

Overpulling the bend and then returning to 90° will release the tension between the spring and the pipe wall and make it easier to remove the spring.

Key point

Bending copper pipe using a spring is something you can try and practice at home. Use this material as a guide, and have a go at setting out and forming 90° bends. You can also try forming an offset. Use an internal spring, and start off with 15 mm copper tube and work up to 22 mm.

Machine bending

This is the most common method used for bending copper tube. Bending machines can be either hand held, or free standing, and

Figure 3.12 Examples of hand held bending machines (Reproduced with permission of Ridgid Tools)

Figure 3.13 Free-standing pipe benders

Figure 3.14 Student using a hand held bender in a centre

Key point

We would strongly recommend that you find out more about machine benders, try and obtain catalogues about the various items shown here, search on websites or better still visit tools shops and take a look at the equipment for yourself.

they work on the principle of leverage. Here are a couple of examples of bending machines.

The small hand held bender is used for pipe sizes of 15 mm and 22 mm, and is light and portable. The free-standing bender can handle pipes up to 35 mm and uses a range of sizes for the back guide and former.

It's important that the machines are set up properly. If the roller is adjusted so that it's too loose it will cause rippling on the inside of the radius of the pipe. If too tight, it will reduce the pipe diameter at the bend, an effect called throating.

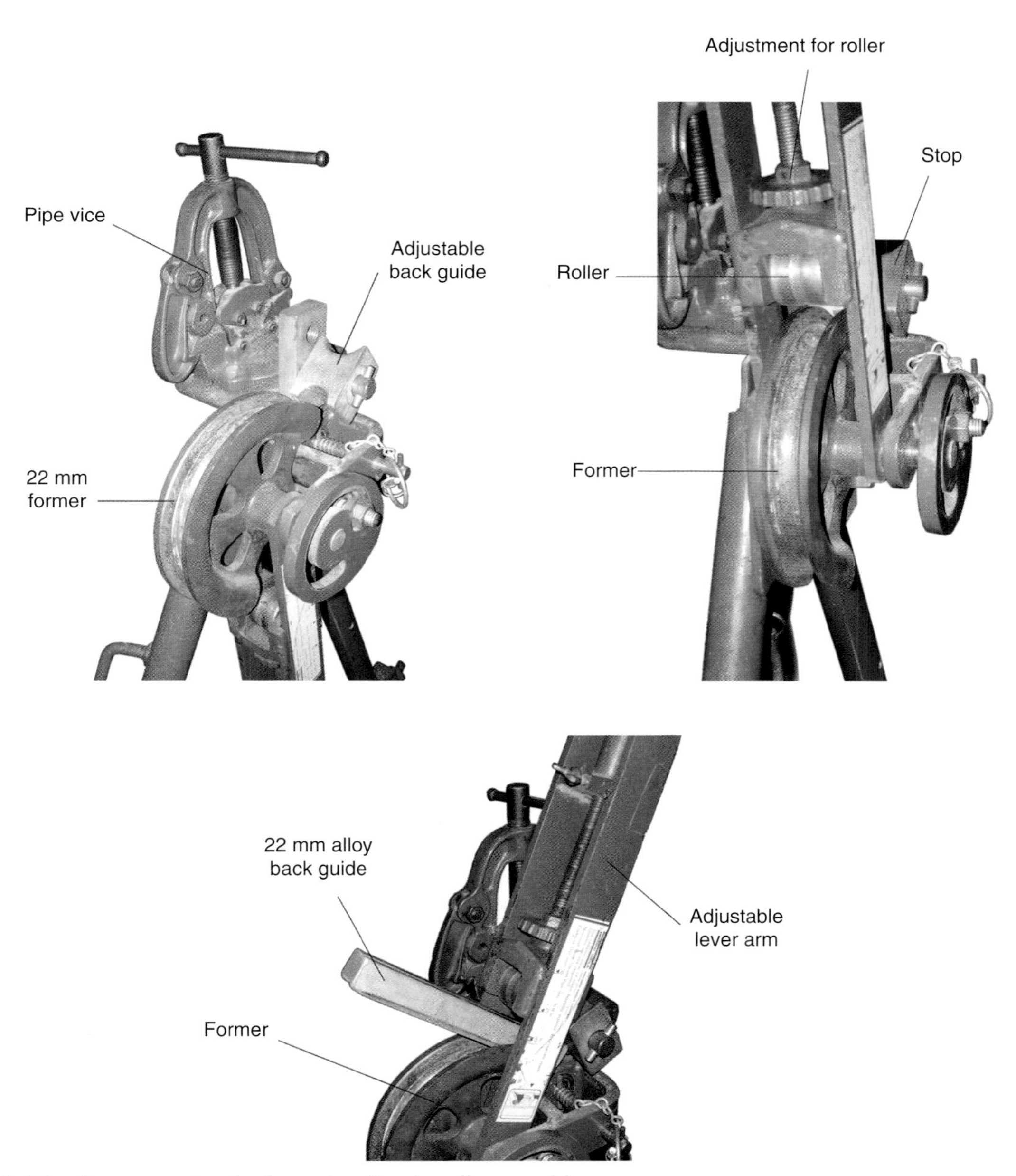

Figure 3.15 Components of a free-standing bending machine

Setting out for pipe bends using a machine bender

Bending machines produce a much tighter bend (approximately three times pipe diameter) than is possible using a spring. This method is by far the most widely used for bending copper tubes in the UK.

Here we'll look at three types of machine bends:

- 90° or square bend
- Offset
- Passover sets.

90° or square bends

When forming square bends the bend can be set to either the inside of the bend and inside of the former, or outside of the bend and outside of the former; the procedures are almost identical, both methods are equally correct and will produce perfect bends. Here's what to do when setting the bend to the outside of the former:

- Mark the pipe to the required measurement using a pencil, this should be taken from the back of the bend to the fixed point
- Make sure that the pipe is pushed fully into the former and is also inserted in the stop
- Place the alloy guide over the pipe (don't use the steel version, as these are used for bending low carbon steel tubes)

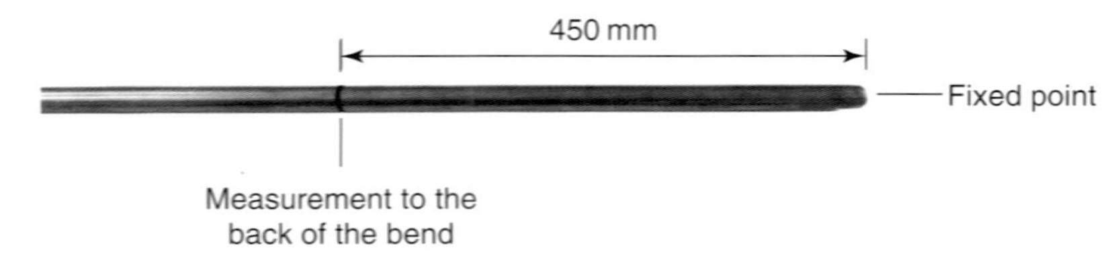

Figure 3.16 Machine Bend Measurement

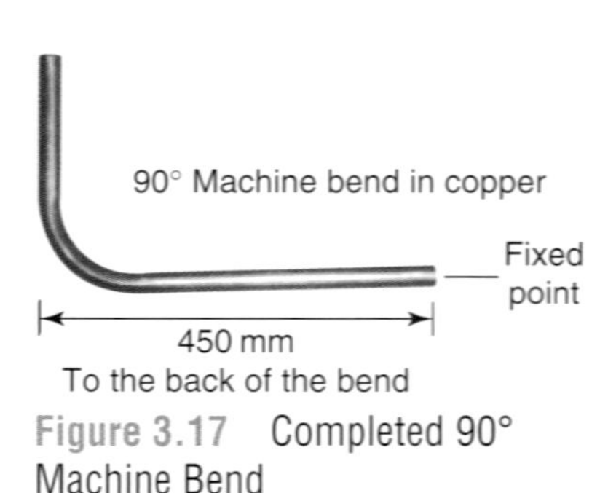

Figure 3.17 Completed 90° Machine Bend

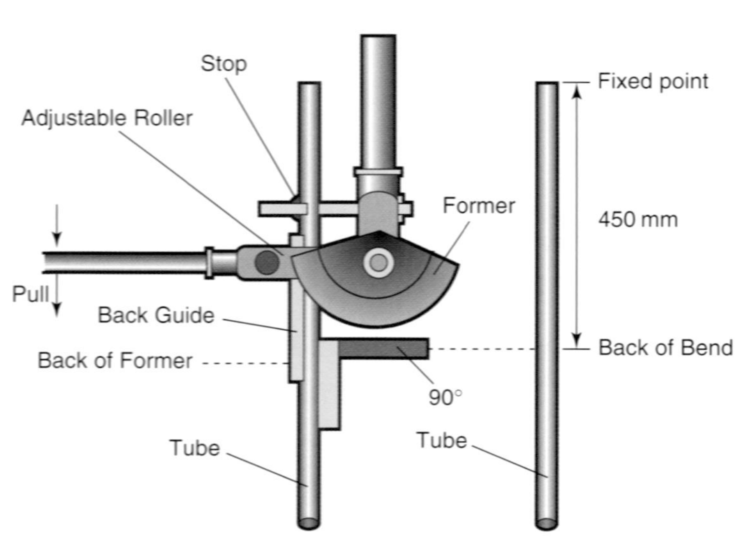

Figure 3.18 Tube Position in Former

- Adjust the pressure onto the guide enough to hold the pipe in position (if adjustable type)
- The square is then placed against the mark on the pipe and adjusted until the square touches the outside of the former
- Make a final adjustment to ensure the correct bending position
- Pull the lever arm to bend the pipe slightly over the required 90° angle as this will counteract the spring back in the bend

Making a return bend, or bending the same pipe again in a different position is achieved using the same technique, only now the first bend becomes the fixed point.

Offsets

You may hear offset being referred to as a double set, with an ordinary single bend being known as a set. The machine is set up in the same way as that of the 90° bend. There are a couple of variations for producing an offset, it can be produced by measurements from the site, or by producing a template, which could be made from strong wire, or drawn on the floor in chalk or on a piece of sheet timber.

- The first bend or set on the pipe is made in the required position which will have been marked on the pipe

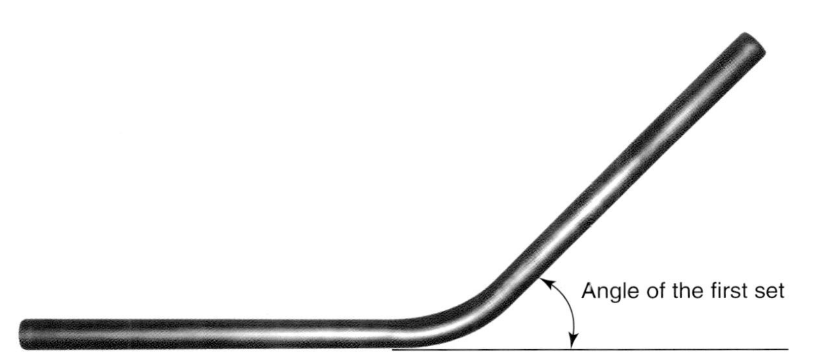

Figure 3.19 Angle of the first set

- The angle of the first bend is usually 45°, but this is not essential. However, where the angle is critical, it can be taken from the actual job using a bevel, and then the bevel angles are used to produce a template
- As before, the bends are made from the back of the former, and the pipe is adjusted in the machine, holding a slight pressure on the lever to hold the pipe in place
- A straight edge is positioned against the outside of the former and parallel with the pipe

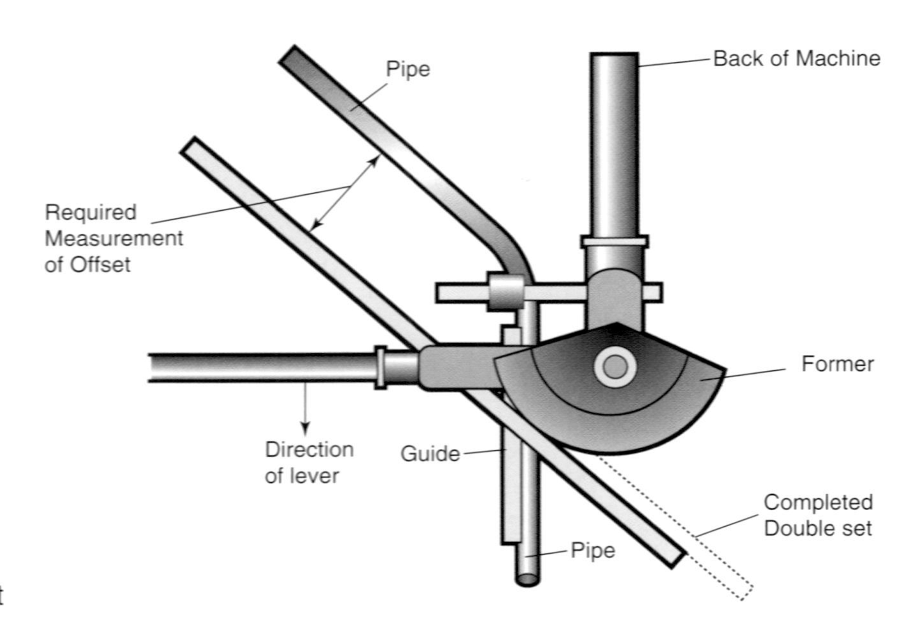

Figure 3.20 Pulling the offset

- The pipe in the machine is adjusted until the required measurement for the offset has been achieved. Alternatively, if using a template, when the first bend has been pulled it is removed from the machine, placed on the template, and marked off in the position of the second bend
- The second bend is now pulled by applying pressure to the lever arm and bending the pipe until the pipe legs are parallel.

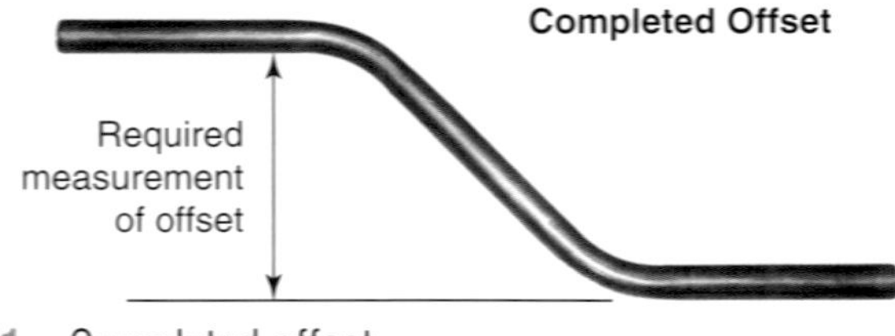

Figure 3.21 Completed offset

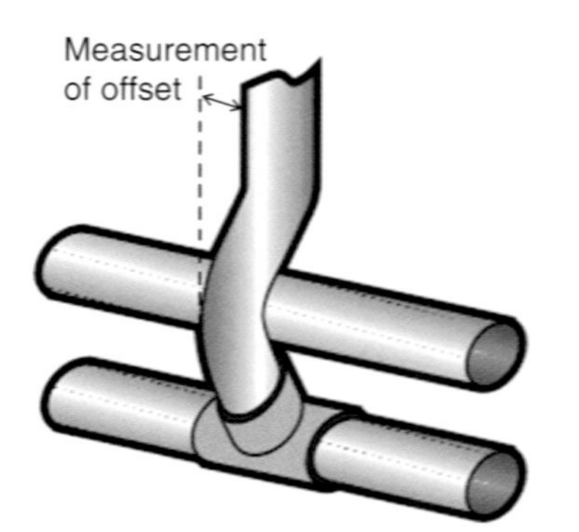

Figure 3.22 Pass-over offset

Pass-over bends

These are used to clear other obstacles such as other pipes, and can be either pass-over offsets (see Figure 3.22) or crank pass-over bends (see Figure 3.23).

The measurements for a pass-over bend are taken in the same way as an ordinary offset. The angle of the first pull will be governed by the size of the obstacle it has to 'pass over'. It is to be made sure that the first bend is not too sharp or it will be difficult when pulling the offset bend (Figure 3.24).

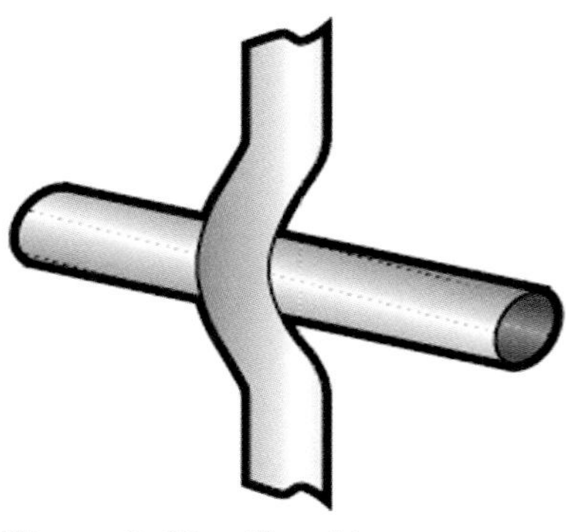

Figure 3.23 'Crank' pass-over bend

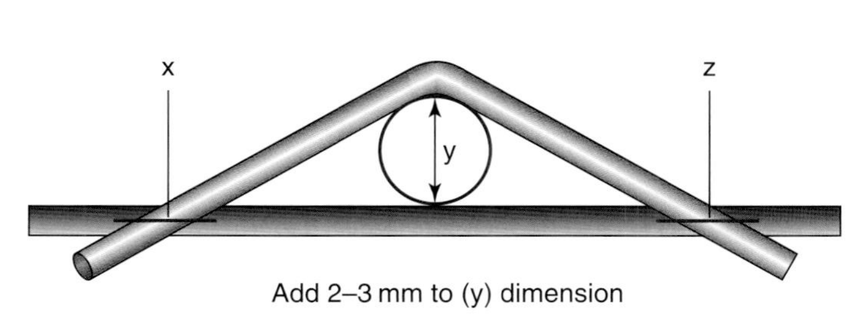

Measurement of crank passover

Figure 3.24

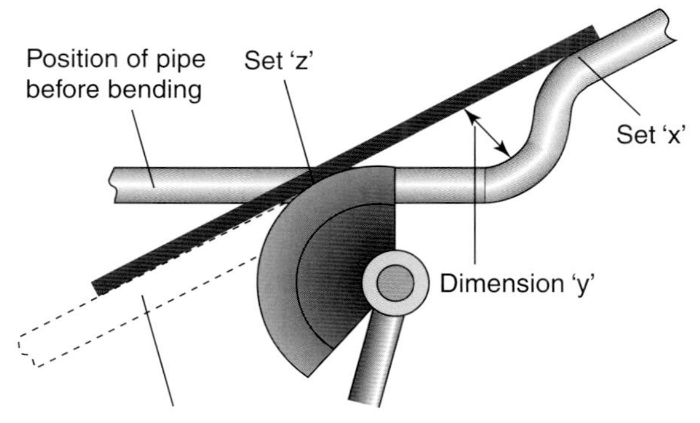

Dotted line indicates tube after bending. i.e. until in line with straight edge

Figure 3.25

> **Key point**
>
> We make every effort to explain how to bend copper pipe in these pages, but the best way to learn is by practice. Follow this information carefully, particularly when working on site.

A straight edge is placed over the bend at the distance of the obstacle and the pipe is marked. These will be the back of the finished offsets as shown in Figure 3.25.

The pipe is then returned to the machine, and when the first mark lines up with the former, the first pipe is turned around in the machine; the second mark is lined up in the former and pulled to complete the pass-over.

Low carbon steel (LCS) pipe bending

In domestic plumbing applications, steel pipes are bent using a hydraulic pipe bender.

Use of hydraulic machines

Hydraulic machines are needed to bend LCS tubes. This is due to the strength of the material, and the thickness of the pipe. Because of this you do not need to fully support the pipe with a back guard, as with copper pipe.

Hydraulic bending machines are used to form all bends, including 90° and offsets.

The hydraulic mechanism is usually oil based, and because liquids are incompressible, once under pressure it can exert considerable force on the pipe.

A typical hydraulic press bender in use is shown in Figure 3.26.

Setting out the offset:

- Mark off the required measurement for the first set onto the pipe
- Place pipe in machine, but do not make any deduction
- The measurement X mm is from the fixed end of the pipe to the centre of the set
- Pull the first set to the required angle
- Take the pipe from the machine and place a straight edge against the back of the tube. Mark the measurement of the offset as point A.
- Replace the tube back in the machine and line the mark up with the centre of the former. Pull the second set and check against the wire template. Again, allow a 5° over pull to allow for it to spring back.

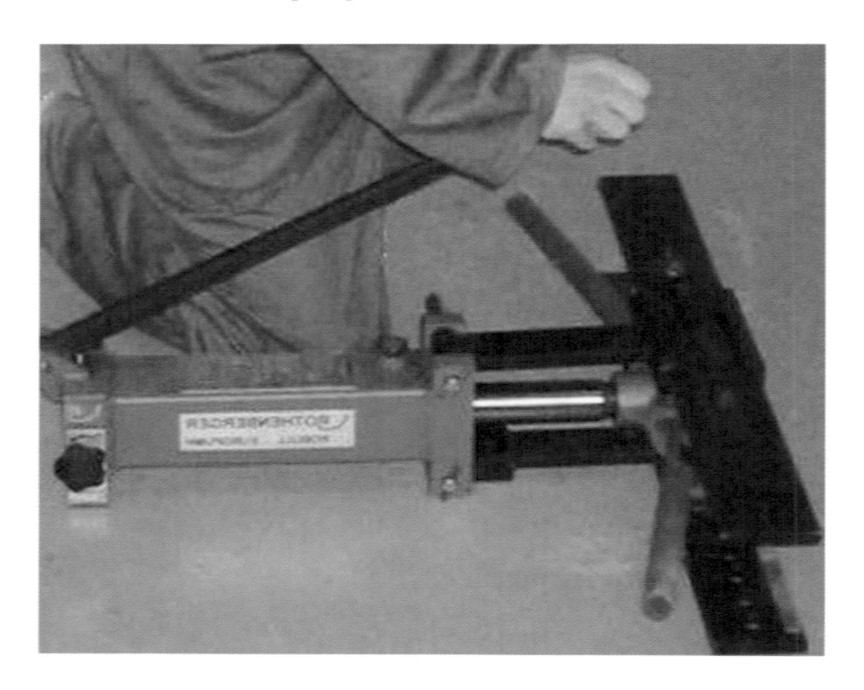

Figure 3.26 Hydraulic press bender

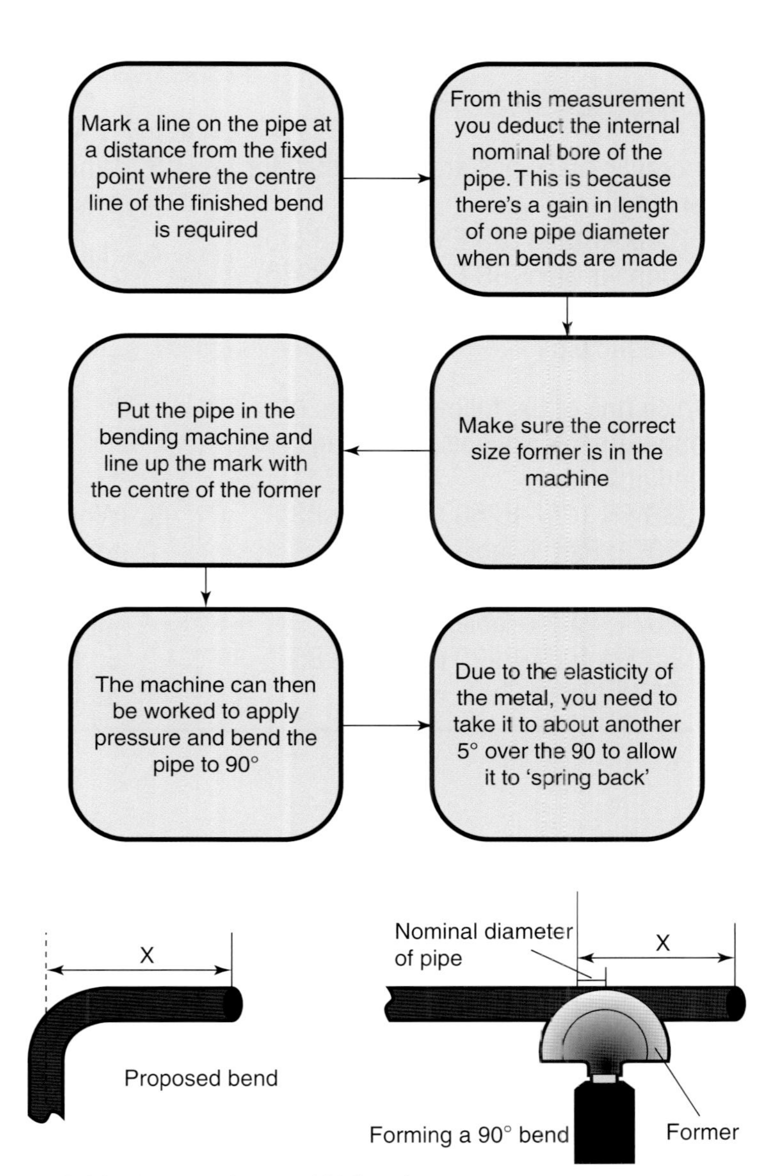

Figure 3.27 Steps to form a 90° bend

Figure 3.28 How to form a 90° bend

Key point

It is a good idea to make a template from steel wire bent to the required angle to help you achieve the required offset profile (see Figures 3.29(a) and 3.29(b)).

How to form an offset

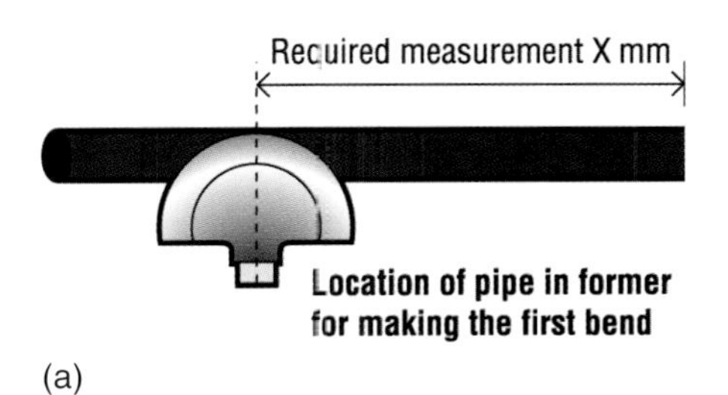

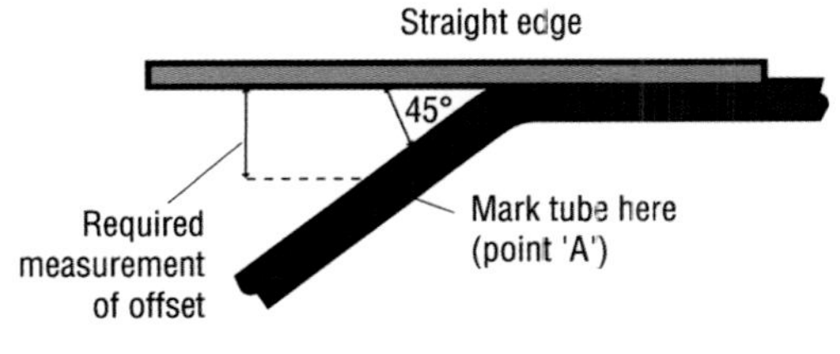

Figure 3.29 Low carbon steel offset

Test yourself 3.3

1. Hydraulic machines are best suited for bending:
   a. Copper tube (BS EN 1057 – R250) ☐
   b. Copper tube (BS EN 1057 – R220) ☐
   c. Low carbon steel tube ☐
   d. Plastic pipe ☐

2. Which one of the following types of copper tube is suitable for spring bending?
   a. BS EN 1057-R250 (formerly BS 2871; Part 1 Table x) ☐
   b. BS EN 1057-R220 (formerly BS 2871; Part 1 Table y) ☐
   c. BS EN 1057-R220 (formerly BS 2871; Part 1 Table w) ☐

3. When bending copper pipe using springs, briefly explain how the spring works?

4. Hand held bending machines are usually limited to bend copper tubes with diameters of up to:
   a. 22 mm ☐
   b. 32 mm ☐
   c. 42 mm ☐
   d. 52 mm ☐

5. Which of the following tools can be used to help determine suitable bend angles?
   a. Internal bending spring ☐
   b. External bending spring ☐
   c. 600 mm straight edge ☐
   d. 600 mm folding rule ☐

Pipework material and jointing processes

Introduction

In domestic plumbing installations, the main pipework materials you will work with will be:

- Copper
- Low carbon steel and malleable iron fittings
- Plastic.

Pipes, fittings and jointing materials acceptable for water regulation purposes are listed in the Water Regulations Advisory Scheme (WRAS) Water Fittings and Materials Directory.

A WRAS symbol shows that a product has been tested for approval and is listed in the Directory. Further information can be obtained from www.wras.co.uk.

Copper tube and fittings

Copper tube development

The first recorded use of copper for conveying water goes back to a conduit that has been dated back to 2,750 B.C., discovered at

Abusir in Egypt. Copper water pipes and cisterns were also widely used by the Romans and good examples of copper plumbing can still be seen at the archaeological site of Herculaneum, Italy, which was uniquely preserved by the eruption of Vesuvius in 79 A.D.

Historically copper tubing was expensive and only installed in prestige buildings. It was not until the development of modern types of fittings in the 1930s, which led to the introduction of light gauge copper tubes, that copper plumbing systems became highly competitive with other materials. In 1996 the latest specification for copper tubes, EN 1057, was adopted across Europe.

In the UK this specification was published as BS EN 1057:1996, '*Copper and copper alloys – Seamless, round copper tubes for water and gas in sanitary and heating applications*'. It replaced the previously familiar standard BS 2871 Part 1: 1971, '*Copper and Copper Alloys Tubes – Copper tubes for water, gas and sanitation*'.

In drawing up this standard, the opportunity was taken to rationalise tube sizes across Europe. At first glance the changes to the BS 2871 Part 1 standard must have seemed quite extensive and the available options confusing.

Yorkshire Copper Tube simplified this process by branding their products as Yorkex, Kuterlon and Minibore in line with Tables X, Y and W in BS 2871 Part 1. Under BS EN 1057, temper condition (material strength) is designated with an 'R' number, the higher the number indicating a stronger material. Tables 3.1–3.5 show the relationship of the current Yorkshire range with BS EN 1057 and BS 2871 Part 1. These tables have been reproduced with the kind permission of Yorkshire Copper Tube.

Key point

- R250 (Table X) is widely used for domestic installations but should not be used underground
- R220 (Table Y) is used for external underground installations, and can be supplied in either a blue or green (water service) or yellow (gas) plastic coating
- R220 (Table W), like Table Y, is fully annealed and is used for microbore heating

Soft condition is denoted R220, half hard R250 and hard R290. Because of the variety of sizes, both diameter and thickness should be specified when ordering to BS EN 1057. For example, when ordering half hard copper tube with an outside diameter of 15 mm and a thickness of 0.7 mm (formerly 15 mm in BS 2871 Part 1 Table X tubing) the official designation is EN 1057 − R250 − 15 × 0.7 mm. More simply, it can be ordered as 15 mm Yorkex.

BS EN 1057 – R250 half hard straight lengths are also available in chromium plate. They are used where pipework is exposed to the eye, and an attractive finish is required.

Table 3.1

Yorkex – Half Hard Range		
Size mm (od × wall)	**EN 1057 Designation**	**BS 2871 Part 1 Designation**
6 × 0.6	6 × 0.6 mm – R250	6 mm Table X
8 × 0.6	8 × 0.6 mm – R250	8 mm Table X
10 × 0.6	10 × 0.6 mm – R250	10 mm Table X
12 × 0.6	12 × 0.6 mm – R250	12 mm Table X
15 × 0.7	15 × 0.7 mm – R250	15 mm Table X
22 × 0.9	22 × 0.9 mm – R250	22 mm Table X
28 × 0.9	28 × 0.9 mm – R250	28 mm Table X
35 × 1.2	35 × 1.2 mm – R250	35 mm Table X
42 × 1.2	42 × 1.2 mm – R250	42 mm Table X
54 × 1.2	54 × 1.2 mm – R250	54 mm Table X

Table 3.2

Yorkex – Hard Range		
Size mm (od × wall)	**EN 1057 Designation**	**BS 2871 Part 1 Designation**
35 × 1.0	35 × 1.0 mm – R290	New size
35 × 1.2	35 × 1.2 mm – R290	33 mm Table X
42 × 1.0	42 × 1.0 mm – R290	New size
42 × 1.2	42 × 1.2 mm – R290	42 mm Table X
54 × 1.0	54 × 1.0 mm – R290	New size
54 × 1.2	54 × 1.2 mm – R290	54 mm Table X
66.7 × 1.2	66.7 × 1.2 mm – R290	66.7 mm Table X
76.1 × 1.2	76.1 × 1.2 mm – R290	76.1 mm Table X
108 × 1.5	108 × 1.5 mm – R290	108 mm Table X
133 × 1.5	133 × 1.5 mm – R290	133 mm Table X
159 × 2.0	159 × 2.0 mm – R290	159 mm Table X

Table 3.3

Kuterion – Straight Tube Range		
Size mm (od × wall)	**EN 1057 Designation**	**BS 2871 Part 1 Designation**
6 × 0.8	6 × 0.8 mm – R250	6 mm Table Y
8 × 0.8	8 × 0.8 mm – R250	8 mm Table Y
10 × 0.8	10 × 0.8 mm – R250	10 mm Table Y
12 × 0.8	12 × 0.8 mm – R250	12 mm Table Y
15 × 1.0	15 × 1.0 mm – R250	15 mm Table Y
22 × 1.2	22 × 1.2 mm – R250	22 mm Table Y
28 × 1.2	28 × 1.2 mm – R250	28 mm Table Y
35 × 1.5	35 × 1.5 mm – R290	35 mm Table Y
42 × 1.5	42 × 1.5 mm – R290	42 mm Table Y
54 × 2.0	54 × 2.0 mm – R290	54 mm Table Y
66.7 × 2.0	66.7 × 2.0 mm – R290	66.7 mm Table Y
76.1 × 2.0	76.1 × 2.0 mm – R290	76.1 mm Table Y
108 × 2.5	108 × 2.5 mm – R290	108 mm Table Y

Table 3.4

Minibore – Coil Range		
Size mm (od × wall)	**EN 1057 Designation**	**BS 2871 Part 1 Designation**
6 × 0.6	6 × 0.6 mm – R220	6 mm Table W
8 × 0.6	8 × 0.6 mm – R220	8 mm Table W
10 × 0.7	10 × 0.7 mm – R220	10 mm Table W

Table 3.5

Kuterion – Coil Range		
Size mm (od × wall)	**EN 1057 Designation**	**BS 2871 Part 1 Designation**
12 × 0.8	12 × 0.8 mm – R220	12 mm Table Y coil
15 × 1.0	15 × 1.0 mm – R220	15 mm Table Y coil
22 × 1.2	22 × 1.2 mm – R220	22 mm Table Y coil
18 × 1.2	18 × 1.2 mm – R220	28 mm Table Y coil

Key point

Where might you see chromium-plated pipework on a domestic plumbing installation? Probably the most likely place you would find chrome-plated pipe used is for exposed pipework runs to stand alone gas fires; other examples would be instantaneous showers in bath or shower rooms.

Methods of jointing copper tube

This falls under three main headings:

- Compression joints
- Soldered joints
- Push fit joints.

Compression joints

Can be subdivided into:

- Manipulative
- Non-manipulative.

The term manipulative, as used here, means to work or form the end of the tube.

Manipulative joint (Type B)

Figure 3.30 below shows a typical manipulative fitting.

Type B or manipulative fittings are used with soft copper tube and require the plumber to flare the tube end before the joint is assembled. An adaptor fits between the end of the pipe and fitting as shown. Type B compression fittings can be used to join pipes above or below ground.

Non-manipulative joint (Type A)

Figures 3.31(a)–(d) are reproduced with the kind permission of Pegler Limited. Tel: 01302 560560; Website: www.pegler.co.uk

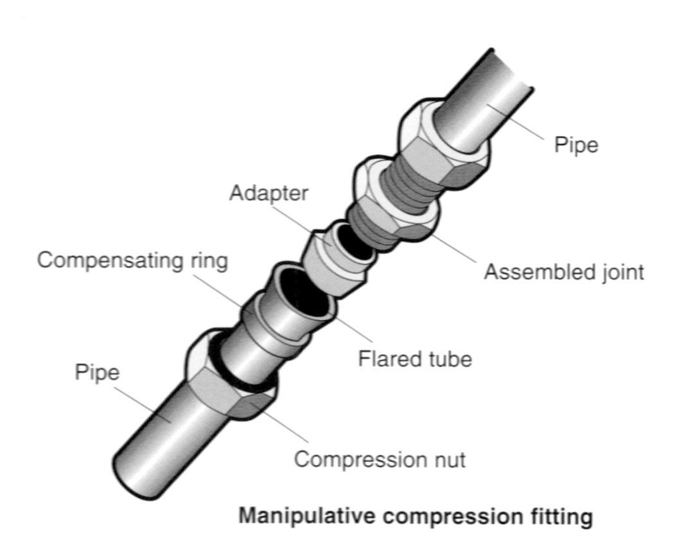

Figure 3.30 Manipulative joint

Key point

What do you think is the difference between these types of solder jointing methods? The difference between the two jointing methods, between the two fittings is that: the solder ring fitting already contains solder in the raised ring within the fitting (integral). When the fitting is heated up, the solder melts (at between 180 and 230°C) and the solder is drawn into the fitting by capillary attraction. It is the same principle for the end feed fitting, but this time the solder is end fed separately from a spool of wire solder.

The following diagrams (Figures 3.31(a)–(d)) show the basic steps required to form a non-manipulative type A joint.

Type A, or non-manipulative fittings enable the plumber to make a compression joint without carrying out any work on the tube ends other than ensuring that they are clean and cut squarely. Type A compression fittings can be used to join pipes above ground only.

A range of brass, gunmetal, cz-resistant to dezincification, and chromium-plated brass, fitting designs, compatible with pipe sizes, are available.

Soldered capillary joints

Soldered joints can be classified as soft soldered, on which we will concentrate here, and hard soldered, such as silver and silver alloys, which is not a domestic application (Figure 3.32).

Soft soldered joints are made using two types of fittings:

- Integral solder ring
- End feed solder.

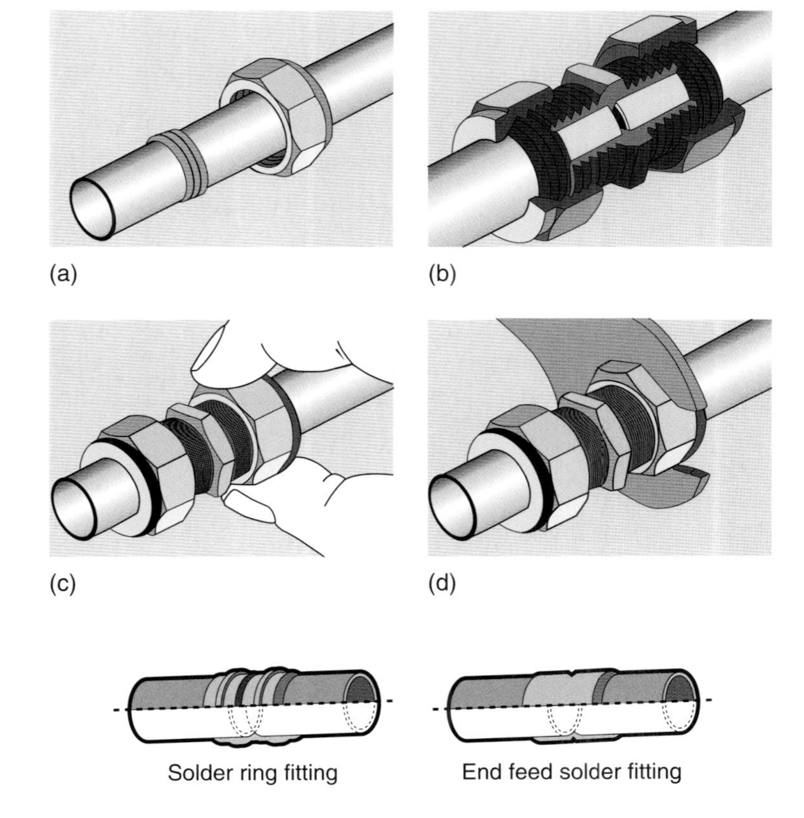

Figure 3.31 Non-manipulative jointing process (a) The pipe is cut to length and de-burred, and a nut and olive fitted, (b) Both pipe ends and olives are pushed home, (c) The nuts are hand tightened, (d) The joint is completed by tightening using a spanner or adjustable grips (Reproduced with permission of Pegler Limited)

Figure 3.32 Solder fittings

The jointing process

Diagrams in Figure 3.33 are reproduced with the kind permission of Yorkshire Fittings Ltd.

Tel: 0113 270 1104 Website: www.yorkshirefittings.co.uk/

Figure 3.33 shows the jointing of an integral solder ring fitting.

- Clean and de-burr the pipe
- Clean the fitting
- Apply the flux
- Apply heat until you see the solder appear.

There are a number of fluxes on the market that are heat activated. This means a cleaning action takes place during the heating process, so you do not have to pre-clean the pipe end or fitting. When using these fluxes, you should make sure they are non-acidic, non-toxic and WRAS-approved for use on hot and cold pipework installations.

Push fit joints for copper pipes

A number of types of push fit joints are available for use on hot and cold water supplies. Here is a typical example, illustration supplied by Hepworth from their Hep2O range (Figure 3.34).

They are made from plastic. A grab ring is used to lock the pipe in place and a neoprene 'O' ring makes it water-tight. The fittings are

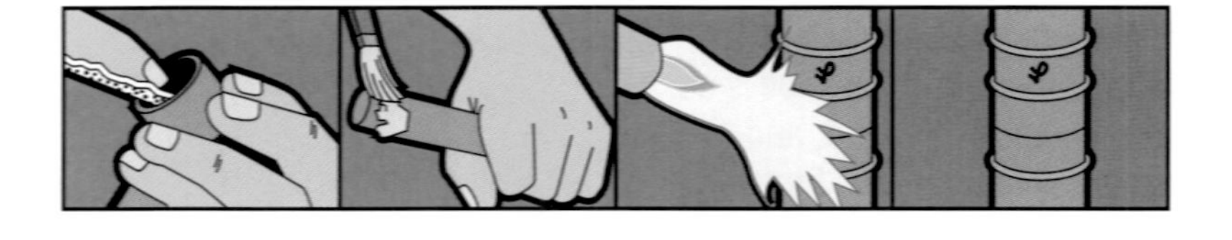

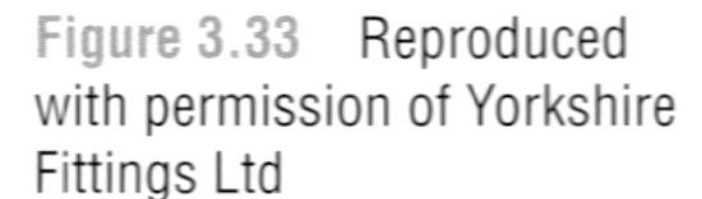

Figure 3.33 Reproduced with permission of Yorkshire Fittings Ltd

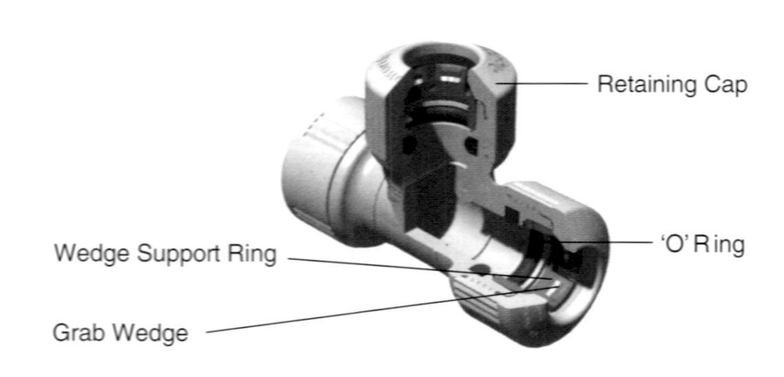

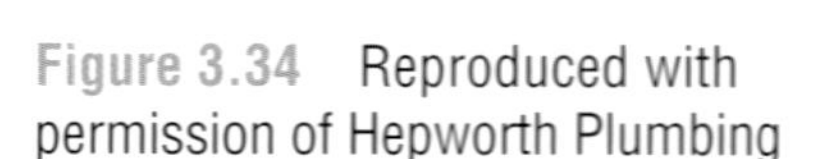

Figure 3.34 Reproduced with permission of Hepworth Plumbing

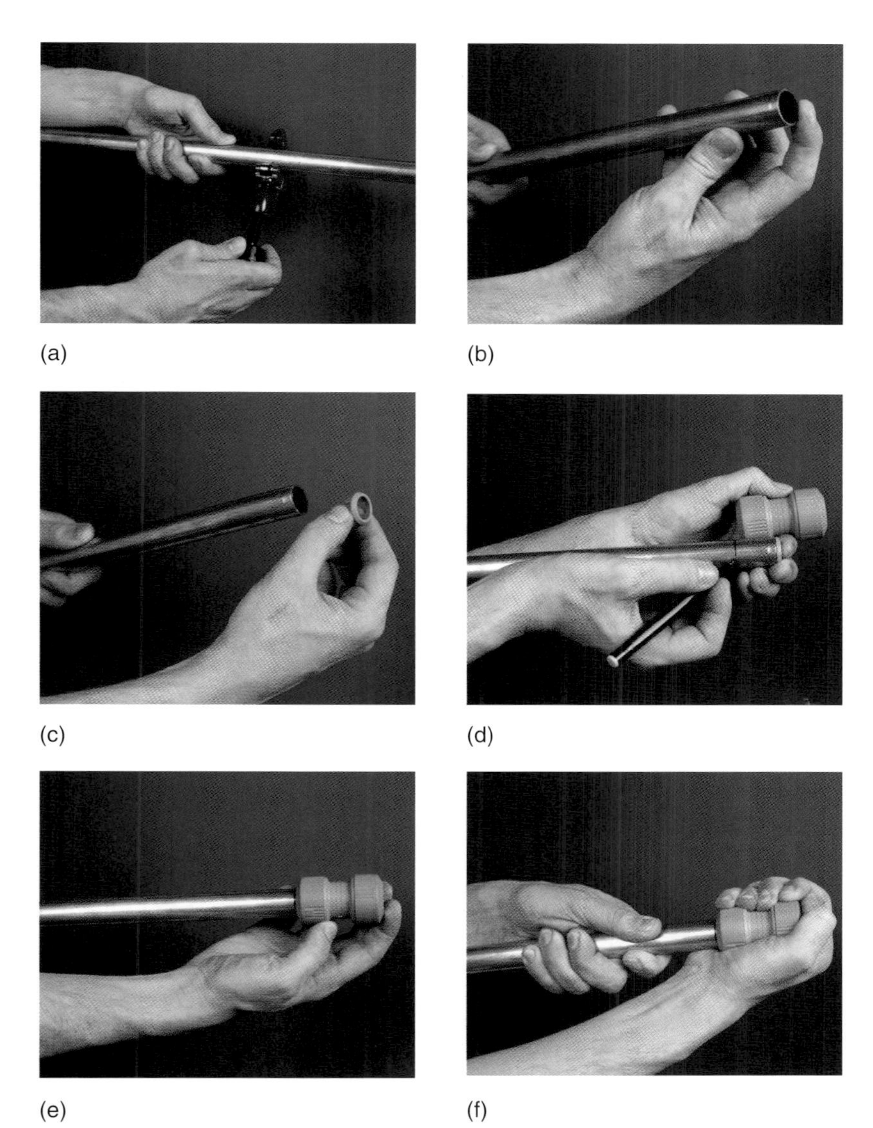
(a) (b) (c) (d) (e) (f)

Figure 3.35 (a) Cut tube to length, (b) Clean and de-burr the pipe, (c) Insert ring to end of pipe, (d) Check pipe depth, (e) Fully push fitting into place, (f) Complete joint (Reproduced with permission of Hepworth Plumbing)

quite bulky, so do not look too aesthetically pleasing in exposed locations.

How to make the joint? (See Figure 3.35).

Copper to copper push fit systems

Yorkshire Fittings produce copper to copper push fit fittings (also available in stainless steel and for use on barrier pipe). The trade name is Tectite, and the fittings work on the same principle as the 'grab ring'.

Example of a Tectite push fit tee fitting.

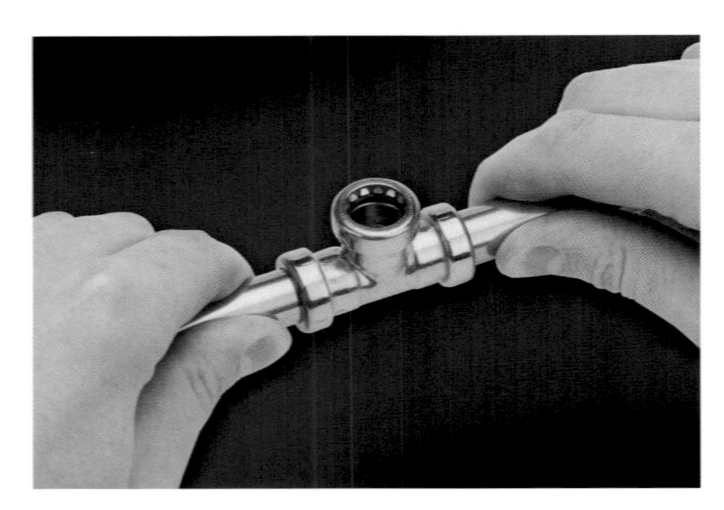

Figure 3.36

Yorkshire Fittings produce a range of press fit fittings known as the XPress system. The jointing procedure works on the principle of exerting pressure from a press fit tool around the end of a fitting which encompasses a purpose made 'o' ring, thus forming a perfect seal.

Ensuring that the tube is fully inserted into the fitting, the jaws of the press-fit tool are placed around the collar of the fitting, which contains a butyl or EPDM 'O' ring (the XPress Gas range contains NBR 'O' rings). With the jaws at a 90° angle to the fitting, the press-fit tool is activated and the jaws compress the 'O' ring tightly onto the tube creating a strong and reliable joint.

Key point

Yorkshire Fittings produce catalogues explaining about Tectile and XPress fittings, including jointing techniques. Try to obtain copies of these catalogues to see how it's done.

Images reproduced with the kind permission of Yorkshire Fittings contact www.yorkshirefittings.co.uk

Example of a XPress push fit reducing tee fitting.

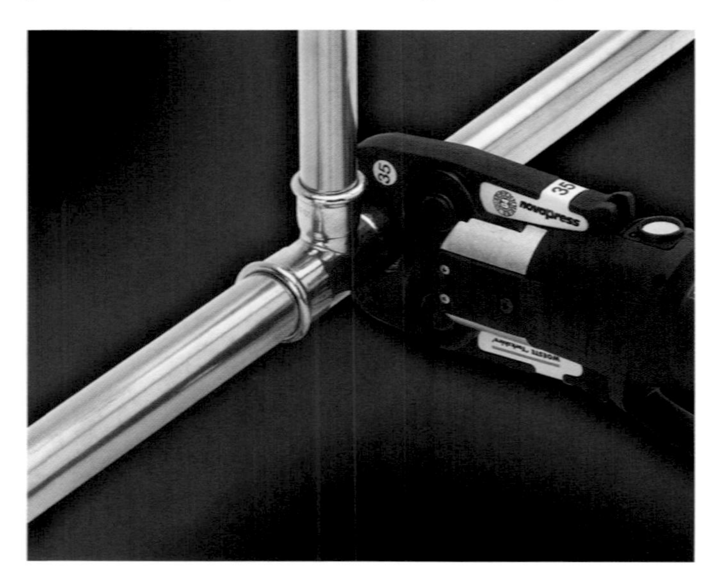

Figure 3.37

Key point

Black painted LCS must only be used on wet heating systems, oil and gas supply pipework. It must not be used on cold and hot water supplies. Why do you think this is so?

Low carbon steel (LCS) pipe and fittings

Often referred to as mild steel, low carbon steel pipe is supplied in three grades:

- Light – which is identified by the colour code brown
- Medium – colour code blue
- Heavy – colour code red.

It can be supplied either in a painted black finish, or with galvanised coating.

Generally speaking, light-grade tube is not used on plumbing pipework. You are most likely to work on medium-grade pipes, and occasionally heavy grade.

Key point

Nominal bore means that it is not the actual bore of the pipe as this will vary depending on the thickness of the pipe wall, which in turn will be determined by the grade.

Medium and heavy grades are available in 6 m lengths, ranging from 6 to 150 mm diameter, specified as nominal bore.

Methods of jointing

For domestic installations, there are two main jointing methods:

- Threaded joints
- Compression joints.

Threaded joints

This is shown in Figure 3.38.

Jointing LCS pipe can be done by cutting threads into the end of the LCS pipe to give a British Standard Pipe Thread (BSPT), then jointing them together with a range of female threaded fittings

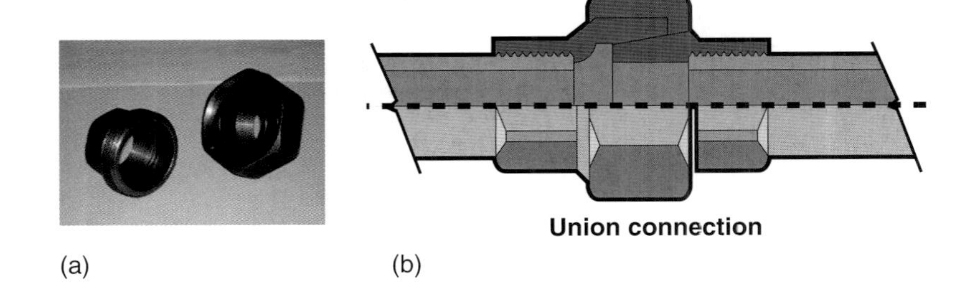

Figure 3.38 (a) Section through a male/female union (b) Section through union connection showing pipework

Try this

You will cover pipe threading using stocks and dies and pipe jointing as part of your practical training at your college or centre. When you have completed this exercise write up a report of what you did. Include in the report:

- Tools and equipment used
- Materials and fittings
- Safety precautions
- How you tackled the job step-by-step including how you made the joint.

Discuss the report with your tutor.

made from steel or malleable iron (Figure 3.39). The threads are cut using stocks and dies; the stocks being the body and handle of the tool, the dies being the actual cutter.

Use of pipe threading machine

Pipe threading machines, like the one shown here (or even smaller portable versions) provide a quicker and easier method of forming threads for LCS pipes. The machine is an 'all in one' combined pipe cutter, de-burring reamer, also comprising stock head and dies (Figure 3.40).

Threaded pipe fittings for LCS

As mentioned earlier, these can be made of steel or malleable iron. Steel fittings can withstand higher pressure, but are more expensive than malleable iron. They are manufactured to BS 1740 for steel and BS 1256 for malleable iron.

(a)

(b)

Figure 3.39 (a) Head gear (b) Ratchet threader (Reproduced with permission of Ridgid Tool)

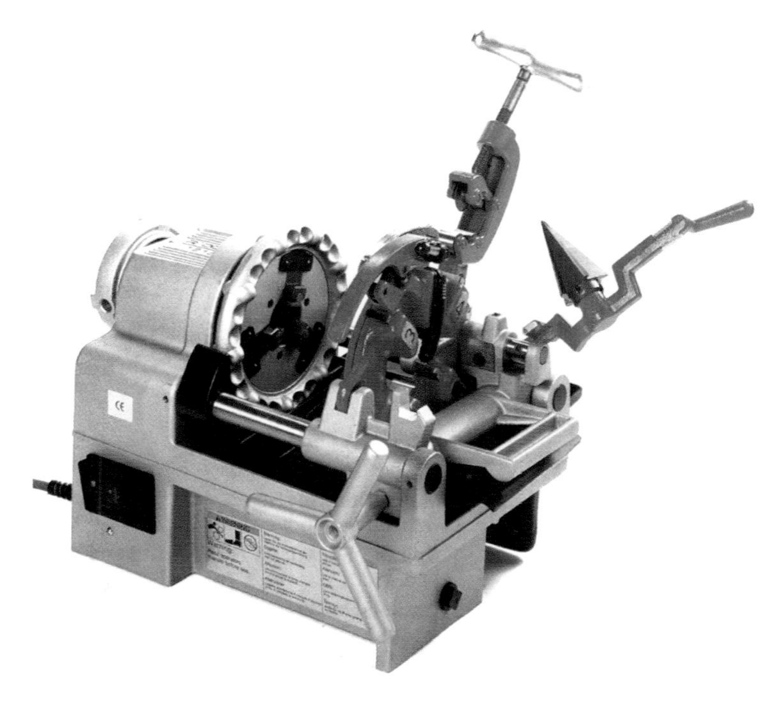

Figure 3.40 Pipe threading machine (Reproduced with permission of Ridgid Tool)

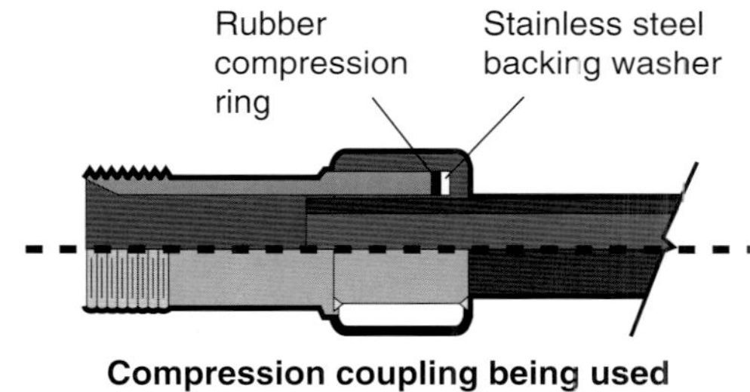

Figure 3.41 Compression fitting

Key point

Due to the way in which screwed joints are made and installed (rotated in the pipe), it is sometimes difficult to remove or assemble lengths of pipework. Where this is the case, a union connector like the one shown in Figures 3.38(a) and 3.38(b) which allows the pipework joint to be 'broken', should be used.

Malleable iron fittings are adequate for domestic installations and, again like copper tube fittings, there is a wide range available.

Compression joints

There are a number of manufacturers' designs for compression joints.

Here is a typical example (Figure 3.41).

The fitting is designed to enable steel pipes to be joined without threading. Made of malleable iron, they use locking rings and seals which are tightened onto the pipe. They can be used on water and gas supplies, and although more expensive than threaded joints, they do save time on installation.

Plastic tube and fittings

Plastic pipes and fittings fall into a number of categories: Polythene, Propylene (pp) and Polyethylene (MDPE), all of which are by-products of polymerisation of ethers:

- BS 4991: Propylene copolymer for pressure pipe should not be used in installations where the working temperature exceeds 20°C
- BS 6572: Blue polythene pipes up to a nominal size of 63 mm for below ground supply of wholesome water
- BS 6730: Black polythene pipes up to a nominal size of 63 mm for above ground supply of wholesome water.

Other plastic pipes used for hot and cold water installation, wastes and overflows are:

- BS 7291 Part 1: polybutylene (PB) pipes (10–35 mm)
- BS 7291 Part 3: Cross-linked polyethylene (PE-X) pipes (10–35 mm)
- BS 7291 Part 4: Chlorinated polyvinyl; (PVC-C) pipes (12–63 mm) and unplasticised polyvinyl (UPVC)
- ABS – Acrylonitrile butadiene styrene – No British standard is available for the material and it is not suitable for use on hot water services.

Before we go into pipe jointing in further detail, here is an overview of typical jointing methods for plastic pipes (Figure 3.42).

Key point

The black colouring is actually added to prevent the ultraviolet rays from the sun degrading and hardening the pipe, causing it to crack, and to prevent the transmission of light, protecting the water quality in wholesome water pipes – from the growth of microorganisms, e.g. Legionella.

Jointing methods

Fusion welding: Polythene and polypropylene (MDPE and PP). The jointing process requires the use of special equipment and fittings and is mainly used on water and gas distribution main installations. So, we will not go into it in further detail here.

Types of plastic	Mechanical joints	Solvent welding	Fusion welding	Push fit 'O' ring Discharge pipes/ overflows only
Polythene	✓	×	✓	×
Polyethylene	✓	×	×	×
Polypropylene	✓	×	✓	✓
PVC-U	✓	✓	×	✓
ABS	✓	✓	×	✓

Figure 3.42 Overview of jointing methods

Mechanical jointing:

This applies to the jointing of:

- Polythene/Polyethylene pipework
- ABS, PVC-U and PVC-C.

Polythene/Polyethylene pipe. It is used for underground water mains, and is identified by its blue colouring. It can also be found on internal cold water services, and is coloured black.

The joints are made using metallic or plastic (e.g. polypropylene) compression fittings. The jointing process involves:

- Cutting the pipe to the required length
- De-burring the pipe inside and out
- Sliding the cap nut and compression ring onto the pipe and inserting the approved liner into the pipe
- Making sure the pipe is fully inserted into the fitting and hand tightened
- Completing the tightening process using adjustable grips or spanners.

Key point

Inserting the approved liner into the pipe stops the plastic pipe from squashing when the nut is tightened onto the body of the fitting and pressure is applied by the compression ring.

ABS, PVC-U and PVC-C. Solvent-welded jointing is used to join ABS, PVC-U and PVC-C pipe materials using approved solvent cement. The cement temporarily dissolves the surface of the pipe and fitting, causing two surfaces to fuse together. It is used for joints on soil and vent systems, waste pipes, overflows, and some cold water pressure pipe installations.

Here is a section (Figure 3.43) through a typical solvent-welded joint used on a waste system.

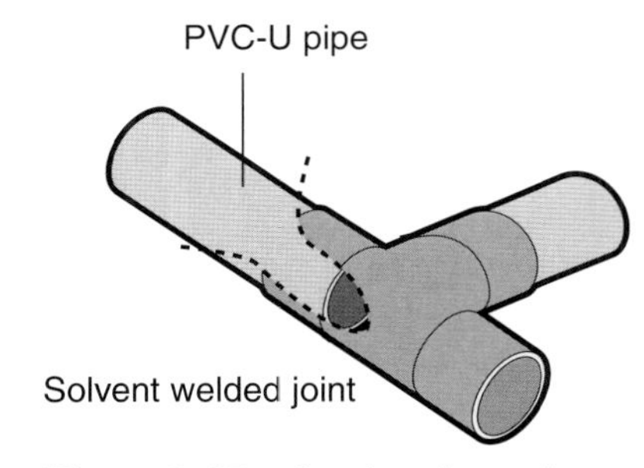

Figure 3.43 Section through solvent-welded joint

Push fit joints – used on domestic and heating overflow pipework. These are used mostly on PVC-U and PVC-C overflow pipework applications. The pipe is cut to length, making sure it is square and de-burred. The outside edge of the pipe is chamfered to give it a leading edge to make it easier when pushing it into the fitting.

Push fit connectors:

These are termed as flexible push fit plumbing systems for 'use only' on hot and cold water installations and central heating circuits. A number of manufacturers produce pipe and fittings for these systems, and the fittings can be used either on plastic or copper. Plastic pipes are available in various diameters ranging from 10 to 22 mm and in lengths of 3 m and 6 m, or in coils of 25, 50 and 100 m.

Key point

Hepworth produce an 'Installer Guide' for their Hep_2o products, (details are given at the end of this unit). They also produce self-adhesive metal tape, which, when applied to their plastic pipe, enables them to be detected when installed in walls.

Many flexible push fit plumbing systems are manufactured from polybutylene (this is part of the polyolefin family of plastics) and allow the permeation of oxygen through the pipe wall. Therefore, polybutylene pipe is also available with a protective barrier wall to prevent occurrence of permeation, this is appropriately named 'barrier pipe'.

Barrier pipes should ideally be used in vented and sealed heating systems, reducing the risk of system corrosion, especially instances where an inhibiter is not used. Barrier pipe conforms to the requirements of BS 7291 Parts 1 and 2. The type of fitting used is the same as in Figure 3.34.

How to make the joint

The instructions are as follows and as shown in Figure 3.44.

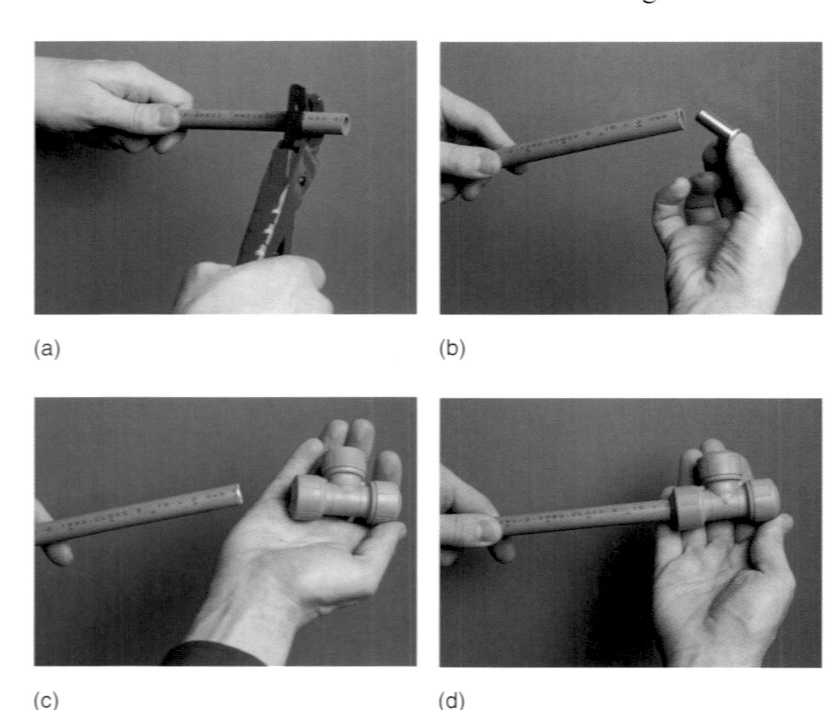

Figure 3.44 (a) Cut pipe to length and de-burr (b) Insert approved support liner (c) Position push fit fitting (d) Complete the joint (Reproduced with permission of Hepworth Plumbing)

Jointing different metals

Fittings that are designed to join copper to plastic are available. These are of mechanical type and are available in either metallic or plastic finish. Fittings for jointing copper to LCS are also available.

Lead pipework is no longer allowed for new installations, but it may be necessary to join onto an existing supply in order to extend a system where a customer cannot afford to replace long runs of lead pipework, or where the joint is at the end of an underground service pipe.

There are various types of fittings available in the market for jointing plastic to lead, for below ground use, and lead to copper. These fittings should be WRAS-approved products.

Here are some examples, as shown in Figure 3.45.

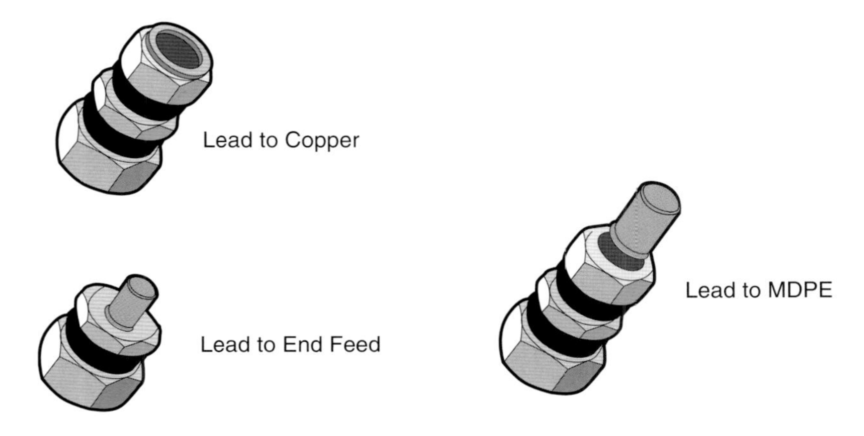

Figure 3.45 Adaptor fittings

Key point

Make sure you include these activities in your portfolio and discuss your findings with your tutor.

Try this

There is a vast range of different types of fittings used for joining copper, low carbon steel and plastic pipes. We strongly recommend that you get hold of manufacturers' fitting catalogues so that you can see what is available in the market. Use the contact details that are included in this unit. Once you have obtained the catalogues, have a look at the next activities.

Try this

Figure 3.46 shows a number of sections from an installation drawing for a domestic dwelling. You are required to produce a fitting schedule for each. Using this support material and manufacturers' catalogues you are required to specify a suitable plastic material for the underground service pipework from the external stop valve to the internal stop tap and drain valve, and then for the rising main to the cistern. Include pipe sizes and fittings.

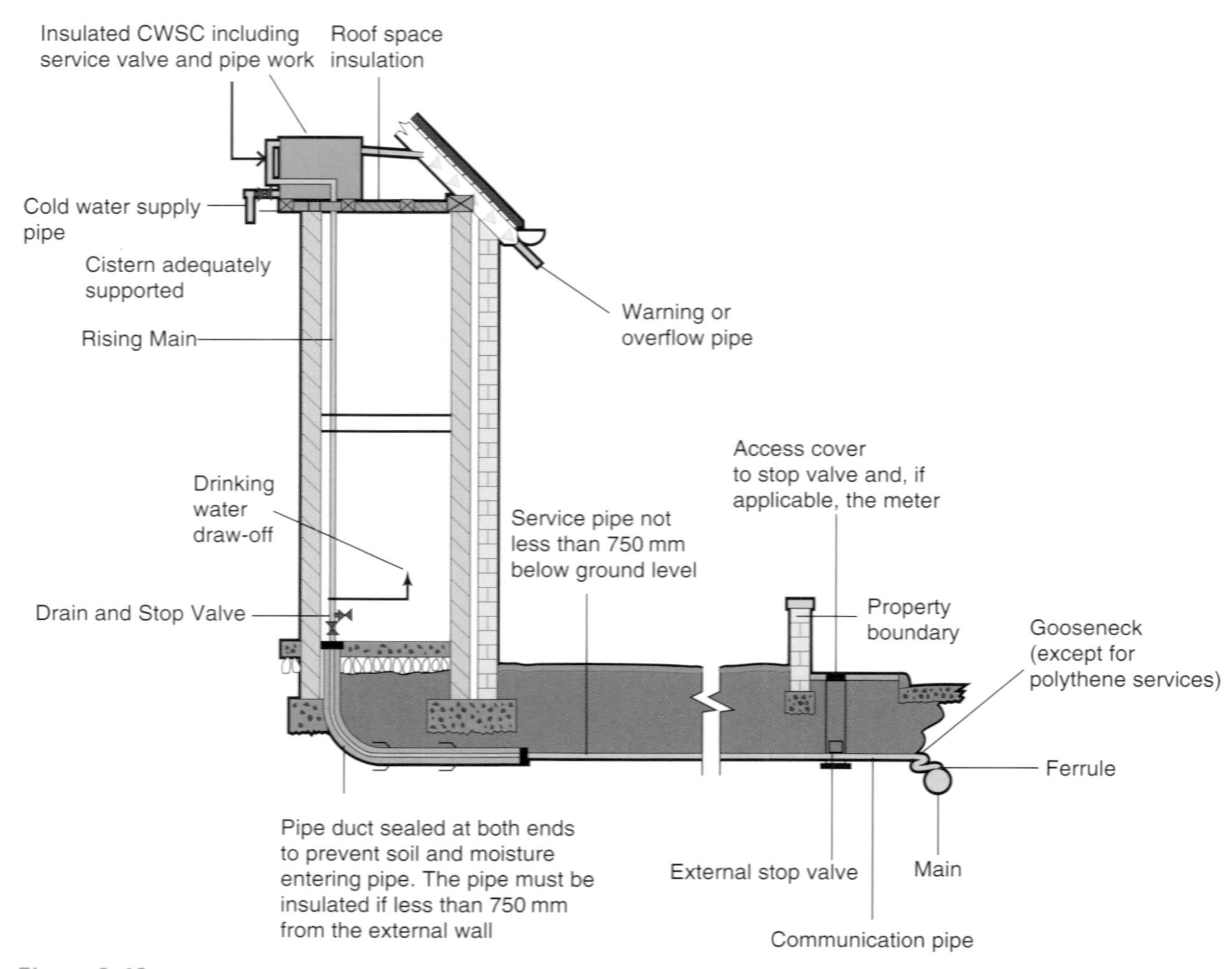

Figure 3.46

Try this

This installation is to be in copper (Figure 3.47). Here you need to:

1. Specify the grade of copper to be used
2. Specify the type of jointing method to be used and why
3. Using this support text and manufacturers' catalogues produce a fitting schedule only for the pipework installation. Include the crossover.

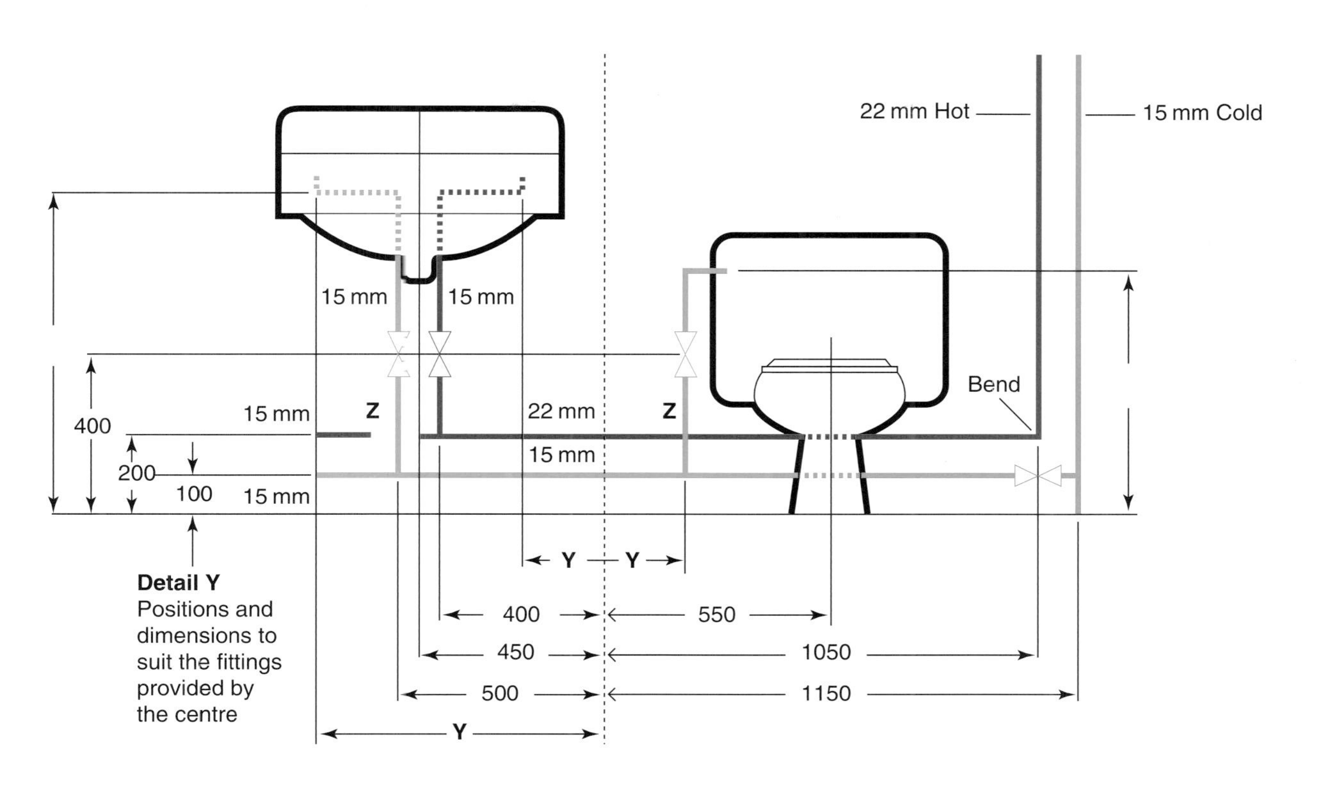

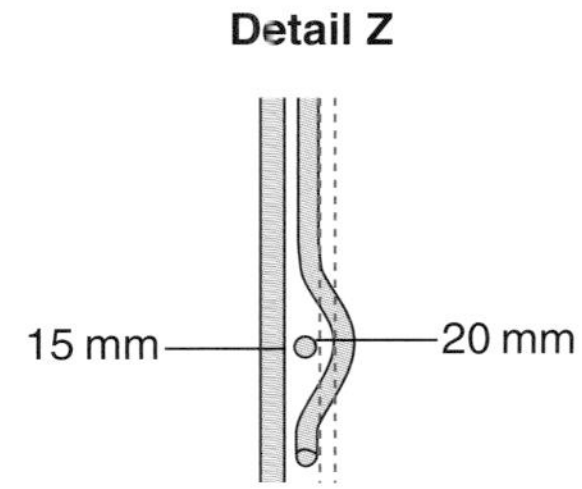

Figure 3.47

Try this

This heating installation is to be in LCS (Figure 3.48).

1. Specify the grade of LCS to be used
2. Using this support material and manufacturers' catalogues, produce a fitting schedule only for the pipework installation. Include the crossover.

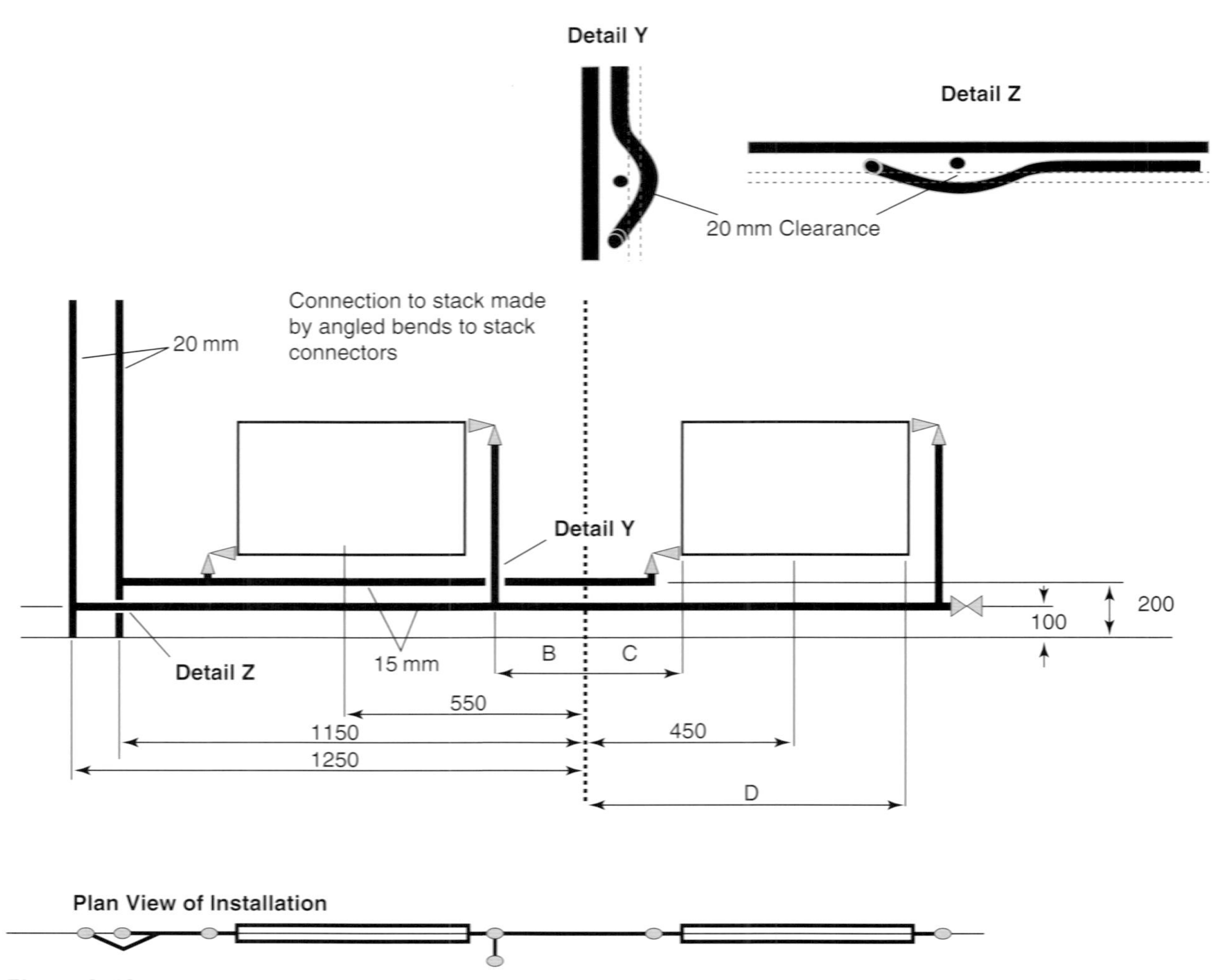

Figure 3.48

Test yourself 3.4

1. What are the three main materials used for new domestic plumbing installations?
2. From the list below, tick two answers which you think are correct.

   The two main types of material used for fittings to joint LCS are:
   - Steel
   - Stainless steel
   - Copper
   - Malleable iron
3. You have been asked to make a joint at a given length of LCS tube to an elbow. Describe how you would tackle the job, including tools and equipment used. Oh, you do not have access to a powered threading machine.

Fixing devices, pipe supports and brackets

Introduction

In this section, we will take a look at the various types of fixing devices, such as screws and plugs. Different types of clips and brackets are used for securing the pipework, so that it looks neat, and is kept in the position it should be. Fixings should also provide sufficient support to the pipework or fittings so that they withstand possible accidental damages from people treading on it, children pulling at it and so on.

As a plumber you will be required to fix the pipework, sanitary appliances, boilers and radiators to various surfaces. You will also need to know how to refit boards and access traps in timber floors.

Try this

We expect you to do some additional work in this section. There are so many types of fixings, supports and brackets that you need to obtain manufacturers' catalogues to find out information for yourself.

Fixing devices

These include:

- Brass wood screws
- Self-tapping screws
- Turn threaded wood screws
- Steel countersunk screws
- Chipboard screws
- Mirror screws
- Plastic wall plugs
- Plastic board fixings
- Cavity fixings
- Nails
- Corrosion-resistant (plated) screws.

Try this

The range of fixing devices is vast, and too detailed to describe fully here. You should try to get hold of a hardware catalogue so you can have a look at the range for yourself. Try www.screwfix.com

Key point

The condition of the wall you are fixing it to might not be very good, so you might have to use longer screws and thicker gauge. Trial, error and experience are the factors here!

Screws

Screws are specified in length in inches or millimetres, and gauge. A few examples used in domestic plumbing installations include:

- 3/4″ (20 mm) × No8 for fixing saddle clips
- 1¾ 10sb to b2 (50 mm) to 2½ (65 mm) 12s for fixing radiator brackets.

This is only a rough guide and you often have to make a decision about the appropriate length and gauge for a particular situation.

Brass, alloy or zinc-plated screws are used internally where they may be affected by the moisture. This could be in situations where they are fitted close to boilers and towel rail radiators. They are also used externally for soil and vent stack installations, gutter and rainwater systems. Self-tapping screws are required when fixing into sheet metal.

Steel countersunk screws are used for general tasks, such as fixing clips and radiator brackets.

Try this

Here are a few screw types in Figure 3.49. Can you state what they are? Use catalogues to find out.

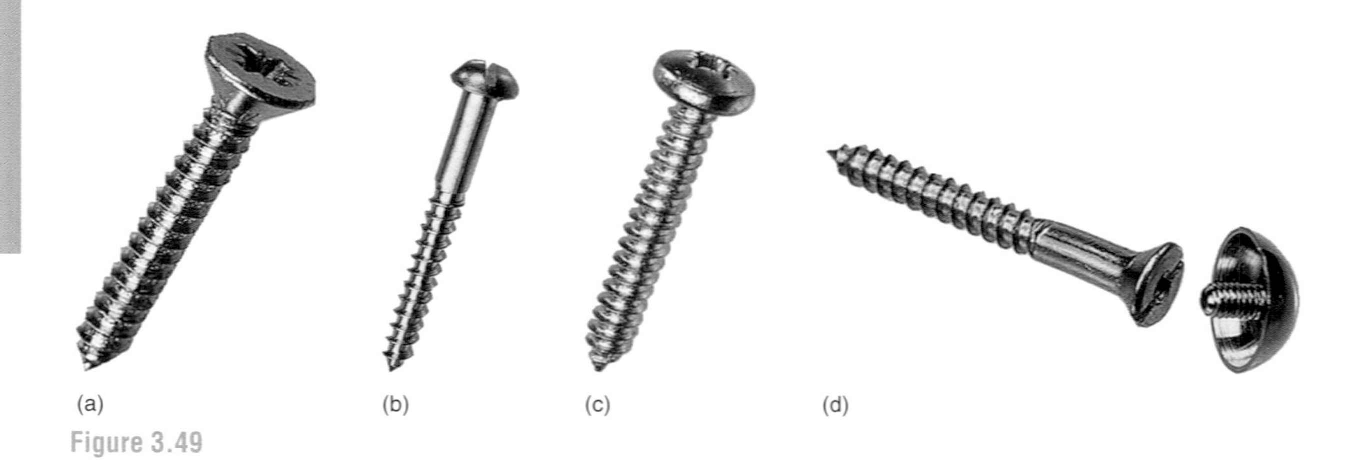

Figure 3.49

Plastic wall plugs

These come in a range of gauges that are appropriate to the gauge of the screw. They are colour-coded for ease of selection.

Selecting the drill, plug and screw

Here is a helpful chart (Figure 3.50). Figure 3.51 gives the details of the fixing operation using plug and screw.

Screw size (gauge)	Drill size (mm)	Plug colour code
6–8	5	Yellow
8–10	6	Red
10–14	7–8	Brown
14–18	10	Blue

Figure 3.50 Screw size, drill size and plug colour code

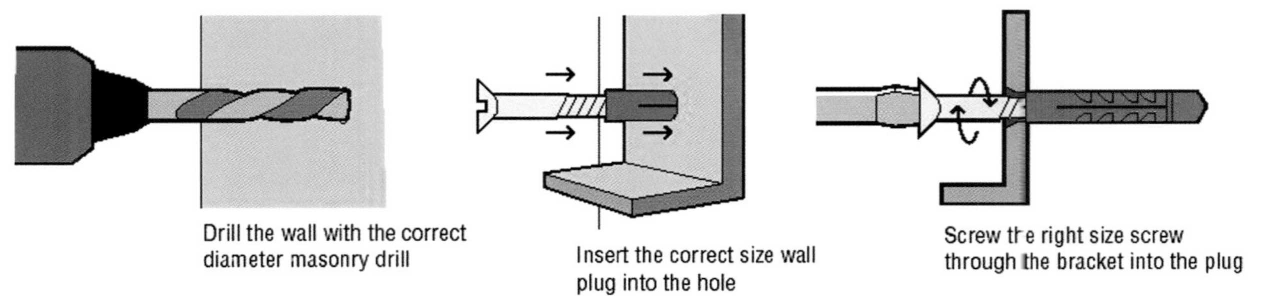

Figure 3.51 Fixing operation using plug and screw

Plaster board fittings

Several types are available from the manufacturers. In the case of the spring toggle, a hole is drilled in the plaster board, big enough to take the toggle when folded. This is inserted through the hole and once through to the space behind the board, the toggle is pushed open by the spring.

The rubber nut fixing works on the principle of drawing the nut mounted in the rubber towards the screw head. As it tightens, the rubber is squashed to form a flange to the back of the board (see Figure 3.52).

Metal plaster board fixing

A cavity fixing works on the same principle as rubber nut fixing, but here the aluminium body is 'squashed' to form the flange. With an all metal plaster board fixing, a small pilot hole is drilled into the plaster board and the complete fixing is screwed into the plaster board (Figure 3.53). The screw is then removed leaving a fixing point similar to a wall plug.

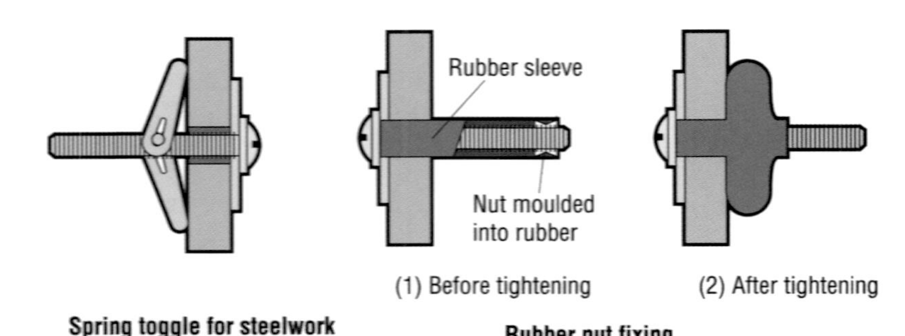

Figure 3.52 Spring toggle and rubber nut fixing

(a)

(b)

Figure 3.53 (a) Metal plaster board and (b) Cavity fixing

Other fixings

The fixings described so far should be adequate for your day-to-day work.

There will be occasions, however, where you may be fixing a heavy appliance, or where the fixing surface is in a poor condition. Fixings of this type would include:

- Coach screws
- Wall bolts.

Nails

A plumber will use a range of nails, particularly on maintenance, repair and refurbishment work.

Activity 3.4

What type of jobs do you think a plumber might require the use of nails

Please check out the answer given at the end of this book.

Try this

Again, use manufacturers' catalogues to see what clips and brackets are available.

Try this

Once you have completed these questions, mark the answers yourself. Not all the answers are in the text, so you may have to find out from the manufacturers' catalogues! This is a good revision exercise. Check your answers with your tutor.

Clips and brackets

It is likely that the bulk of your work will take place in domestic dwellings. In this case, the use of copper or plastic clips is adequate for supporting copper and plastic pipework. Like fixings, there is a range of clips available for this purpose.

However, there may be times when you may have to work in buildings other than domestic dwellings. This could include schools, hospitals or small industrial units, in which case the clips or brackets need to be strong and robust.

Table 3.6 shows recommended spacings for internal pipework fixings.

Table 3.6

Pipe size		Copper pipe		Steel size		Plastic pipe	
(mm)	(in)	Horizontal	Vertical	Horizontal	Vertical	Horizontal	Vertical
15	1/2	1.2	1.8	1.8	2.4	0.6	1.2
22	3/4	1.8	2.4	2.4	3.0	0.7	1.4
28	1	1.8	2.4	2.4	3.0	0.8	1.5
35	11/4	2.4	3.0	2.7	3.0	0.8	1.7
42	11/2	2.4	3.0	3.0	3.6	0.9	1.8
54	2	2.7	3.0	3.0	3.6	1.0	2.1

Test yourself 3.5

1. List four common types of fixings used in plumbing installations.
2. How are screws specified – tick the answer which you think is correct:
   a. Length and gauge ☐
   b. Length and width ☐
   c. Width and gauge ☐
   d. Strength and gauge ☐
3. Where would you be most likely to use a brass or alloy or zinc-plated screw?
4. What type of screw would be used to fix a radiator bracket to a concrete block wall? Tick the answer which you think is correct:
   a. Steel countersunk ☐
   b. Self-tapping ☐
   c. Mirror screw ☐
   d. Chipboard screw ☐
5. State three types of fixings suitable for plaster board stud partitioning.

Associated trade skills

Introduction

Throughout this book, we have made reference as to how the work of a plumber can vary considerably from working on small repair jobs in people's homes to working on large multidwelling housing developments. On the larger type of developments, most of the associated building work will be carried out by the relevant trades, e.g. preparing the fabric of the building to run pipework. But on some jobs you may have to do the work yourself.

Hence, you will need to learn some additional skills to carry out your work effectively. You will need to run pipes under timber floors, which will involve notching joists.

In existing properties, gaining access to work under the floor will mean lifting floor surfaces, so you will need to know how to do that. Often, both on new work and maintenance work, you will be required to cut or drill holes in brickwork, blockwork, concrete and timber.

Key point

Always remember that whether you are cutting through floorboards or chopping through walls there may be existing water or gas pipes or electrical cables lurking to give you a nasty shock. Take care when doing this type of work, and where possible, try to determine if services already exist. This can be done using pipe and cable tracing equipment. Experienced plumbers can usually work out if a pipe or cable may be present by the proximity of the work to appliances, components and fittings.

Lifting floor surfaces

Lifting a length of floorboard to run the pipework through joists using hand tools

This usually involves lifting a single length of floorboard. If it is a full length one, it will be easier because you will not have to cut across the board. Here is what you have to do:

- Using the hammer and sharp bolster, carefully cut the tongue and groove joint down on either side of the floorboard. A pad saw could also be used for this purpose
- Nails should be punched down to enable the board to be removed
- Alternatively you may use a nail bar, sometimes called a wrecking bar or draw bar, to carefully prize up the floorboard and nails
- Once the board is partially lifted, if it is pushed down slightly, the nail heads will be revealed, allowing their removal with the claw hammer
- If you need to lift only part of the board, you will have to make one, possibly two cuts across a joist, so that when the board goes back it has a firm fixing point; if you cannot locate a joist you will have to insert timber cleats

Key point

You can locate the joist by finding the floorboard nails.

- The cross-cuts on the floorboard can be made using an extremely sharp wood chisel or a purpose made floorboard saw.

Key point

Do not forget the information we provided earlier in the unit about the safe use of power tools.

Lifting floorboards using power tools

This is done using a circular saw.

The saw can be used to cut down the full length of the tongue and groove on each side of the board. The cross-cut is made, again over the joist, making sure that the blade does not hit the nails. Setting the saw blade at a slight angle will help to avoid hitting the nails.

Cutting traps in floorboards

This uses similar methods like removing a single floorboard, either using hand or power tools. It is just that there are more boards to cut, although in shorter lengths in order to make an access point to a section of pipe or component such as a central heating pump.

Replacing floorboards and traps

The floorboard lengths and traps should be screwed back into position to make future inspection or maintenance easier, this should be done using countersunk wood screws. When refixed over the pipework, the board surface should be marked accordingly, e.g. water, heating or gas pipework.

Where it has not been possible to find a joist to refit the board or trap, cleats must be used to support the board end. The following illustration (Figure 3.54) shows a trap or board replacement over joists and using cleats.

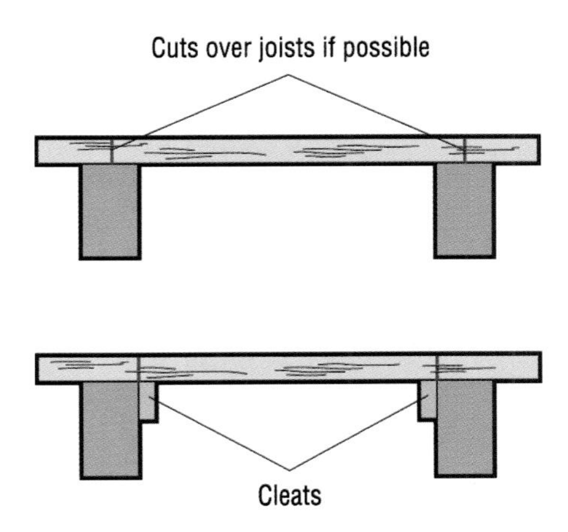

Figure 3.54 Trap or board replacement over joists and using cleats

Removing and replacing chipboard

This is a more difficult job than floorboards, as it is laid in wider sheets. The best way to remove it is by using a circular saw – as for floorboards, the same rules apply. If a power supply is not available, a section of board can be cut using a floorboard saw. Here is what you have to do:

- The section of the board to be removed needs to be marked out across the board so that you have the guidelines to follow for the cut
- If a pad saw is used to make the cut, it is helpful to drill holes in each corner of the area to be lifted in order to start the cutting process
- Using this method would mean replacing the section removed with a new piece of chipboard.

Figure 3.55 Typical pipe guard

Replacing the chipboard

Like the floorboards, the chipboard should be screwed back in position.

Use of pipe guards

As we mentioned earlier, the position of the piperuns under the floors should be marked. A more effective way of protecting the pipes, which pass through the joists, is the use of pipe guards. Figure 3.55 shows a typical example.

Timber joists

It is inevitable that a joist will have to be drilled or notched to permit piperuns under timber floors. The preferred method would be to drill the joist in the centre of its depth as this is the point of least stress. In practice, apart from when using plastic hot and cold water supply pipe, it tends to be impractical.

The main point to remember, for either notches or holes, is that the joist is not weakened. This also applies to the distance from the wall where the joist is notched or drilled.

Think about the consequences for the customer of having joists that have been significantly weakened.

The Building Regulations set out the requirement for notching or drilling joists.

Activity 3.5

Which Building Regulation covers the notching and drilling of joists?

Check out the Building Regulations at www.odpm.gov.uk and then check your answer with the one at the end of the unit.

Building Regulations in practice

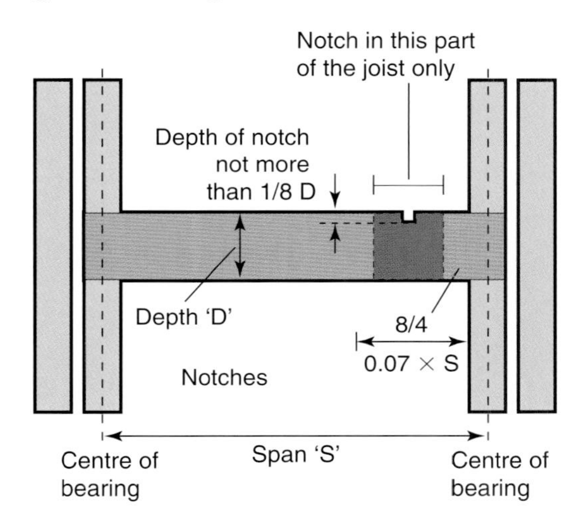

Figure 3.56

Worked out example

For a joist 200 mm deep and 2.5 m long, any notch must have a maximum depth of D ÷ 8. So the depth of the notch is 200 ÷ 8 = 25 mm. The minimum length is 7 × 2500 ÷ 100 = 175 mm. The maximum length is L ÷ 4 from its bearing, so maximum length of 2500 ÷ 4 = 625 mm.

The joists are normally cut using a hand or floorboard saw. They are cut to the required depth and width, and the timber notch is removed using a hammer and a sharp wood chisel.

The width of the notch should be large enough to give freedom of movement, in order to allow expansion and contraction, particularly for hot water supply pipework.

Cutting holes in the building fabric

Again, you can use hand or power tools for cutting the holes. This time we will start with power tools. The selection of a drill will depend on the job at hand. Drills are available with various watt ratings, e.g. 500, 850 and 1400 W and have varying specifications for the different materials as shown in Table 3.7.

Key point

By building fabric we mean:

- Brickwork
- Blockwork
- Concrete
- Timber.

Table 3.7

	500 W	850 W	1400 W
Brick/block (mm)	30	42	152
Concrete (mm)	24	30	120
Steel (mm)	13	13	13
Wood (mm)	32	32	30

The drill bits used will be designed for a specific task and will be purpose made for brick, block, concrete, steel or wood. The diamond core drill will use either diamond- or tungsten-tipped bits.

Key point

A core drill has the advantage over hand tools in providing a much neater finish to the job, and means less 'making good'.

Core drills are excellent for drilling through brickwork, blockwork or concrete where you need to pass a large diameter pipe, or flue pipework.

Wood bits or wood boring bits

These come in the following sizes: 12, 16, 18, 20, 22 and 25 mm and are useful for drilling holes in joists, or other timber constructions for the passage of pipes.

Key point

All pipes that pass through wall structures must be sleeved, with an approved material.

Hole saws

These come in the following sizes: 19, 22, 29, 38, 44 and 57 mm. These are handy while drilling through kitchen units in order to pass installation pipework or waste pipes through the side or back of cabinets. (They are also used to drill holes in plastic cold water storage cisterns.)

Making good

This will involve making good to brickwork, blockwork and concrete. For most jobs, a mortar mixture of 4 parts sand to 1 part

cement will be adequate for jointing any brickwork joints disturbed whilst doing the job, and for making good the gap around the pipe penetration.
Mastic sealant, clear, or close to the colour of the pipework can be used as an alternative.

Test yourself 3.6

1. What are the two main types of wooden floor surfaces?
2. A timber joist measures 200 mm deep by 4 m long. What is the maximum depth of the notch?
3. What is the minimum and maximum distance from the wall that the above joist can be notched?

Check your learning Unit 3

Time available to complete answering all questions: 30 minutes

Please tick the answer that you think is correct.

1. If a central heating system cannot be installed as per the client's original specification, what should a plumber's first action be?
   a. Advise their immediate supervisor or employer about the problem ☐
   b. Obtain the client's verbal agreement to carry out the alterations ☐
   c. Obtain the client's written agreement to carry out the alterations ☐
   d. Proceed when the client's Quantity Surveyor issues a variation order ☐
2. Where can guidance on drilling of timber floor joists to accommodate water systems pipework be found?
   a. Water Bylaws ☐
   b. Construction Regulations ☐
   c. Building Regulations ☐
   d. Water Regulations ☐
3. What should be in place when working on uncovered first floor joists?
   a. Lay duck boards over the joists ☐
   b. Knee pads and safety glove protection ☐
   c. Safety harnesses ☐
   d. Scaffold beneath the joists ☐
4. What PPE would a plumber require when working in a roof space insulated with mineral wool?
   a. Extension lamp ☐
   b. Fire extinguisher ☐
   c. Respirator ☐
   d. Safety mask ☐

(Continued)

UNIT 3

Check your learning Unit 3 (Continued)

5. What is the statutory requirement for cutting the depth of notches in timber floor joists, where D = Depth?
   a. ⅛th D ☐
   b. ⅙th D ☐
   c. ¼th D ☐
   d. ⅒th D ☐
6. What information should be left with the customer after installation of a new boiler?
   a. DIY servicing information ☐
   b. Boiler installation template ☐
   c. Material delivery note ☐
   d. User instructions ☐
7. What is provided with materials to confirm the number and type of items delivered?
   a. Materials schedule ☐
   b. Materials invoice ☐
   c. Delivery advice note ☐
   d. Copy of the order ☐
8. On a large multidwelling building site the progress of other trades can be checked by referring to the:
   a. Bill of Quantities ☐
   b. Work programme ☐
   c. Materials schedule ☐
   d. Site plan ☐
9. Upon delivery of sanitary ware to the site, the plumber should first check if it is:
   a. Adequately protected ☐
   b. Water regulation approved ☐
   c. The correct specification ☐
   d. Free from damage/defects ☐
10. What action should a plumber take if a fault on a gas meter is suspected?
   a. Advise the customer ☐
   b. Repair the fault ☐
   c. Contact the gas supplier ☐
   d. Turn off the supply ☐
11. When jointing copper tube, the purpose of the grab ring on a push-fit plastic fitting is to:
   a. Prevent the pipe damaging the 'O' ring ☐
   b. Ensure the pipe is locked in place ☐
   c. Ensure the fitting is watertight ☐
   d. Prevent the pipe from flattening ☐
12. Sleeves must be provided when a pipe:
   a. Is located in the roof space ☐
   b. Comes into contact with lead ☐
   c. Passes through a brick wall ☐
   d. Is laid in timber floor joists ☐
13. For what type of fitting would the application of solder wire during the jointing process be required?
   a. Integral ring ☐
   b. Capillary ☐
   c. Type A ☐
   d. Type B ☐
14. What grade of copper tube is best suited to being bent using an internal bending spring?
   a. X (R250) ☐
   b. Y (R220/250) ☐
   c. Z ☐
   d. W ☐

Check your learning Unit 3 (Continued)

15. What piece of equipment would be the most appropriate for bending 20 mm low carbon steel pipe?
    a. External bending spring ☐
    b. Stand bender ☐
    c. Hydraulic press bender ☐
    d. Internal bending spring ☐

16. On which tool is 'mushrooming' most likely to occur?
    a. Wood chisel ☐
    b. Screw driver ☐
    c. Cold chisel ☐
    d. Claw hammer ☐

17. What document would be needed when maintaining a water softener?
    a. Maintenance schedule ☐
    b. Installation programme ☐
    c. Manufactures' catalogue ☐
    d. Manufacturers' instructions ☐

18. What type of document would be used to plan the work activities on a long-term maintenance contract?
    a. Bill of Quantities ☐
    b. Programme ☐
    c. Materials schedule ☐
    d. Installation plan ☐

19. On a job to replace a bathroom suite in a one bathroom occupied dwelling, which of the following should a plumber install first in order to minimise disruption to the customer?
    a. Bidet ☐
    b. WC suite ☐
    c. Shower tray ☐
    d. Bath ☐

20. What information should be issued to a customer upon completion of a long-term planned maintenance contract?
    a. Bill of Quantities ☐
    b. Material invoices ☐
    c. Maintenance records ☐
    d. Maintenance programme ☐

Sources of Information

We have included references to information sources at the relevant points in the text; here are some additional contacts that may be helpful.

- Pegler Limited
  St Catherine's Avenue
  Doncaster
  South Yorkshire
  DN4 8DF
  Tel: 01302 560560
  Website:www.pegler.co.uk
- Yorkshire Fittings
  Head Office
  PO Box 66
  Leeds
  LS10 1NA
  Tel: 0113 270 1104
  e-mail: info@yorkshirefittings.co.uk
  Website:www.yorkshirefittings.co.uk
- Copper Development Association
  Grovelands Business Centre
  Boundary Way
  Hemel Hempstead
  Herts HP2 7TE
  United Kingdom
  Tel: 01442 275700
  e-mail: helpline@copperdev.co.uk
  Website:www.cda.org.uk
  (CDA provides some excellent general information on copper tube)
- Hepworth Plumbing Products
  Head Office
  Edlington Lane
  Edlington
  Doncaster
  DN12 1BY
  Tel: 01709 856400
  Website:www.hepworthplumbing.co.uk
- Screwfix
  Screwfix Direct
  FREEPOST
  Yeovil
  BA22 8BF
  Tel: 0500 41 41 41
  e-mail: online@screwfix.com
  Website:www.screwfix.com
- Draper Tools Ltd
  Hursley Road
  Chandler's Ford
  Eastleigh
  Hants
  SO53 1YF
  Tel: 023 8026 6355
  e-mail: sales@draper.co.uk
  Website:www.draper.co.uk
- Ridgid Tool
  Arden Place House
  Prixmore Avenue
  Letchworth
  Herts
  SG6 1LH
  Tel: 01462 485335
  e-mail: sales.uk@ridgid.com
- Water Regulations Advisory Scheme
  Fern Close
  Pen-Y-Fan Industrial Estate
  Oakdale
  Gwent
  NP11 3EH
  Tel: 01495 248454
  e-mail: info@wras.co.uk
  Website:www.wras.co.uk

You may also find the following websites useful: www.piping.georgfischer.com, www.plumbworld.co.uk

UNIT 4

Key Plumbing Principles

Summary

The Key Plumbing Principles Unit covers the 'maths and science' aspect of plumbing. It covers the basic materials, theories and concepts that underpin the plumbing work that you will encounter on a day-to-day basis. Here is what you will cover:

To most people maths and science are not the most exciting topics, but as any practising plumber will tell you, these subjects underpin the bulk of the practical work you are likely to cover.

- Plumbing science
  - Gravity, mass and weight
  - Density
  - Pressure.
- Properties of water and related principles
  - Water
  - Water hardness
  - Corrosion
  - Heat and temperature
  - Capillary attraction
  - Siphonage
  - Properties of heating gases.

- Plumbing materials
  - Standards for plumbing material
  - Properties of materials
  - Metals commonly used in the plumbing industry
  - Plastics commonly used in the plumbing industry.

Before progressing to the first part of the unit, there are two topics that we need to mention here, that whilst not be covered specifically in the Technical Certificate are something you need to be aware of as a plumbing student as they have a bearing on some of the work that you will do. You should also cover these topics at your college or centre.

Units of measurement

Metric units of measurement, such as millimetres, metres and kilograms are used frequently in the plumbing trade.

You will still hear reference to what are known as imperial measurements, such as feet, inches and pounds; because some plumbers still refer to imperial terms and some materials, such as low carbon steel fittings are actually specified in imperial units.

There is a standard international measurement system commonly known as SI units.

Try this

Find out the SI units and their imperial equivalents that are used for:

Length	Area	Volume
Pressure	Velocity	Temperature
Capacity	Mass/Weight	Force
Power	Heat energy	

Write up your findings on separate sheets and include them in your portfolio.

Maths

Again, this should be studied during your time at college or centre, and certainly if you are undertaking an apprenticeship. One of the key skill areas is 'application of number'.

A list of terms that are often used in plumbing applications is given below.

- Length
- Area
- Volume
- Mass
- Temperature.

Activity 4.1

Understand these terms first, then find out the formulas for calculating:

- the area of a Rectangle or Square
- the area of a Triangle
- the area of a Circle
- the surface area of Cylinder

Now find out the formulas for calculating the volume of:

- a Cuboid
- a Prism
- a Cylinder

Once you have completed your research/revision, carry out these calculations and check your answers with those given at the end of the book:

1. Work out the area of the shapes below.

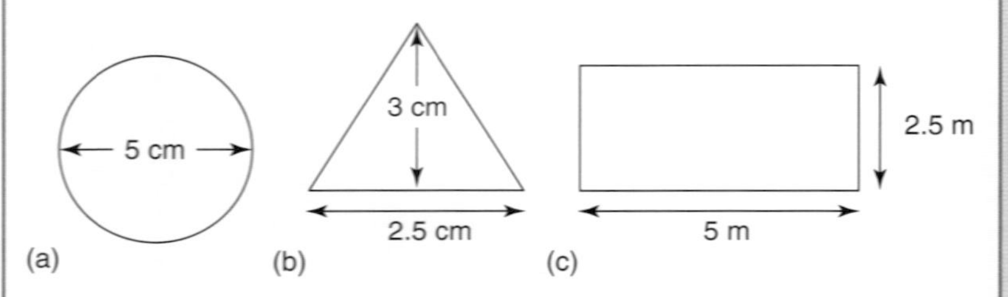

2. Calculate the volume of the following:

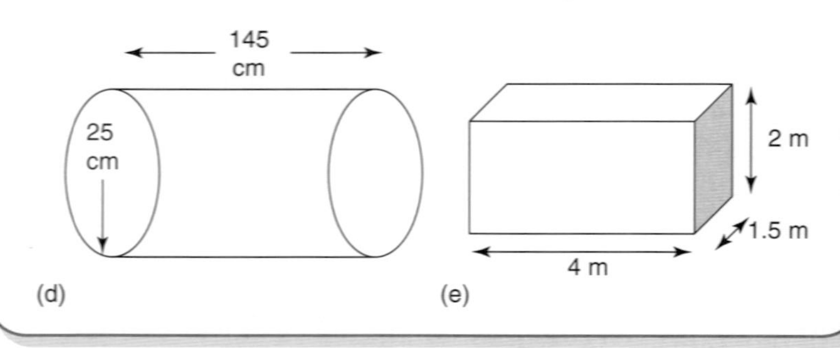

Plumbing science – general principles

Introduction

This unit contains the scientific principles and knowledge that underpin much of the work done by plumbers. Whereas other units concentrate on the 'how to' aspect of plumbing work, this unit explores a few of the 'why' aspects.

- How and why does capillary attraction affect plumbing systems?
- Why is siphonic theory so important to plumbing work?

We will begin this unit by looking at some of the 'physics-related' aspects of plumbing work related to gravity, mass and weight. All these have an effect on how plumbing systems perform as they will have an influence on how substances such as water and air behave under different conditions.

Try this

What is your understanding of the terms gravity, mass and weight? You do not have to write anything down, but see how the definitions given in the next section compare with your initial thoughts.

Gravity

The idea of gravity is not easy to explain. It is a force that each and every object exerts on every other object. When the object is small, the force of gravity it has is also small. Only when the object is large (and has a large mass), like planet Earth, is the force really important.

Key point

Point of interest: Sir Isaac Newton was the first man to successfully explain the concept of gravity. His theories have underpinned many subsequent scientific discoveries, and it is in recognition of his achievements that the unit of measurement used to describe force is 'newton'.

Children obviously learn very early on that things fall to the ground, but it took the genius Newton to explain why this happens. Basically, on Earth, all objects are accelerated towards the centre of the planet by the Earth's gravitational pull.

The two factors that determine 'how much gravity acts' are the masses of the two objects and the distance between them. You may have seen the film of astronauts on the moon. On the moon the astronauts weigh less than they do on Earth, why do you think this is so? Think about the mass of the moon compared to the mass of the Earth. The moon is smaller and therefore, the force of gravity is smaller.

One other thing about the gravity of the Earth is that it acts equally on all objects. If you drop two objects, say a brick and a ball then they will fall at the same rate and hit the ground at the same time. This does not happen when you use two objects with considerably different shapes – say a feather and a brick. The fall of the feather is slowed down because it experiences air resistance, which acts against the direction of travel. Where there is no air resistance (like on the moon), the two would fall at the same rate and hit the ground simultaneously.

That brings us to an important difference between mass and weight. For an object to have weight there must be a force of gravity acting upon it. Therefore, the weight of an object is defined as the force exerted upon it by the gravity of the planet.

Mass

Mass on the other hand is a measure of the matter contained by the object. The astronaut weighs less on the moon but still has the same mass. Just to complicate things further, mass is measured in kilograms whilst weight is measured in newtons, but it is important to understand that mass and weight are two different things.

Key point

Mass is given the symbol M or m and is measured in grams (g) or kilograms (kg).

Imagine looking at a piece of ordinary chocolate and a piece of Aero. If the two pieces were exactly the same size and you were to predict which would be the heavier you would probably go for the ordinary chocolate. This is because, for the same size, the matter in it is more closely packed. You could say the same about a piece of polystyrene and a block of wood of exactly the same volume, or any number of other materials.

When we want to measure mass we do so using grams or kilograms. For some, this can cause confusion, because we normally associate grams or kilograms with weight. However, the newton is used to measure force, and since weight is a downward force caused by gravity, weight is measured in newtons.

It should also be remembered that the mass of an object is constant, whereas the weight of an object can change with gravity, and this will be affected by where the object is in relation to other objects. For instance, your weight on the moon would be approximately one-sixth of your weight on Earth due to the effect of gravity, but your mass remains the same!

Key point

Weight is given the symbol W and is measured in newtons (N).

Weight

The weight of an object is defined as the force exerted on that object by the gravity of a planet. So, for an object to have weight it must experience a gravitational force, where no gravity exists then the object has no weight – but it will still have a mass. As we have already said, astronauts in space experience weightlessness because the gravitational pull of the Earth is reduced, but they still have mass.

All this talk about astronauts; fancy doing some plumbing in space? What follows is taken from a BBC news article on this very subject.

Plumbing work 390 km above the Earth

'Astronauts Michael Lopez-Alegria and John Herrington completed their second stint outside the International Space Station (ISS) on Thursday, continuing construction work on the orbiting outpost.

The two men, who went up to the platform in the Endeavour shuttle, were walking in the vacuum of space for just over six hours.

Their tasks involved plumbing work on the platform's newest addition, a 13.5-metre-long (45 feet) hi-tech beam.

The Port One (P1) truss, as it is known, contains piping that is part of a cooling system for the ISS.

The most dramatic moment in the spacewalk came when Herrington was carried high above the space station on a robotic arm so he could guide a 270-kilogram (600 pounds) rail cart from one side of the platform to the other'.

Density

Density of solids

Remember, solid materials, which have the same size and shape, can frequently have a completely different mass. This relative lightness or heaviness is referred to as **density**. In practical terms, the density of an object or material is a measure of its mass (g) compared to its volume (cm^3) and can be worked out using the following formula:

$$\text{density} = \frac{\text{mass}}{\text{volume}}$$

The densities of all common materials you will come into contact with during your plumbing career have been determined. Lists of comparative densities can usually be found in science text books, available in a library.

A few examples are shown in Table 4.1.

Table 4.1

Material (Solids and liquids)	Relative density
Water	1.0
Linseed oil	0.95
Aluminium	2.7
Cast iron	7.2
Mild steel	7.7
Copper	8.9
Lead	11.3

Density of liquids

Like solids, liquids also have differing densities. The difference in densities of all the substances (solid, liquid or gas) depends upon the number of molecules that are present within the volume of a particular substance.

As a plumber it is extremely important to know the density of water and understand that the density changes depending upon the temperature of water. Water is less dense when it is heated:

- 1 m^3 of water at 4°C has a mass of 1,000 kg
- 1 m^3 of water at 82°C has a mass of 967 kg.

This is because as water is heated, its molecules move further apart due to an increase in their energy and thus water becomes less dense. Water reaches its maximum density at 4°C.

Key point

Propane and Butane are commonly known as Liquefied Petroleum Gas or LPG. Together with natural gas, coal and oil they make up the main fuel types used to power domestic central heating systems throughout the United Kingdom. As a plumber you will probably come into contact with LPG on an almost daily basis, as many blow torches used for soldering and other plumbing-related work activities are fuelled by propane.

Density of gases

When considering the density of gases, comparisons are usually made in relation to the density of air, and on whether other gases are lighter or heavier than air. Table 4.2 gives a quick indication as to the densities of some of the gases you are most likely to come across in the course of domestic plumbing work. Other properties associated with these gases will be covered later in this unit.

Relative density

Relative density (this is sometimes also referred to as specific gravity) is an effective way of measuring the density of a substance

Table 4.2

Gas	Relative density
Air	1.0
Natural gas (methane)	0.6
Propane	1.5
Butane	2.0

or an object by comparing its weight per volume to an equal volume of water, e.g. 1 m^3 of water has a mass of 1,000 kg, 1 m^3 of copper has a mass of 8,900 kg.

1 m^3 of copper is 8.9 times heavier than 1 m^3 of water and therefore has a relative density of 8.9. Water has a relative density (specific gravity) of 1.0. Any substance with a relative density less than 1.0 will float on water; substances with a higher relative density will sink.

As water has a relative density of 1.0, enabling comparisons of relative density with solids and liquids, air has a relative density of 1.0 to enable comparisons between the gases.

Pressure

Pressure is exerted by solid objects in a downward direction only; however, liquids exert pressure downwards and sideways. Figure 4.1

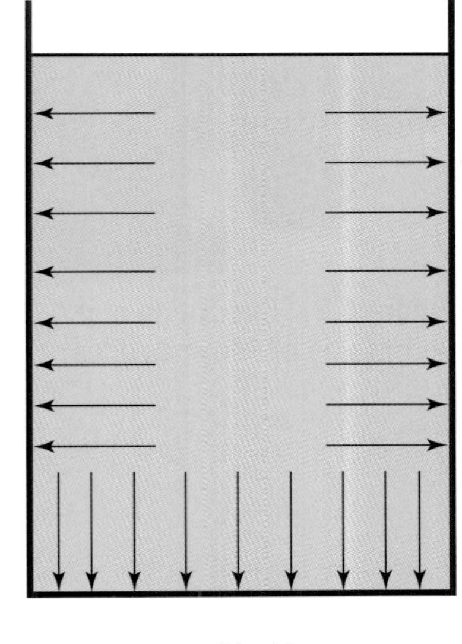

Figure 4.1 Pressure exerted by liquids and solids

shows the difference between pressures exerted by solids and liquids.

As almost every aspect of a plumber's job is linked in some way to the shape or form of water, understanding the intricacies of water pressure is very important.

Water pressure

Basically, there are two methods of creating pressure in plumbing systems.

- By connecting a pump, like the one in Figure 4.2, into the system pipework
- By using the weight of the water itself.

The higher the column of water (referred to by plumbers as the 'head of water'), the greater the pressure that is exerted at its lowest point (see Figure 4.3).

Figure 4.2 Circulating pump (Reproduced with permission of Pegler Limited)

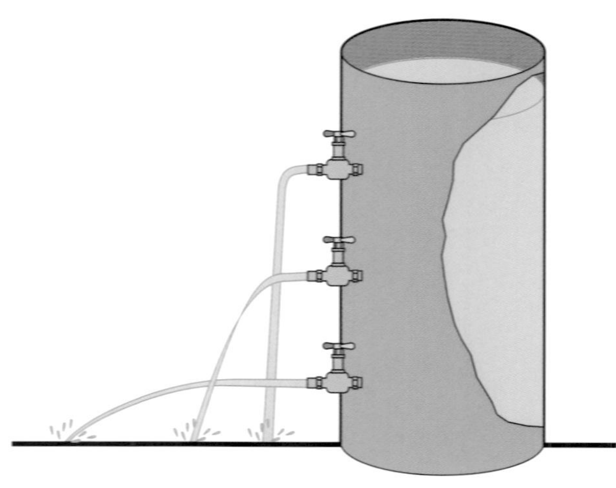

Figure 4.3 Water pressure

Practical applications of water pressure are evident in systems that utilise a cold water storage cistern (CWSC). The general rule is that the higher the location of a CWSC, the better the pressure will be at the draw off points.

May the force be with you

We looked at gravitational force earlier. Pressure can be defined as the force applied per unit area and is measured in newtons per square metre (N/m^2), a unit also known as pascal (Pa). One newton is the force that gives a mass of 1 kg an acceleration of 1 metre per second squared ($1\ N = 1\ kg \times 1\ m/s^2$).

Here is an example:

> **Key point**
>
> The force produced by 1 kg mass = 9.81 N, and this figure is used in calculations to determine water pressure.

A container having a water capacity of $1\ m^3$ would contain 1,000 l of water and would therefore weigh 1,000 kg, so this would exert a pressure of:

$1{,}000 \times 9.81 = 9{,}810$ newtons per square metre or $9.81\ kN/m^2$ at its base.

If the container is extended to 3 m height, then the pressure is increased three times to $3 \times 9.81 = 29.43\ kN/m^2$.

Total pressure is a measurement of the intensity of pressure acting on a given area. If you need to calculate the total pressure for an area, then you will have to find the intensity of pressure first and then multiply it by the area it will be acting upon.

Here is an example, as shown in Figure 4.4:

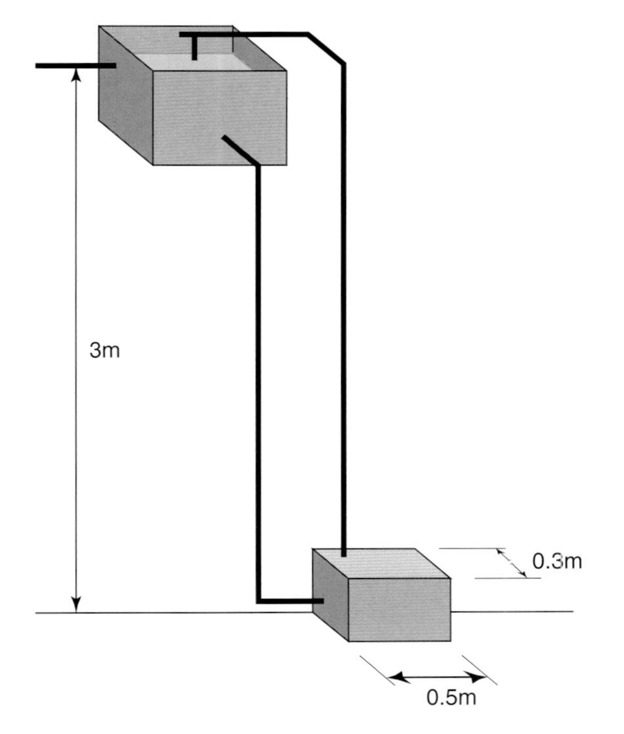

Figure 4.4 Total pressure

You need to find the total pressure at the base of a central heating boiler which measures 450 × 300 mm at the base and is installed 3 m below the water level in a CWSC, so:

Total pressure = (intensity of pressure) × (area acted upon)

Therefore: (head × 9.81) × (L × W of base)

Therefore: 3 m × 9.81 × 0.45 × 0.3 = 3.97 kN

You will probably come across other terms used to identify pressure such as the 'bar' or 'pounds per square inch' ($lbf/in.^2$). These can be expressed as:

- 1 bar = 100,000 N/m^2
- 1 $lbf/in.^2$ = 6,894 N/m^2

More pressure points

Pressure, therefore, is a measurement of a concentration of force. The effect of concentration of pressure can be seen if water flowing through a pipe is forced through a smaller gap by reducing the diameter of the pipe. Think about a garden hosepipe with an adjustable nozzle and how you can maximise the force of the jet of water by reducing the pipe bore. The force is not being increased, it remains the same. What happens is that by reducing the area through which the water comes out, the pressure of the jet of water is increased.

As a plumber, you will need to have a good understanding of the effects that pressure can have on pipes and fittings that you install. The internal pressure a pipe or vessel will be subjected to will be dependent on what it contains (water or gas); this factor must be taken into consideration when deciding which material is most suitable for its construction.

Key point

Remember: Head gives pressure not volume.

Look at Figure 4.5; at which tap, A or B, do you think there is the greatest pressure?

If you said A, you would be correct.

Atmospheric pressure

The Earth is surrounded by an envelope of air, which is held to its surface by gravity. The weight of air creates a pressure on the Earth's surface and the pressure exerted by the weight of air

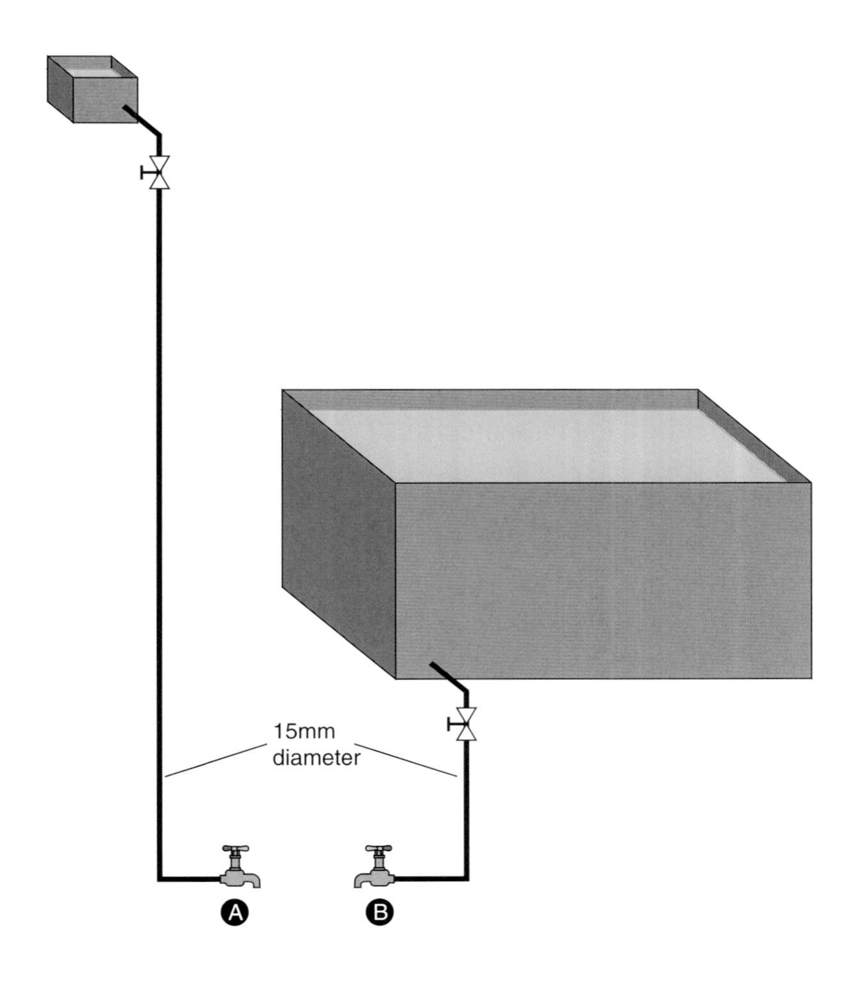

Figure 4.5 Pressure and volume

pressing down on the ground below will vary depending on the ground's height above sea level. For example, the pressure at the top of Mount Everest would not be as great as the pressure at the bottom of a valley below sea level.

The pressure created at sea level is $101{,}325\,N/m^2$. This is approximately equal to 1 bar.

Plumbers must be aware of the effects of atmospheric pressure to ensure that they avoid creation of 'negative' pressure or vacuum situations within domestic water systems.

Time to check your progress. Have a look at the following exercises which relate to plumbing science.

Test yourself 4.1

1. The measure of matter contained by the object is referred to as its:
   a. Mass ☐
   b. Weight ☐
   c. Pressure ☐
   d. Force ☐

2. Which of the following examples shows the correct unit measurements?
   a. Pressure → Grams ☐
   b. Weight → Joules ☐
   c. Mass → Grams ☐
   d. Mass → Newtons ☐

3. Consider the formula below:
   Weight = mass × acceleration due to gravity (9.8 m/s)
   What would be the weight of a 3 kg block of mild steel?
   a. 9.8 N ☐
   b. 15.4 N ☐
   c. 19.6 N ☐
   d. 29.4 N ☐

4. Water reaches its maximum density at a temperature of:
   a. 0°C ☐
   b. 4°C ☐
   c. 20°C ☐
   d. 60°C ☐

5. If 1 m^3 of lead has a mass of 11,400 kg, on this principle what is its relative density?
   a. 11.4 ☐
   b. 9.8 ☐
   c. 7.2 ☐
   d. 22.4 ☐

6. In the diagram below (Figure 4.6), at which point A, B or C is the pressure the greatest?

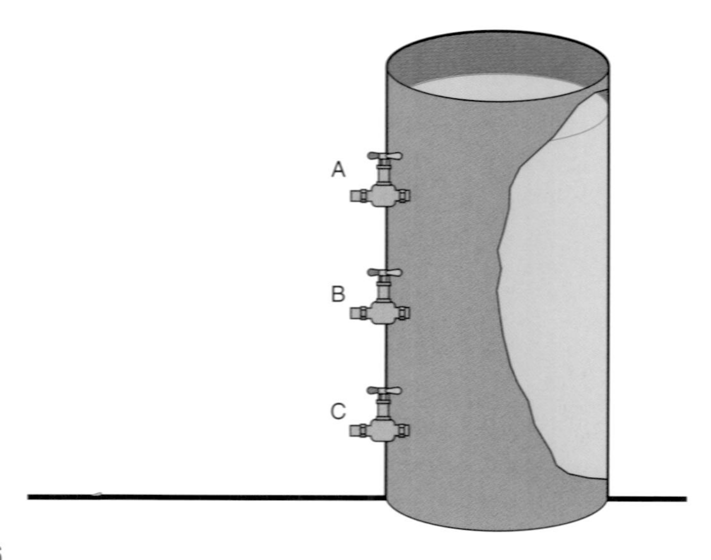

Figure 4.6

Properties of water and related principles

Some of the problems that occur in water supply systems can be attributed to the properties of water. The corrosion or biodegrading of system materials and components due to water that is slightly acidic can result in sludge and silt formation in cold water storage systems.

Similarly, in hard water districts, limescale deposits on pipework and in hot water storage vessels can also harbour harmful bacteria.

This section is not about turning you into a water treatment chemist, but is intended to provide a basic understanding of the properties of water in the context of plumbing systems.

Water

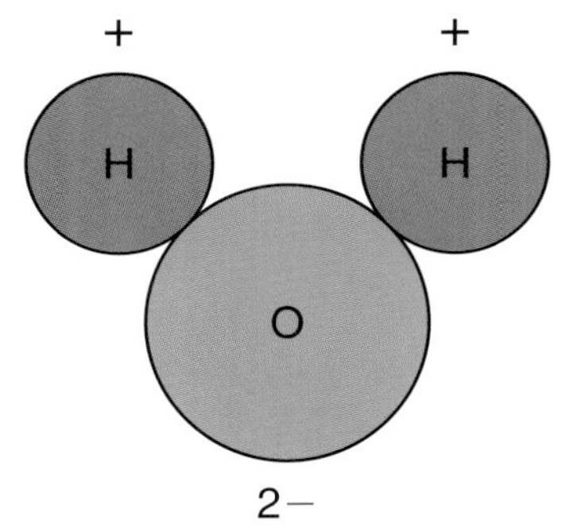

Figure 4.7 Composition of a water molecule

Key point

Legionella bacteria is the pathogen that causes Legionnaire's Disease, one of many dangerous diseases that can be spread through badly designed and/or maintained water systems. Plumbers need to be aware of the conditions (including the properties of water) that can promote growth of the Legionella bacteria so as to minimise this risk.*

**Pathogen means an 'agent causing disease'*

Pure water consists of two atoms of hydrogen and one atom of oxygen. Electronic attraction between the hydrogen atoms gives water some unique and peculiar properties. Water has a high surface tension; it falls as drops and forms a puddle on a flat surface. It has a capillary action that allows it to run through the tightest of spaces, and it is lighter as a solid than it is as a liquid, hence ice floats on water.

Figure 4.7 shows a water molecule; H_2O.

Water is a universal solvent. It is capable of dissolving most of the materials and carrying that dissolved material in solution. An example of this is when rain water dissolves limestone rock, resulting in high concentrations of calcium in the water supply, and of course if it is not treated it will cause limescale deposits in hot water systems, reducing efficiency, and providing a hiding place for Legionella bacteria.

pH value

pH is a logarithmic scale that measures the acidity or alkalinity of water. Pure water has a pH of 7.0. If water dissolves acidic materials, the pH falls and if it dissolves alkaline materials, the pH rises. Demineralised water as used in the semiconductor industry contains absolutely no salts. If a sample is taken and its pH measured, the result would be between 5.0 and 6.0, because the water absorbs carbon dioxide from the air.

Acidic and alkaline water can both damage the materials used in plumbing systems, by causing corrosion. Metals in particular are at risk from the corrosive effects of acids and alkalis. Slightly acidic water will also break down materials in plumbing systems, resulting in the build up of silt and other debris in the system.

All water has what is known as a 'pH value', and Figure 4.8 shows how this indicates the level of acidity or alkalinity. Water in its most natural form (rainwater) has a slightly acidic value, due to the levels of carbon dioxide and sulphur dioxide in the atmosphere.

These levels can be significantly increased in areas affected badly by pollution, where the local atmosphere contains high concentrations of either sulphur dioxide or carbon dioxide. The pH value of water is also affected by the different ground strata that water passes through after it falls as rain.

The different basic categories are shown in Table 4.3.

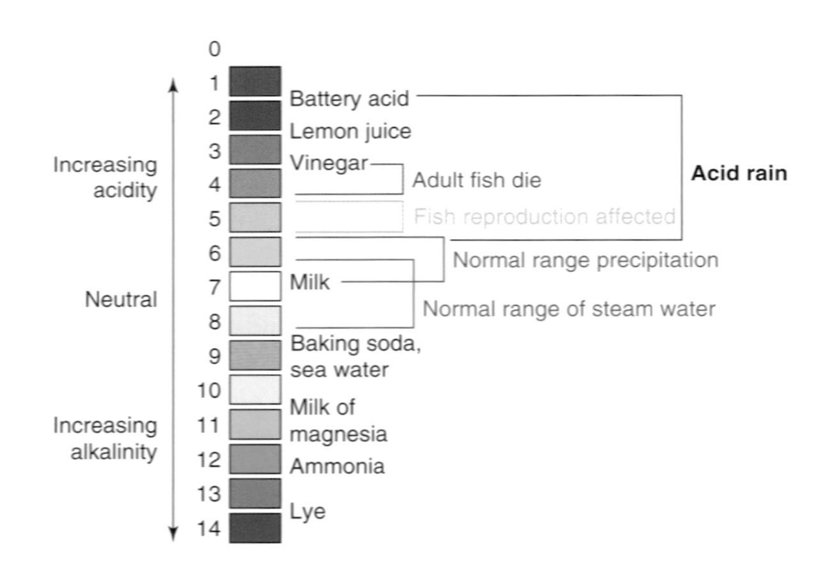

Figure 4.8 How various additives affect the pH of water

Table 4.3

Type of ground condition	Chemical added to water	pH
Moorland	Carbon dioxide added	Acidic
Sandstone	Carbon dioxide retained	Acidic
Salt	Calcium sulphate	Alkaline
Limestone and chalk	Calcium carbonate	Alkaline

Calcium hardness

This is most commonly known as limescale. Water contains different levels of calcium, and water that comes from a limestone source will be particularly high in calcium. High temperature, high pH and high conductivity can cause calcium to come out of water and produce limescale (calcium carbonate). As already mentioned, limescale deposits can cause problems in hot water systems.

Water can be classified as having varying degrees of softness or hardness, or can in fact be neutral. Water hardness is measured in parts per million (ppm).

Here is a quick summary detailing the main characteristics of soft and hard water.

Soft water

Water is classified as soft, when it easily produces a lather with soap. This is due to the absence of dissolved salts, such as calcium carbonates or sulphates (chalk and limestone). Soft water is classified to be slightly acidic because it absorbs carbon dioxide from the atmosphere.

Hard water

Water is classified as hard, when it is difficult to produce a lather with soap. This is because the rain water has fallen on ground containing calcium carbonates or sulphates, dissolving them and taking them into solution as it passes through.

Water hardness

Water hardness can be described as temporary or permanent:

- *Permanent hardness* occurs as a result of the natural solvency of pure water, which enables it to dissolve the sulphates of limestone. It is the cause for the normal problems associated with hard water (difficult to form a lather with soap and detergents).
- *Temporary hardness* is responsible for the hard scale which can accumulate on the inside of boilers, circulating pipes and hot water storage vessels, restricting the flow of water, reducing the efficiency of appliances and components and ultimately causing damage and eventual system failure.

Hard water is undesirable in domestic installations as it produces limescale in pipework, heating equipment and sanitary appliances. This can result in higher maintenance costs. Hard water also requires the use of a lot more soap and detergent for washing purposes as the 'hardness' makes producing a lather far more difficult.

Turbidity (suspended solids)

Turbidity is a measure of materials that are in suspension in water, as water not only dissolves, but also acts as a carrier for other materials.

In domestic systems, the turbidity of the water can be affected by the presence of products of corrosion, iron, copper, scale particles and biofilm.

The particles will deposit in low flow areas in the system, these are areas where there is an insufficient flow to keep the particles moving. These are ideal conditions for the development of Legionella bacteria.

Dissolved oxygen

Oxygen is present in all water, at about 8 ppm. This makes water corrosive, as metals use oxygen to take them back to their natural state. Steel would rather exist as rust so it grabs the oxygen in water. Steel domestic water tanks are generally galvanised to protect them from this oxygen exchange.

Galvanised tanks corrode in the presence of oxygen to form zinc oxide, which normally appears in the water as a white powder.

Here is an example of what we mean and this is shown in Figure 4.9.

There are probably over 300 tests that could be performed on a water sample to measure all its possible contaminants. The examples given above are an indication of some of the most common (and of most relevance to the plumber).

Corrosion

Corrosion has been mentioned throughout the text, it is essential to consider the effects that corrosion can have on various types of

Figure 4.9 Galvanised tank corrosion

Key point

The term ferrous refers to all materials that contain iron (from the metal's chemical symbol 'Fe').

Key point

Plumbosolvency refers to the process by which acidic water corrodes lead supply pipework, thus adding an amount of lead to the water that is supplied.

materials. Of all the materials used in the plumbing trade, metals are most likely to be more at risk from corrosion.

The main causes of corrosion are:

- The effects of water and air
- The direct effects of acids
- Electrolytic action (galvanic action).

Corrosion by water

Ferrous metals, such as iron and steel, are particularly vulnerable to the effects of corrosion caused by water. The effects of ferrous corrosion are commonly seen in the central heating systems as black ferrous oxide and red rust (haematite and magnetite) build up in radiators. A by-product formed from this action is hydrogen gas which accumulates in radiators. This build up can be released by 'bleeding'.

In areas where the water has a high acidity (soft water) the internal wall of the copper pipe may become slightly discoloured. This action will not affect the safety or quality of the drinking water. However, in situations where lead pipework is still in use there is a risk of the water becoming contaminated with lead due to the plumbosolvency of the acidic water. This situation creates a potential health risk to consumers; however, most water suppliers now add phosphates to the water supply to combat the risks caused by plumbosolvency in acidic water areas.

Atmospheric corrosion

Pure air and pure water have little corrosive effects, but together in the form of moist air (oxygen + water vapour), they can attack ferrous metals, such as iron and steel very quickly to form the common oxide known as rust. The corrosive effects of rusting can completely destroy the metal.

Other chemicals, such as carbon dioxide, sulphur dioxide and sulphur trioxide, are present in the atmosphere, and tend to be more abundant in industrial areas. They increase the corrosive effect air can have on particular metals especially iron, steel and zinc.

Coastal areas suffer from increased atmospheric corrosion due to the amount of sodium chloride (salt) from the sea which is dissolved into the local atmosphere.

Non-ferrous metals, such as copper and lead, generate protection against atmospheric corrosion in the form of barriers (usually carbonates and sulphates), which form on the outer surface of these metals to prevent further corrosion. This protection is also known as *patina*. Patina will be mentioned again in the section on lead weathering unit.

Corrosive effects of building materials and underground conditions

Some types of wood (such as oak) have a corrosive effect on lead and latex cement and foamed concrete will often affect copper materials.

Various types of soil can have an adverse effect on underground pipework. Heavy clay soils may contain damaging sulphates which can corrode lead, steel and copper. Ground containing ash and cinders is also very corrosive; if pipes are to be laid in grounds of this type they should be wrapped in protective material. Copper pipework for underground use is provided with a plastic coating. Cold water pipework is colour-coded in either blue or green.

Electrolytic (Galvanic) action and corrosion

Electrolytic action is caused when two very dissimilar metals, e.g. a galvanised tank and a copper fitting, come into contact in water. This is demonstrated in Figure 4.10. It is an experiment showing

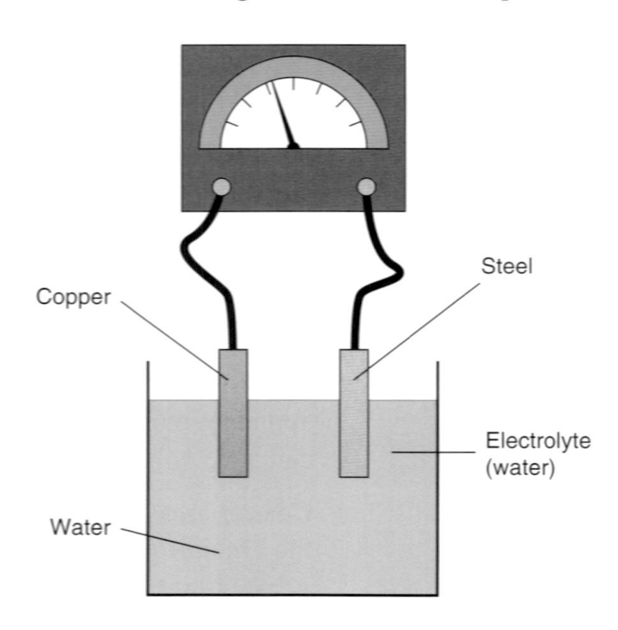

Figure 4.10 Electrolytic action

how electrically charged ions flow from an **anode** (+) to a **cathode** (−) through a medium known as the **electrolyte** (usually water).

The cathode in the figure is copper and the anode is zinc. Zinc would be the subject of corrosion in this experiment. The voltmeter demonstrates the presence of an electrical current flow.

Electrolytic corrosion takes place when the process of electrolysis leads to the destruction of the anode. The length of time it takes for the anode to be destroyed will depend on:

- The properties of water which acts as the electrolyte; if water is hot, acidic, or contains impurities, the rate of corrosion will be increased
- The position of the metals which make up the anode and the cathode in the electromotive series.

The electromotive series

The list below shows common elements used in the plumbing industry, the order in which they appear indicates their electromotive properties.

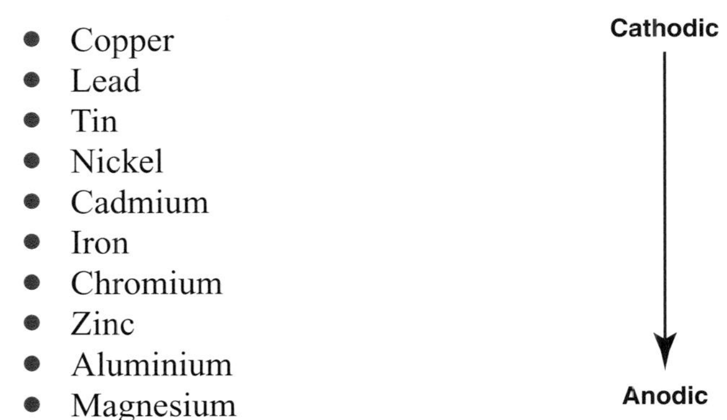

- Copper
- Lead
- Tin
- Nickel
- Cadmium
- Iron
- Chromium
- Zinc
- Aluminium
- Magnesium

The relevance of this list is that the elements higher up in the list will destroy those lower down through the process of electrolytic corrosion. The further apart in the list the materials appear, the faster the erosion will take place. For example, copper will destroy zinc at a quicker rate than it will destroy lead.

In some circumstances, cathodic protection can give protection against electrolytic action. This involves using what is termed as a sacrificial anode. These can be fitted in cylinders, cisterns, tanks and to pipelines, e.g. a magnesium ring fitted at the base of a copper cylinder would act as the sacrificial anode and would corrode before the cylinder that it protects.

> **Try this**
>
> Look for examples of electrolyte corrosion in practice, particularly on older systems. If you get a chance, take a photo for your portfolio or records.

Electrolytic corrosion in practice

Plumbers need to be aware of situations where electrolytic corrosion could take place. If two dissimilar metals are placed in direct metallic contact with each other, and both the metallic elements are then surrounded by water or damp ground, the likelihood is that electrolytic corrosion will occur.

The combination of these factors creates what is effectively a basic electrical cell through which a very small current will pass.

A problem you may encounter if working on older systems is 'dezincification corrosion' of brass fittings, causing blockage or failure of the fitting. This is caused by the deterioration of the zinc content in brass, while in contact with copper pipework, for example. Fitting manufacturers now produce 'DZR fittings' which are not affected by dezincification.

Heat and temperature

Latent and sensible heat

Heat which brings about a change in state with no change in temperature is called latent (hidden) heat. All pure substances are able to change their state, solids become liquids, liquids become gas. These changes occur at the same temperature and pressure combinations for any given substance. It takes the addition of heat or removal of heat to produce these changes.

Heat which causes a change in temperature in a substance is called sensible heat. When a substance is heated (heat added) and the temperature rises as the heat is added, the increase in heat is called sensible heat. Likewise, heat may be removed from a substance. If the temperature falls and the heat is removed, this again, is sensible heat.

In summary:

- Latent heat = heat which produces a change of state without a change in temperature, e.g. heat that converts ice to water
- Sensible Heat = heat which increases the temperature of a substance or material without changing its state.

Difference between heat and temperature

The principal difference between heat and temperature is that heat is recognised as a unit of energy measured in Joules (J).

- Temperature is the degree of hotness of a substance
- Heat is the amount of heat energy (J) that is contained within a substance.

For example, imagine an intensely heated short length copper pipe and a large container of hot water.

The pipe has a temperature of 250°C
The water has a temperature of 70°C

The pipe is far hotter, but actually contains less heat energy.

Specific heat capacity

Key point

Specific heat values vary as the temperature changes.

The specific heat capacity is the amount of heat required to raise the temperature of 1 kg of material by 1°C. The heat required differs from material to material. For example, it requires 4.186 kJ to raise the temperature of water by 1°C (for calculation purposes this is usually rounded off to 4.2), but only 0.385 kJ would be needed to raise the temperature of copper by 1°C. See Table 4.4 for specific heat values.

Table 4.4

Material	kJ/kg to increase temperature by 1°C
Water	4.186
Aluminium	0.887
Cast iron	0.554
Zinc	0.397
Lead	0.125
Copper	0.385
Mercury	0.125

Measuring temperature

You would have investigated the units used for measuring temperature at the beginning of this unit. Although the SI unit of temperature measurement is Kelvin, the unit you will deal with most frequently is degree Celsius (or Centigrade) written as °C.

Temperature is normally measured using a thermometer, of which there are various types. Most plumbers use digital thermometer kits which enable them to measure the temperature of flow and return pipework in systems, and the temperature of water at draw off points to ensure they meet the design requirements. An example of a typical kit is shown in Figure 4.11.

Key point

One of the main failures in the installation of PVC soil and waste installations, particularly in multi-storey buildings, is damage caused due to thermal expansion because measures haven't been taken to make sufficient allowances for it.

Thermal expansion and contraction

Most materials will expand when heated. This is because all substances are made up of *molecules* (*groups of atoms*) which move about more vigorously when heated. This in turn leads to the molecules moving further apart from each other resulting in the materials becoming larger. As the material cools, the molecules slow down and move closer; thus the material gets smaller or contracts.

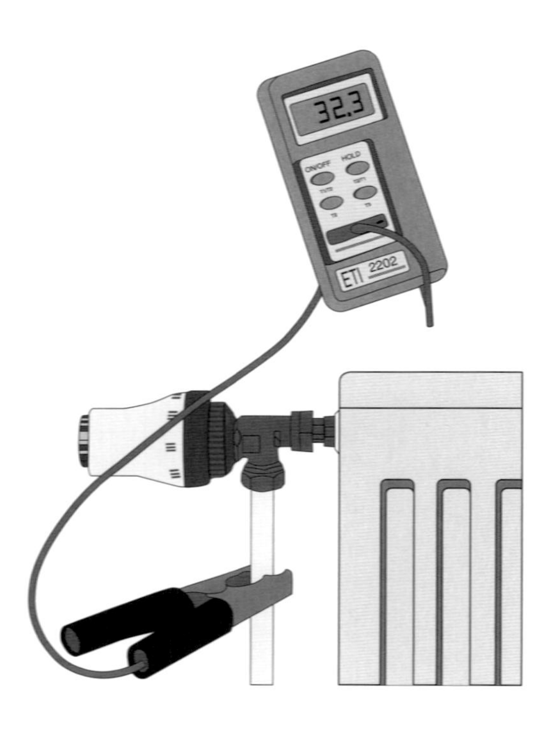

Figure 4.11 Digital thermometer kit

Table 4.5

Material	Coefficient (°C)
Plastic	0.00018
Zinc	0.000029
Lead	0.000029
Aluminium	0.000026
Tin	0.000021
Copper	0.000016
Cast iron	0.000011
Mild steel	0.000011
Invar	0.0000009

The amount a material expands in length when heated, termed the coefficient of linear expansion and can be measured by a simple calculation using the formula below:

$$\text{Length} \times \text{temperature rise} \times \text{coefficient}$$

Table 4.5 shows the coefficient values for some of the most common materials used in the plumbing industry.

Example: Find how much a 4 m long plastic discharge stack will expand due to a temperature rise of 21°C.

$$4 \times 21 \times 0.00018 = 0.01512\,\text{m or } 15{:}12\,\text{mm}$$

Activity 4.2

Jot down possible situations in which you think it will be important to consider thermal expansion and contraction, check your thoughts with the suggested answers at the end of this book.

Heat transfer

As a plumber you will need to have an understanding of the methods of heat transfer because you will be dealing with the effects of this process on a daily basis. There are *three* methods of heat transfer. (You will probably remember some of the methods from your school science lessons, but do not worry if you do not). We will also touch on these again in Unit 6, which is about hot water systems.

Conduction

Conduction is the transfer of heat energy through a material. It takes place as a result of increased vibration of molecules, which occurs when materials are heated. The vibrations from the heated material are then passed on to the adjoining material, which then heats up in turn. Some materials are better at conducting heat than others, for example:

- Metals tend to be good conductors of heat
- Wood is a poor conductor of heat.

Gases and liquids also conduct heat, but poorly, air and water are especially poor conductors of heat. When considering materials used in the plumbing industry, copper has a higher rate of conductivity than steel, iron or lead. Wood, ceramic and plastic, which are poor conductors of heat, are known as *thermal insulators*.

Convection

Convection is the transfer of heat by means of the movement of a locally heated fluid substance (usually air or water). As a fluid is heated, the process causes expansion, which in turn causes a lowering of density. The less dense warm fluid begins to rise, and is replaced by cooler, denser fluid from below. Eventually, convection currents are set up which allow for a continuous flow of heat upwards from the source. Examples of systems which use convection currents for heat transfer include:

- Electric convector heaters warm the air at one place in a room, and the resulting convection currents transport the heat around the room
- A domestic hot water vessel heated via an immersion heater for the supply of hot water depends on convection currents to transfer heat from an immersion heater (similar to the 'element' in an electric kettle) to the rest of the water in the storage vessel.

It is very easy to demonstrate the thermal effects of convection currents – this can be done by hanging a piece of light material such as paper above a convector heater. The movement of the hanging material will clearly show the existence of rising lighter currents of warm air.

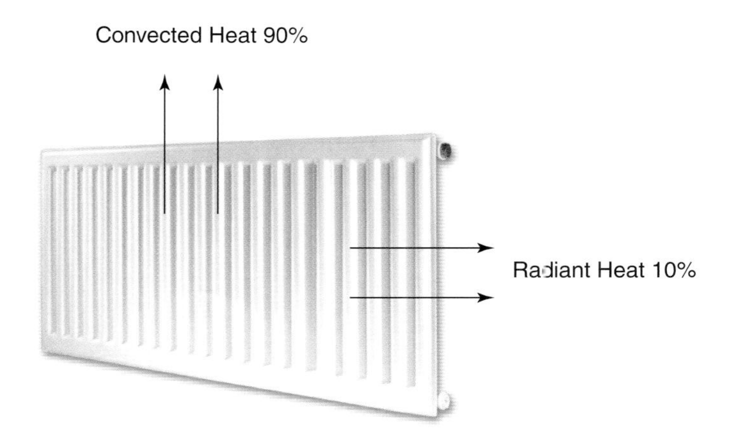

Figure 4.12 Domestic radiator

Radiation

Heat transfer by conduction and convection requires a medium through which to travel. Radiation or radiant heat requires none – it transfers heat from a hot body to a cooler one, by means of 'heat' waves. Radiant heat can be felt as the 'glow' from a fire, heat from a radiator, or as heat from the Sun.

Heat radiation is better absorbed by some materials than others; in these instances colour is an important factor. Dull matt surfaces such as black will absorb radiated heat more efficiently than shiny polished surfaces.

It is interesting to remember that the majority of domestic central heating 'radiators' only give off approximately 10% surface heat by **radiation** – the remaining 90% heat energy warms the room by **convection**. This is also the case with modern radiant/convector heaters. Figure 4.12 shows a domestic radiator.

Capillary attraction

Capillary attraction is the process by which a liquid (99% of the time this will be water) is drawn or rises up the surface of a solid material. The relevance of this action is especially important to plumbers as capillary attraction can affect plumbing installations in domestic properties.

Surface tension – what is the attraction?

Surface tension describes the way in which water molecules 'cling' together to form what is effectively a very thin 'skin'. This

Figure 4.13 Surface tension

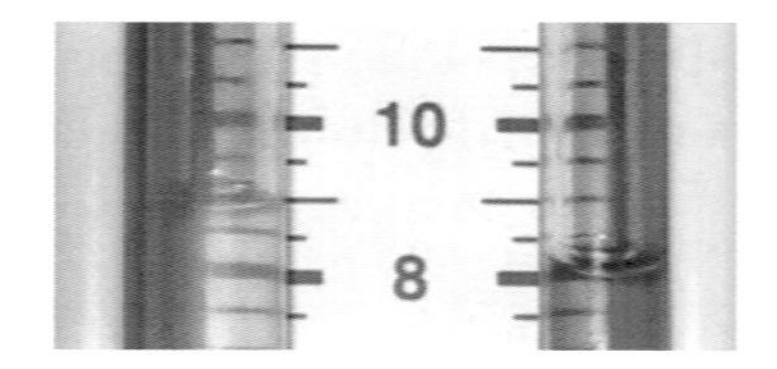

Figure 4.14 The meniscus

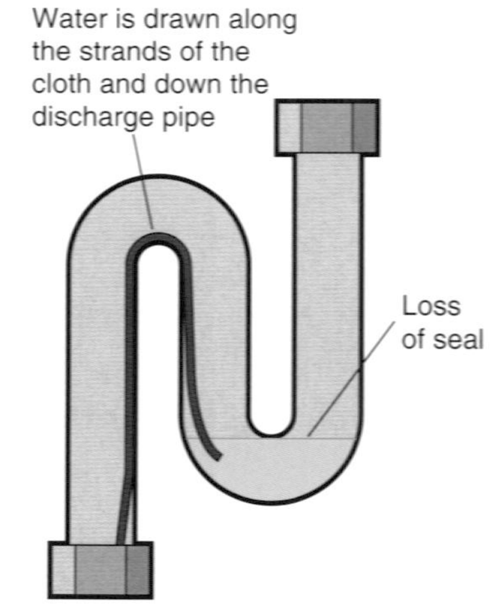

Figure 4.15 Loss of trap seal caused by capillary action

can be demonstrated by completely filling a drinking glass with water. When it is full, examine the glass with your eye at the same level as the top of the glass. The water will appear above the upper limit of the glass with a slight curvature. Why does not it spill over the side?

> *The answer is because of the cohesion (water molecules 'sticking together') as a result of surface tension (see Figure 4.13).*

Adhesion is the force of attraction between water molecules and the side of the vessel that the water is contained in. This leads to the slightly curved 'skin' that appears when water is held in a vessel, as shown in Figure 4.14.

The correct terminology for this 'skin' is the **meniscus**, as shown in Figure 4.14. An example of the effect of adhesion and cohesion can be seen in capillary attraction.

Practical examples of capillary attraction

Plumbers must consider the possible effect of capillary attraction while planning and installing lead roof work and flashings. Water could find its way through tiny gaps between lapped roof weatherings. This can be eliminated by fabricating a small anti-capillary groove in the face of the sheet lead work that is underneath the lap, thus breaking the path of the capillary action.

Capillary action can also take place in the waste pipe 'S bend' drainage traps. If a length of waste material such as a strand of mop material or a length of dishcloth becomes lodged in the 'S bend', water will rise up the material by capillary action dropping down the waste discharge pipe causing loss of trap seal, permitting foul drainage smells to enter back into the property (see Figure 4.15).

Siphonage

We touched on atmospheric pressure in Unit 2. Although we do not often notice it, the atmosphere exerts a great pressure upon all materials on Earth (approximately equivalent to a metre square column of water 10 m high).

Atmospheric pressure is the key to siphonage. The principle of siphonic action is used in several plumbing applications from siphoning the water out of old hot water cylinders that do not have

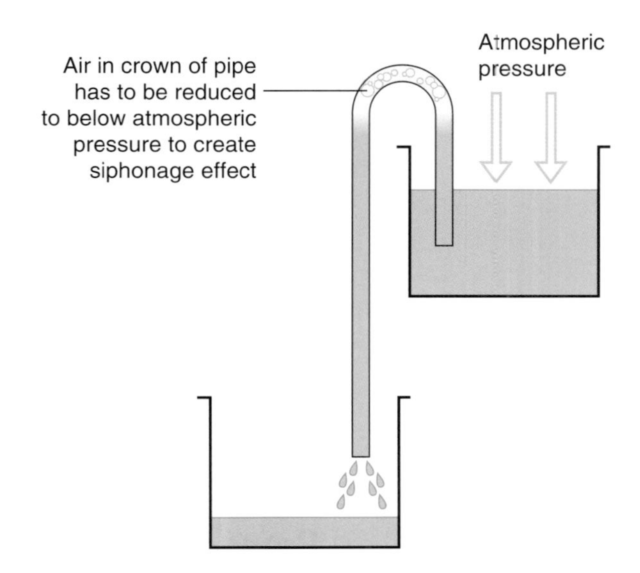

Figure 4.16 Principles of siphonic action

a drain tap fitted to siphoning the waste contents out of a double trap siphonic WC pan.

The illustration in Figure 4.16 will help you understand how siphonage occurs.

Basically, siphonage works when atmospheric pressure is able to force water through a channel such as a length of hose, or a run of pipe often seemingly against the pull of gravity.

For siphonage to take place, the air in the channel must be reduced to below that of atmospheric pressure. This can be done by forcibly 'sucking' the water through the open end of the pipe using something like a hand or stirrup pump, not your mouth!

Water flow

The rate of water flow through a pipe is affected by frictional resistance caused by the water 'rubbing' against the inner wall of the pipe. This action should be an important consideration when designing pipework installations. Think about dropping a small ball through a large diameter pipe, it will fall through without resistance, if a ball that touches the side of the pipe is used, the resistance will slow its movement considerably. This rubbing effect or friction is further increased if the internal surface of the pipe wall is rough. The effect of this is known as **frictional resistance** or frictional loss.

Comparison of flow in smooth and rough pipework

- Water flowing through a 10 m length of 25 mm diameter pipe would flow at approximately 22 l/min for copper and plastic pipes having smooth internal surfaces
- Water flowing through a 10 m length of 25 mm diameter pipe would flow at approximately 18 l/min for galvanised low carbon steel pipe, due to its rougher internal surface.

Frictional resistance will also be increased by the number and types of fittings used in a pipework run. Elbow fittings give about 50% more frictional loss than fabricated pipe bends.

Properties of heating gases

Although gas work is not covered in the plumbing syllabus until Level 3, you will still come across gas-fired installations in the course of your day-to-day work. As this is the case, you should understand the basic properties of the heating gases used in domestic properties across the UK.

As suggested earlier in this unit, there are three main heating gases used in domestic properties:

Key point

The main constituent of natural gas is methane, which accounts for approximately 90% of its composition.

- Natural gas
- Propane
- Butane.

Key point

Remember: Natural gas is lighter than air, which means that any escaped gas will collect at high levels/roof spaces. Propane and butane are heavier than air which means that escaped gas will collect at low levels – especially important if you are working in a trench or basement.

Of these fuels, natural gas is by far the most common and is used for fire heating and hot water systems in the vast majority of domestic properties. Propane and butane (LPG gases) are occasionally utilised as a fuel in domestic situations (particularly in rural areas), but are more commonly used for mobile heating and catering applications (caravans, residential park homes, boats and leisure crafts, etc).

Whilst it is not vital for you to understand all the characteristics of these gases at this stage of your development, it is useful for you to have some general knowledge. Table 4.6 gives some general information on natural gas, propane and butane. The main thing to remember about heating gases is that they are explosive when mixed with air.

Table 4.6 Gas characteristics chart

Characteristic at standard temperature and pressure	Natural gas	Propane	Butane
Specific gravity (SG) (relative density) (SG of air = 1)	0.6	1.5	2.0
Gross calorific value CV (mJ/m^3)	38.76	93.1	121.8
Flame speed (m/s)	0.36	0.46	0.45
Flammability limits (% gas in air)	5–15	2–10	2–9
Ignition temperature (°C)	704	530	408

Note: These figures are approximate due to slight variations in gases

Test yourself 4.2

1. If you know that the water in a domestic property is free from dissolved salts such as calcium carbonate and sulphates, you would expect the water to be:
   - a. Hard ☐
   - b. Soft ☐
   - c. Neutral ☐
   - d. Mineral rich ☐
2. Soft water is usually classified as being slightly:
   - a. Neutral ☐
   - b. Alkaline ☐
   - c. Acidic ☐
   - d. Ferrous ☐
3. The pH scale is used to measure:
   - a. Atmospheric pressure ☐
   - b. Specific gravity ☐
   - c. Acidity and alkalinity ☐
   - d. Heat loss ☐
4. External iron pipework such as gutters and above ground discharge systems in an industrial town are likely to be at a greater risk from corrosion, due to higher atmospheric levels of what chemical?
   - a. Sulphur dioxide ☐
   - b. Calcium carbonate ☐
   - c. Copper sulphate ☐
   - d. Potassium permanganate ☐
5. When can electrolytic corrosion (galvanic action) take place?
   - a. Dissimilar plastics are placed in direct contact with each other ☐
   - b. Dissimilar metals are placed in direct contact with each other ☐
   - c. Metals of the same type are placed in direct contact with each other ☐
   - d. The materials are installed in coastal areas ☐

(*continued*)

Test yourself 4.2 (*Continued*)

6. Electrolytic corrosion can be reduced by using a sacrificial:
   a. Electrolyte ☐
   b. Cathode ☐
   c. Inhibitor ☐
   d. Anode ☐
7. Kelvin is the SI unit for measuring
   a. Linear expansion ☐
   b. Temperature ☐
   c. Pressure ☐
   d. Force ☐
8. Rearrange the chart below so that the process matches the correct description.

Process	Description
Conduction	Molecules in a material are heated and move further apart
Specific heat capacity	Movement of a fluid substance as a result of heating
Convection	The transfer of heat through waves from a hot body to a cooler one
Radiation	The amount of heat required to raise the temperature of 1 kg of a material by 1°C
Thermal expansion	The transfer of heat through direct contact with a heated up substance

9. An example of the effect of adhesion and cohesion can be seen in:
   a. Siphonage action ☐
   b. Frictional resistance ☐
   c. Stratification ☐
   d. Capillary attraction ☐
10. Capillary attraction can take place on roof weatherings if
    a. The weathering materials are lapped closely together ☐
    b. The weathering materials are kept apart ☐
    c. The water has a high pH value ☐
    d. The weatherings are of dissimilar materials ☐
11. For siphonage to take place what factor is required?
    a. Specific gravity ☐
    b. Water temperature ☐
    c. Atmospheric pressure ☐
    d. Kinetic force ☐
12. Which of the following examples would have the greatest frictional resistance on the flow of water?
    a. 3 m 22 mm copper pipe one bend ☐
    b. 3 m 22 mm copper pipe ☐
    c. 3 m 20 mm low carbon steel one elbow ☐
    d. 3 m 20 mm low carbon steel ☐
13. Which of the following gases is likely to explode if 13% is mixed with air?
    a. Butane ☐
    b. Nitrogen ☐
    c. Propane ☐
    d. Methane ☐

Check your answers at the end of the book.

Plumbing materials

Introduction

Plumbers have to work on a wide range of materials.

Activity 4.3

Can you think of a few examples of materials and where they might be used? Here is one to get you started:

Material	**Usage**
Plastics	Pipework, fittings, cold water storage cisterns

Write down your ideas on a separate sheet of paper and check them out at the end of this book.

You need to have a good understanding of the properties of plumbing materials (a material's properties are its physical and working characteristics and its suitability for the type of work in which it will be used), as this will have a bearing on some of the things that you do while working with them.

Standards for plumbing materials

Plumbing materials have to meet minimum standards of quality and performance in order to make sure they are capable of doing the job they are intended to. As well as minimum standards, there is also a need for 'standardisation'. This is because there are a number of manufacturers of plumbing materials and components, so it is also important that things like sizes and dimensions of fittings and components are the same throughout the industry. To help ensure that minimum standards and standardisation are adhered to, a number of organisations have been established in Britain and Europe.

British Standards

The British Standards Institute (BSI) is the organisation for standards in the UK, and was set up under a Royal Charter as an independent organisation. It was established because of the lack of 'standardisation' in the construction industry, the standards were set up in the UK way back in 1901 to address the problems caused by the fact that there was no set size for items, such as pipes and fittings.

British Standards begin with the letters BS followed by the number of the standard. For example, domestic circulating pumps should conform to BS EN 60335 and BS EN 1151 (The EN refers to European Standards).

Activity 4.4

Materials that meet the requirements of BSI carry the BSI kitemark. Would you be able to recognise the BSI kitemark? Find out what it looks like.

European Standards

Figure 4.17 European safety standard mark

You may have seen the symbol shown in Figure 4.17 on products you have purchased. It signifies that a product is certified to an EN (Euronorm's) standard and means the manufacturer will have taken the product through a series of tests, which are regularly checked under the European Council Quality Control Schemes. These start with the letters EN followed by the number, similar to British Standards.

Some products may show both BS and EN, for example, copper tube should conform to BS EN 1057: Part 1 and fittings to BS EN 1254. This simply means that products showing both these marks have met the requirements of British and European Standards.

Water Regulations Advisory Scheme

Formally known as the Water Byelaws Scheme, the Water Regulations Advisory Scheme (WRAS) has been carrying out the testing of fittings for many years and will continue to advise on Water Regulations in the future. As part of their work they produce a Fittings and Materials Directory, which lists all approved fittings. This directory is an important guide to all those who aim to comply with or enforce Water Regulations.

WRAS is a really good contact and they have a number of useful publications, some of them available free of charge. Their phone number is: 01495 249234. Their website is www.wras.co.uk.

Now, with the regulations and standards out of the way, we will concentrate on the materials themselves.

Properties of materials

What are a material's basic properties?

In general terms, you may say this could be how strong it is, how well it conducts heat or electricity, or how flexible it is.

Scientifically, materials are classified according to a variety of properties and characteristics. The properties can be measured as the materials react to a variety of influences, which include:

- Mechanical properties, such as hardness, strength, elasticity, toughness, stiffness, ductility, malleability
- Thermal properties like conductivity (how well or poorly a material will conduct heat)
- Electrical properties like conductivity (how well or poorly a material will conduct electricity)
- Chemical properties like reactivity and solubility
- Optical properties like transparency, reflectivity, refractivity
- Magnetic properties.

Hardness

Key point

Hardness is measured on a scale, Moh's scale of 1 (talc) to 10 (diamond).

There are many different aspects of materials which could be considered as a measure of hardness. Hardness can mean resistance to permanent or plastic deformation by scratching, indentation, bending, breaking, abrasion or fracture. This is a very important factor in materials which have to resist wear or abrasion – a sink tap for example – and frequently needs to be considered along with the strength of materials.

Strength

The strength of a material is the extent to which it can withstand an applied force or load (stress) without breaking. Load is expressed in terms of force per unit area, and is measured in newtons per square metre (N/m^2). This can be in the form of:

- Compression force, as applied to the piers of a bridge, or a roof support
- Tensile or stretching force, as applied to a guitar string, tow rope or crane cable
- Shear force, as applied by a shearing machine or scissors, or when materials are torn (see below). Materials are therefore described as having compressive, tensile or shear strength.

Elasticity

Almost all materials will stretch to some extent when a tensile force is applied to them. This increase in length on loading, compared to the original length of the material, is known as strain.

As increased loading continues, a point is reached when the material will no longer return to its original shape and size on removal of the load, and permanent deformation has occurred. The material is said to have exceeded its elastic limit or yield stress, beyond which the material is suffering plastic deformation – it is being stretched irreversibly.

Here are how some common materials 'shape up'

- Mild steel has little elasticity, but has a high yield stress and is fairly ductile, i.e. has a large range over which it can sustain plastic deformation. It also has a high tensile strength.
- Cast iron is brittle – it has poor elasticity and has no ability to sustain plastic deformation, although its tensile strength is higher than that of concrete.
- Copper has little elasticity, but is ductile. It has an ultimate tensile strength less than half that of mild steel.
- Concrete has little elasticity, and the lowest tensile strength of the four, but is extremely strong in compression.

Some other important characteristics which must be considered when considering material used in the plumbing trade are:

- *Plasticity* The exact opposite of elasticity: a material which does not return to its original shape when deformed
- *Ductility* Is the ability of a material to withstand distortion without fracture, an example is a metal such as copper that can be drawn out into a fine wire
- *Durability* The material's ability to resist wear and tear
- *Fusibility* The melting point of a material, i.e. when a solid changes to a liquid
- *Malleability* The ability of a metal to be worked without fracture; sheet lead is a very malleable metal
- *Temper* The degree of hardness in a metal

- *Tenacity* A material's ability to resist being pulled apart
- *Thermal expansion* The amount a material expands when heated.

Materials used in plumbing work

By now you should have an idea of the basic properties of material. Next we are going to look at plumbing pipework materials. There is no perfect pipework material that is suitable for all applications; different materials perform better in relation to different factors and conditions which can affect pipework such as:

- Pressure
- Properties of the water
- Cost
- Bending and jointing method
- Corrosion resistance
- Expansion
- Appearance.

There are two basic types of pipework material: metal and plastic.

Metals commonly used in the plumbing industry

Metals used in the plumbing industry include steel, iron, copper, brass, lead, tin, zinc and aluminium.

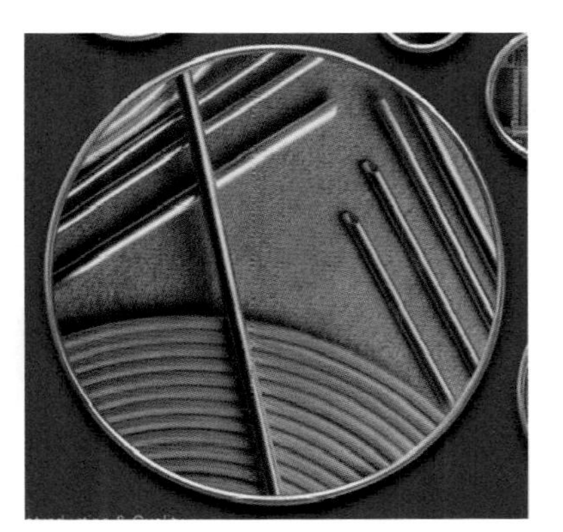

Figure 4.18 Copper tube (Reproduced with the kind permission of IMI Yorkshire Copper Tube Limited)

Copper

Copper is supplied in lengths and coils and in a range of diameters. Copper is a malleable and ductile material which you will use frequently throughout your plumbing career. There are several types of copper tubes (Figure 4.18) manufactured for use in the plumbing industry:

- R220 soft annealed coils – formally designated Table W – is a copper tube used for microbore pipework (usually found on central heating systems) and can be bent by hand or bending machine
- R250 half hard straight length – formally designated Table X – is the copper tube most commonly used above

ground for most plumbing and heating installations. It is fairly rigid and usually needs to be bent using a bending machine or spring.

- A more rigid thin-walled copper tube – formally designated Table Z – is also available. The tube is mainly used for export purposes and cannot be bent.

Tubes complying with the above specifications are not intended to be buried underground.

- R220 thick-walled soft annealed coil – formally designated Table Y copper tube – is most commonly used underground for the conveyance of water.

Key point

Copper is also used in the manufacture of pipework fittings.

The outside diameter of the tube is the same for each type, but there are differences in the internal bore size due to the variations in wall thickness.

Lead

The term plumber is derived from the chemical symbol Pb, and the Latin phrase plumbum, which when translated, means 'worker of lead'. Traditionally, lead was commonly used for water supply, sanitary and rainwater pipework, but it has now been superseded by the use of materials, such as plastics and copper.

These days, its main use in the plumbing industry is for sheet lead weatherings as its use for new water supply pipework is prohibited, although you may come into contact with lead pipework if renewing an old service pipe.

Lead is a very heavy, valuable metal which requires careful handling. It is ideal for sheet roof work as it is extremely malleable, as demonstrated in Figure 4.19 with the 'bossing' of an internal corner.

Cast iron

Cast iron is an alloy of iron and approximately 3% carbon. It has been used in the plumbing industry for many years for above and below ground drainage pipework. Cast iron is very heavy but quite brittle, and can withstand many years of general wear and tear. You will probably come into contact with it on older properties – on newer buildings it has been superseded by plastic-PVC.

Figure 4.19 Sheet lead work (Reproduced with the kind permission of the Lead Sheet Association)

As with lead, cast iron is sometimes used on historic buildings and unlike lead, cast iron is occasionally used on new build public and commercial buildings, where its strength and rigidity are an advantage.

Cast iron was used in the manufacture of baths and is still available in 'Victorian' styles.

Alloys

Alloys can be produced either by mixing different metals or by mixing metals with non-metallic elements, such as carbon. Steel is one of the most common alloys in the world and many plumbing materials and appliances are made of different types of steel.

Other alloys include:

- Brass (alloy of copper and zinc) which is used for various types of plumbing fittings (see Figure 4.20). Taps are chrome-plated for decorative effect.

The image shows an example of brass used in the manufacture of pump isolation valves.

Figure 4.20 Pump isolation valve (brass) (Reproduced with permission of Pegler Limited)

- Bronze (copper and tin), again, used mainly for fittings
- Solder (lead and tin), which is used for soldering capillary fittings, although solder containing lead is no longer permitted for use on water supply pipework, which is a requirement of the Water Regulations.

Steel

Low carbon steel

Low carbon steel (LCS) or mild steel is an alloy made from iron and carbon. It is commonly used in the plumbing and heating industry in larger premises, such as factories and commercial buildings.

In the domestic market, radiators are manufactured from steel, and LCS pipework and fittings are used in some small residential buildings for central heating systems.

LCS pipe is manufactured to BS 1387:1985 and comes in three grades of weight: light, medium and heavy. As with copper tube, the internal bore and wall thickness varies.

- Light LCS tube thin walls larger bore
- Medium LCS tube medium walls medium bore
- Heavy LCS tube thick walls smaller bore

Key point

Galvanised tubes have an outer layer of zinc, which prevents the process of oxidation or rust.

Light LCS is mostly used for electrical conduit. As a plumber, you may come across it occasionally, but are more likely to work with medium and possibly heavy grades. Medium and heavy LCS tubes are used for water supply and heating services, and are capable of sustaining high pressures. Heavy grade LCS tube can be identified by a red band painted towards the end of the tube, and medium by a similarly positioned blue marking.

When LCS tube is used for water supplies it must be galvanised.

Radiators like the one shown in Figure 4.21 are manufactured from mild steel.

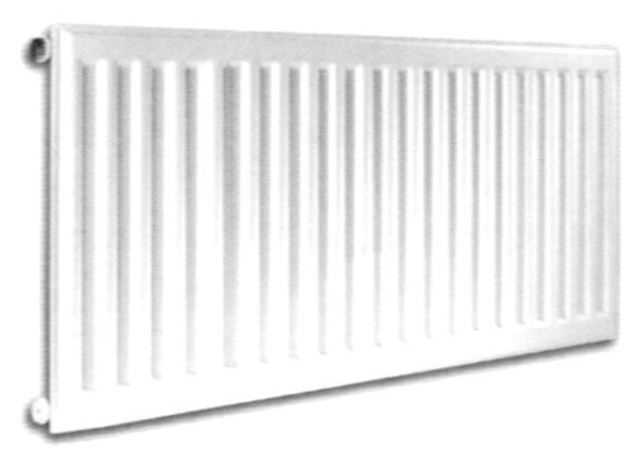

Figure 4.21 Mild steel radiator

Stainless steel

Stainless steel pipework was used extensively in the domestic market during the copper shortage of the 1970s; it is not that common today, although you may come across it while completing maintenance work.

The tube has a shiny appearance due to the chromium and nickel content and is protected from corrosion by a microscopic layer of chromium oxide, which quickly forms around the metal and prevents further oxidisation. This tube is produced with bores of 6–35 mm and has an average wall thickness of 0.7 mm. The outside diameters are similar to those of R250 copper tubes.

UNIT 4

Figure 4.22 Stainless steel WC

Stainless steel pipework is more commonly found where exposed pipework and sanitary appliances are needed, as it is a very strong metal (much stronger than copper) and is easy to clean. Stainless steel is commonly used for:

- Sink units and other sanitary appliances
- Urinal units and supply pipework
- Commercial kitchen or catering installation pipework.

Figure 4.22 shows an example of a WC suite manufactured in stainless steel. Models like this would most likely be installed in public toilets.

Plastics commonly used in the plumbing industry

Introduction

Plastics (polymers) are products of the oil industry. When crude oil (petroleum) is refined by fractional distillation, petroleum vapour is fed into a fractionating column, and different products are allowed to condense out at different temperatures. One of the products from this process is ethene (also called ethylene).

Ethene is the basis for most of the plastics industry. Under conditions of high temperature and pressure, and in the presence of a catalyst, molecules of ethene (monomers), can link together in long chains (polymers) – hence *polythene*. If the ethene monomer is modified by the replacement of one of the hydrogen atoms by another atom or molecule, further monomers result, which lead to the production of other plastics or polymerisation.

After all the chemistry talk, the plastic can be found in shapes like the ones shown in Figure 4.23; these will be used for soil above or below ground drainage.

There are two main categories of synthetic plastics used in the plumbing industry:

1. Thermosetting
2. Thermoplastics

Thermosetting plastics are generally used for mouldings. They soften when first heated which enables them to be moulded, setting

Figure 4.23 Plastic soil and discharge pipes and fittings

hard when cooled. Once hard their shape is fixed and cannot be altered by further heating.

Thermoplastics can be re-softened when heated. Most of the pipework materials you will come into contact with fall into this category. The different types of thermoplastics share many of the same characteristics:

- Strong resistance to acids and alkalis
- Low specific heat (do not absorb as much heat as metallic materials)
- Poor conductor of heat
- Affected by sunlight – leads to embrittlement of plastic, also called degradation.

Types of thermoplastic

The list below gives basic information about the types of thermoplastic pipework materials you may encounter during your plumbing career.

- Low density polythene is a flexible pipe material used to channel chemical waste
- Medium density polyethylene is moderately flexible and is used frequently for potable water supply pipework
- High density polythene is more rigid, and again is used for chemical or laboratory waste

- Polypropylene (PP) is a tough plastic with a relatively high melting temperature, it can be used to channel boiling water for short periods of time
- Polyvinyl chloride (PVC) is one of the most common pipework materials, and is used for discharge and drainage pipework
- Unplasticised polyvinyl chloride (PVC-U) is more rigid than PVC and is used for cold water supply pipework
- Acrylonitrile butadiene styrene (ABS) is able to withstand higher temperatures than PVC, it is used for small diameter waste, discharge and overflow pipework
- Polytetrafluoroethylene (PTFE) can withstand very high temperatures, up to 300°C, and is generally used as a thread sealant
- Polystyrene is brittle and light; it is used generally for insulation purposes, but must have fire-retardant capabilities.

Other materials relevant to the plumbing industry

Ceramics include those products which are made by baking or firing mixtures of clay, sand and other minerals – bricks, tiles, earthenware, pottery, china. There is a sense in which the kiln firing process is creating 'artificial metamorphic rocks' by using heat to fuse together the individual ingredients of the product into a matrix. The main constituent of all these products is silicon, clay is aluminium silicate; sand is silica dioxide.

Figure 4.24 Ceramic WC

This category would also include those products made by 'curing' mixtures of sand, gravel, water, and a setting agent (usually cement) to form concrete, and mortar, a sand, water and cement mixture.

Vitreous china is made from a mixture of white burning clays and finely ground minerals which are mixed and fired at high temperatures. It is used in the manufacture of sanitary appliances like the one shown in Figure 4.24.

Test yourself 4.3

1. What does the abbreviation BSI stand for?
   a. Best Spoke Institution ☐
   b. British Standard Institution ☐
   c. Building Standard Institution ☐
   d. British Specification Institution ☐
2. Load is expressed in terms of force per unit area and is measured in:
   a. Mohs ☐
   b. Watts ☐
   c. Metres ☐
   d. Newtons ☐
3. Low Carbon Steel (LCS) is an alloy made from carbon and?
   a. Zinc ☐
   b. Chromium ☐
   c. Iron ☐
   d. Boron ☐
4. Copper tube most suitable for underground use was formally designated as
   a. Table X ☐
   b. Table W ☐
   c. Table Z ☐
   d. Table Y ☐
5. The following statement describes a thermoplastic and what it is suitable for:
   One of the most common pipework materials, used for discharge and drainage pipework.
   Which thermoplastic does the statement apply to?
   a. PVC ☐
   b. Polythene ☐
   c. Polypropylene ☐
   d. Polystyrene ☐
6. What installation pipe material is now generally prohibited for use?
   a. Stainless steel ☐
   b. Copper ☐
   c. Galvanised iron ☐
   d. Lead ☐

Check your answers at the end of the book.

Check your learning Unit 4

Time available to complete answering all questions: 30 minutes

Please tick the answer that you think is correct.

1. On the Fahrenheit scale water freezes at:
   a. 0 degrees
   b. 16 degrees
   c. 25 degrees
   d. 32 degrees
2. At what temperature does water reaches its maximum density?
   a. 0°C
   b. 2°C
   c. 4°C
   d. 8°C
3. What happens to the molecules when a substance changes state from a liquid to a gas?
   a. They move further apart
   b. They begin to vibrate
   c. They move closer together
   d. They fuse together
4. A water storage tank measures $2 \times 1.5 \times 2\,m^3$ and contains water up to a depth of 1.6 m. Given that the pressure exerted by $1\,m^3$ of water is $9.81\,kN/m^2$, which of the following calculations shows how the total pressure exerted on the base of the tank should be calculated?
   a. $9.81\,kN/m^2 \times 1.6$
   b. $1.6\,m \times 9.81\,kN/m^2 \times 2\,m \times 1.5\,m$
   c. $2\,m \times 1.5\,m \times 2\,m \times 9.81\,kN/m^2$
   d. $2\,m \times 1.5\,m \times 9.81\,kN/m^2$
5. Frictional resistance within a cold water pipework system is influenced most by the:
   a. Pipework material
   b. Pipework diameter
   c. Number of elbows in the pipework
   d. Number of bends in the pipework
6. Which domestic appliance utilises the principle of siphonage to meet its performance requirements?
   a. Thermostatic shower
   b. Instantaneous shower
   c. Corner bath
   d. Double trap WC
7. Copper has a specific heat capacity of 0.385 kJ/kg°C. How much heat energy would be needed to heat 1 kg of copper by 2°C?
   a. 0.385 kJ
   b. 0.77 kJ
   c. 3.85 kJ
   d. 7.7 kJ
8. What plumbing system by-product is most directly linked to hard water?
   a. Carbon dioxide
   b. Hydrogen
   c. Ferrous oxide
   d. Limescale
9. Water from an area which has a limestone and chalk strata is most likely to be
   a. Alkaline
   b. Acidic
   c. Cloudy
   d. Neutral
10. In which of the following combination of metals would electrolytic corrosion be most pronounced?
    a. Copper and lead
    b. Iron and zinc
    c. Copper and aluminium
    d. Aluminium and magnesium

(Continued)

Check your learning Unit 4 (Continued)

11. Which of the following is the correct formula for fuse rating?
    a. Volts × ohms = watts ☐
    b. Volts × amps = ohms ☐
    c. Volts ÷ watts = amps ☐
    d. Watts ÷ volts = amps ☐

12. The SI unit of measurement for electrical resistance is:
    a. Watt ☐
    b. Ampere ☐
    c. Ohm ☐
    d. Volt ☐

Sources of Information

We have included references to information sources at the relevant points in the text; here are some additional contacts that may be helpful.

- BSI Product Services
  Maylands Avenue
  Hemel Hempstead
  Hertfordshire
  HP2 4SQ
  Tel: 01442 230 442
  Website: www.bsi.org.uk

UNIT 5

Cold Water Supply

Summary

One of the basic human needs is a supply of fresh drinking water, and it is your job as a plumber to get the cold water from the mains to the consumer's cold water taps. Refer to Figure 5.1.

To make sure that water is supplied correctly, the Water Supply (Water Fittings) Regulations 1999 set out the installation requirements for cold water systems; one of which is making sure that the water arrives at a temperature less than 20°C, you'll find out why later.

This unit sets out the 'whys and wherefores' of how to get the water from the point of collection and storage to its final distribution points inside people's homes.

The cold water supply unit covers the following topics:

- **Cold water supply and treatment**
  - Regulations
  - Collection and storage of water
  - Water distribution
  - Water treatment
  - Service and mains connection
  - Requirements of service pipework
  - Water meters.
- **Domestic cold water supply in domestic dwellings**
  - Types of systems, direct and indirect
  - Scale inhibitors and conditioners
  - Requirements of cold water storage cisterns
  - Frost protection.

Cold water supply and treatment

Introduction

Have you ever thought of how much water a person uses on average each day? It is approximately 130 l, and if you multiply that by the population, of around 65 million, it is quite a lot of water; all of which has to be collected, treated, distributed and supplied to the end user, me and you!

Figure 5.1

Water Suppliers

Procuring or obtaining water that is fit for human consumption to the consumer is the job of the Water Suppliers or Water Companies as they are also called.

They are required by law, under the **Water Supply (Water Fittings) Regulations 1999** – Schedule 1, to provide sufficient quantities of pure wholesome water to the dwellings.

Water Regulations

Students progressing to NVQ/Technical Certificate Level 3 will cover the Water Regulations as a specific topic, but everything you cover on cold and hot water systems will be influenced by the regulations.

Key point

It is important that you remember the Water Supply (Water Fittings) Regulations 1999, as these make provision for preventing the contamination, waste, misuse, undue consumption and erroneous measurement of water supplied by a water undertaker. In short, they form the basis for how a plumber will install hot and cold water pipework and fittings.

We recommend that you should have access to a copy of the WRAS Guide, your college or centre should have a copy, or if you want your own copy try to find it at your local library or alternatively you can buy one. It is published by WRAS Publications, Fern Close, Pen-y-Fen Industrial Estate, Oakdale, Newport NP11 3EH. They also have a website: www.wras.co.uk. Other contact details:

Tel: 01495 248454
Fax: 01495 249234
E-mail: info@wras.co.uk.

Water supply and installations in England and Wales were previously controlled and regulated by locally made Water Byelaws and Acts. These were replaced by the Government with national Water Regulations, known as Water Supply (Water Fittings) Regulations 1999, which came into force on 1 July 1999. The regulations only apply to England and Wales. Separate arrangements apply in Scotland and Ireland.

Activity 5.1

What do you think are the main objectives of the regulations? There are five listed in the regulations, starting with 'preventing the contamination of a water supply'. Write them down on separate sheets of paper and put them in your portfolio, the answers are at the end of this book.

1. Preventing the contamination of a water supply
2.
3.
4.
5.

Key point

From what we have discussed earlier, who do you think enforces the regulations? Yes, it is the Water Suppliers or Water Companies as they are sometimes known.

Who enforces the regulations?

Water Suppliers are encouraging installers to register as approved contractors. The approved contractor scheme requires that a contractor meets all the requirements of the scheme criteria and has good working knowledge of the Water Regulations. They are then allowed to display a certificate that their own installation work is covered by the Water Regulations.

Currently there is no legal obligation for a person working on a water installation plant to be qualified. However, anyone who carries out such work could be prosecuted for breach of regulations, and if convicted of an offence, may be liable to pay a hefty fine.

Hot tips

If you want to find out more about the regulations and their enforcement, there is information on the WRAS website mentioned earlier, available as downloadable PDF leaflets entitled 'What are the Water Regulations? and 'Information on the Requirements for Notification'. These are also available in hard copy.

Water

You have studied the properties of water in Key Plumbing Principles, so there is no need to go over it again here, but understanding the properties of water will help you to appreciate why systems are designed the way they are, as well as some of the problems that can occur in systems, for example when water freezes or when limescale deposits are formed when water containing calcium carbonate is heated above 65–70°C.

The water that finds its way to our taps is derived from rainfall. Rain water finds its way into rivers, lakes and the sea, where it evaporates forming water vapour and eventually falls as rain again. This continuous succession of events is known as the Water Cycle – more discussion about this can be found later in the unit.

Activity 5.2

Water can be found in three physical forms, what do you think they are? We have given you one answer; jot down the other two in your portfolio. Check your answers at the end of this book.

1.
2. Liquid – as water.
3.

Figure 5.2 will help you to summarise the information from Activity 5.2.

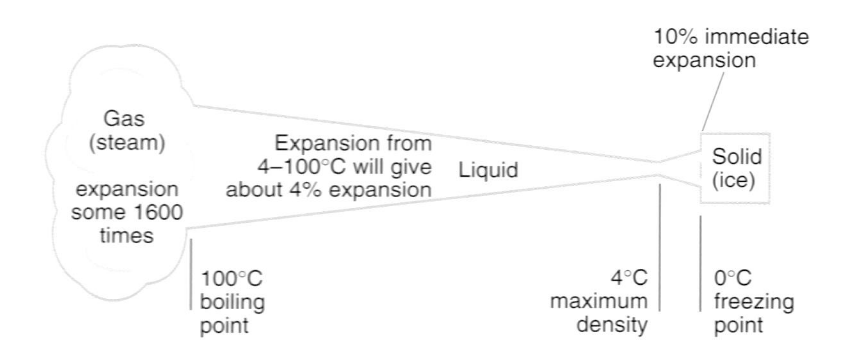

Figure 5.2 Physical forms of water

Key point

Remember, you have already covered the properties of Water in Unit 4, if you need to refresh your memory.

Rain water is relatively pure, but because water is a natural solvent it can dissolve gases in the atmosphere and solids in the ground to form solutions. This means that by the time it eventually reaches natural storage areas, such as reservoirs and lakes, it is unlikely to be pure because it has picked up impurities from either the atmosphere or the ground en route.

Where does water come from?

We mentioned earlier that water is sourced from rainfall. The continuous process of water evaporating and condensing from and to earth is called the rain cycle.

The water cycle

Water evaporates from rivers, lakes, the sea and the ground and forms clouds. Clouds contain water vapour which, under suitable climatic conditions condenses and falls as rain, as shown in Figure 5.3.

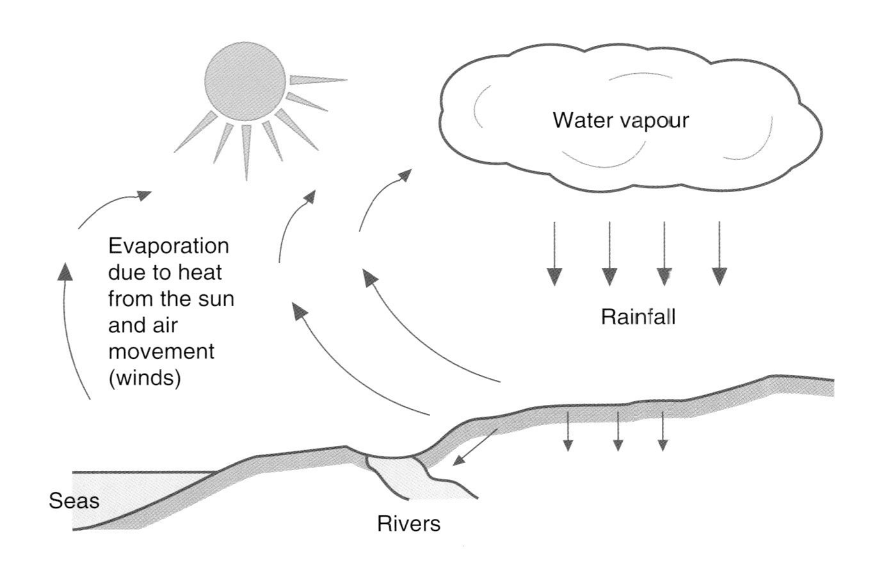

Figure 5.3 The water cycle

When rain falls to the ground, some of it flows into streams, rivers, and lakes, the remainder soaks into the ground, where it collects temporarily and eventually evaporates.

The rain water also percolates through the ground, forming natural springs or water pockets and are referred to as underground sources.

Water Suppliers usually obtain their water for public consumption from two main sources:

- Surface sources such as:
  - Upland surface water
  - Rivers and streams.
- Underground sources such as:
  - Wells
  - Artesian wells
  - Springs.

Water from these sources cannot be supplied directly to the consumer as it is likely to be contaminated, river water being a good example. It is vital that the water is treated first, and that also is the job of the Water Supplier.

Water treatment

The treatment of water is dependent on where it is sourced from and what impurities it contains.

Water that is sourced from springs and wells is naturally purified, and should need little disinfection.

The quality of water that is sourced from reservoirs or rivers (raw water) will determine the level of treatment. Usually this will involve several stages of treatment including settling, filtering and final 'polishing' with activated carbon grains to remove minute traces of impurities and improve water taste.

Water storage

Water Suppliers store water either in its raw state in impounding reservoirs or lakes, or as treated wholesome water in service reservoirs.

Water Suppliers endeavour to store an adequate amount of wholesome water in their service reservoirs for emergency purposes; this should last for about 24 hours. This provision safeguards against mains supply, or pump failure, and allows time for the repair to be carried out before the supplies run out.

Getting water to the tap

After being treated, water is distributed from the Water Supplier to individual buildings and domestic premises through a network of pipes known as *mains*. The mains belong to the Water Supplier and it is their job to maintain them.

The local mains supply provides the 'final leg' for the supply of water to buildings and domestic dwellings such as your home.

Key point

If you would like to know more about the sources of water supply, treatment and distribution, you should try contacting your own local Water Supply Company. You'll find their number on your water bill.

Mains connection

The most commonly used materials for new underground installations and for the replacement of existing mains is PVC-U, conforming to BS 3505 Unplasticised Polyvinyl Chloride. Also, blue polyethylene conforming to BS 6572 up to a diameter of 63 mm is suitable for below ground use.

The Water Supplier is also responsible for the ferrule connection of a new or replacement supply pipe to the distribution main.

The connection between the main and the consumer's stop tap is made using a ferrule. The ferrule is basically a valve that is tapped into the main, with a connection to the supply pipe. The ferrule can be turned off in order to isolate the supply pipe.

Once the ferrule connection is completed, the connection between the supply pipe and the consumer's stop valve can take place. Figure 5.4 shows the completed ferrule installation. Note the gooseneck in the service pipe.

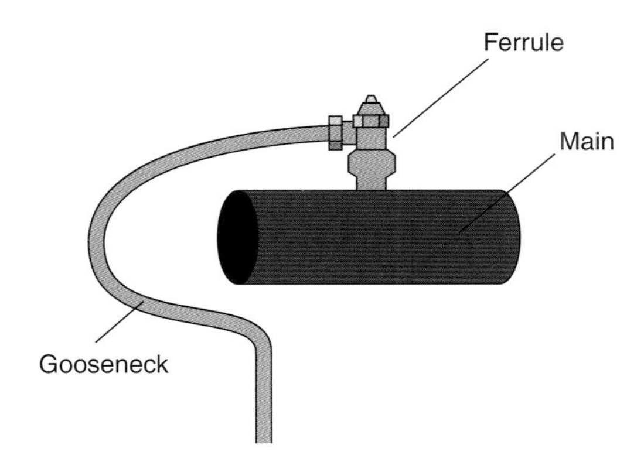

Figure 5.4 Ferrule and goose neck

Try this

What do you think is the purpose of the gooseneck in the service pipe?

The purpose of the gooseneck is to allow for any ground movement, e.g. subsidence or shrinkage that may take place. If the pipe is laid straight, the ground movement can damage the pipe causing it to split or fracture. Some Water Suppliers, while installing polyethylene pipes, do not fit goosenecks, as the pipes are considered to be sufficiently flexible.

Key point

All underground pipework using compression fittings must be type 'B' manipulative compression joints.

From the mains to the premises

It is useful that you understand what is required to get water from its original source to the consumer. What we have covered so far falls within the responsibility of the Water Supplier, but once you get past the external stop tap, also known as the consumer's stop tap, the real work of the plumber begins.

Figure 5.5 shows the pipework layout from the mains to the internal stop valve and drain tap within a building. Study it carefully and have a go at sketching in your portfolio from memory, you may have to work on something like this eventually.

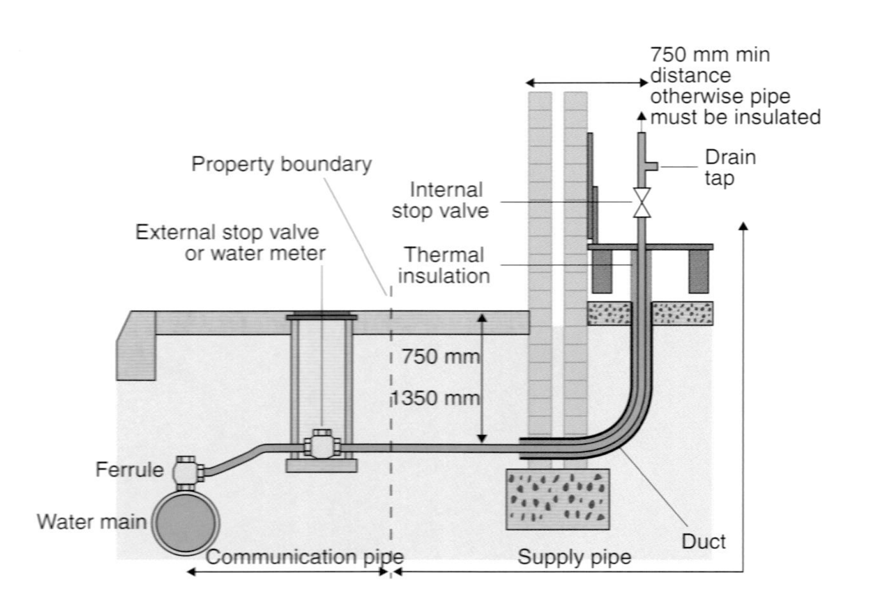

Figure 5.5 Details from the mains to the internal stop tap

Case study 5.1

Leslie has been asked to check that the installation in Figure 5.5 is correct. Leslie consults the regulations and makes the following notes as a cross-check:

- The supply pipe or distributing pipe that is providing water to the individual dwellings must be fitted with a conveniently located stop valve to enable the supply to be shut off. The minimum permitted size for a cold water service into a building is 15 mm diameter.
- Where a cold water supply pipe enters a building via a duct, the pipe duct must be sealed at both the ends; this is to prevent any gases or vermin from entering the building
- In order to protect against frost damage, the mains and service pipes should be at least 750 mm beneath the surface of the ground. This is a requirement made under the Water Regulations Schedule 2, Paragraph 7(4).
- The maximum depth of cover over the pipe should not exceed 1.350 m, as this would prevent ease of access

- Metallic pipes such as copper should be protected against possible corrosion from soil, particularly where acidic soils exist. This can be achieved by:
  - Using plastic-coated pipe
  - Wrapping the pipe suitably to protect from corrosion
  - Installing the pipe inside a duct, sealing the ends of the duct on final completion.

Installation of external stop valves

As you can see in Figure 5.5, the supply pipe to the building can be isolated from the main by installing an external stop valve. This enables the external supply to be isolated for maintenance and emergency reasons, so easy access to the stop valve is very important, as is keeping it in a good working condition.

External stop valves for below ground use should be screw down type, valves complying with BS 5433 or plug cocks conforming to BS 2580. Ideally the valve should be located in a stop valve chamber, constructed from 150 mm PVC, or earthenware pipe sited on a firm base.

The top of the chamber has a robust stop valve cover. A stop valve cover is generally made from steel plate or a combination of steel plate and plastic.

Figure 5.6 Internal stop valve

Try this

Why not see if you can try and locate the **external stop valve** in the dwelling where you live? This is used to isolate the supply to a dwelling, and plumbers may be required to locate it. The stop valve within the box should be accessible, i.e. clean from debris etc. **Just locate it; please do not do anything with it!!**

Try this

Figure 5.6 shows an existing stop valve installation in a domestic dwelling of about two years old. Notice the incoming service is in plastic, and the plastic to copper fitting. What do you think the green and yellow wire is for? What do you think the white substance is on the bottom compression nut?

Installation of internal stop valves

The internal stop valve within a building must be:

- In an accessible position
- Fitted above the floor level
- As close as reasonably practicable to where the supply enters the building
- Installed so that when closed, it will prevent the flow of water to all distribution points within the building
- Well maintained so it's fully operational.

Try this

What type of fitting must be installed immediately above the stop valve and why?

The installation must include a drain tap immediately above the stop valve, so that if it is turned off, the cold water in the pipework can be drained off for repair work, or in instances where a building may be left empty for a long period of time, especially through the winter months.

Water meters

The installation of water meters for domestic premises is becoming more widespread, and from 1989 most local Water Suppliers require them to be installed on new buildings or on properties that have been substantially converted, see Figure 5.7.

Figure 5.7 Typical water meter

They can be installed externally in a purpose-made chamber, or fitted internally on the supply pipe, at the point where it enters the building which is the most preferred practical method.

This type would most likely be fitted externally. Meters are also available that are suitable for installation inside a dwelling.

The meter measures in cubic metres (1 cubic metre = 1000 litres or 220 gallons) the actual quantity of water used by the consumer, and on billing the consumer is charged for the metered amount used.

Prior to working on metallic, e.g. copper, cold supply pipe or where a water meter is to be fitted, it is very important for you to check that there is a permanent main equipotential bonding to the pipework (cross bonding). Equipotential bonding will be covered in greater detail in the electricity unit.

Key point

Where a building or dwelling has a plastic incoming water service (polythene), and metallic internal pipework system (copper), it is extremely important that there is permanent equipotential earth bonding to the metallic pipework.

For the existing water meters, there must be additional earth bonding fitted across the inlet and outlet sides of the meter; this ensures that the earth continuity is maintained and not broken either by meter removal, or the use of fibre/rubber sealing washers.

The British Standards BS 6700 Specification for design, installation, testing and maintenance of service supplying water for domestic use within buildings and their curtilages makes recommendations for the installation of water meters.

Test yourself 5.1

1. What regulations cover the requirements of the cold water supply to a domestic dwelling?
2. The regulations cover five main objectives, state one of these objectives.
3. When water freezes, by what percentage does it expand?
4. Which type of water is most likely to cause the build up of limescale in water systems when heated above 65–70°C?
5. Name two underground sources from where water can be obtained.
6. Briefly describe the term 'the rain cycle'.

Domestic cold water supply

Towards the end of the previous section we followed the path that water takes virtually from its source to where it enters a building, from the ground to the internal stop valve.

In this section, we will concentrate on studying the cold water supply pipework within a domestic dwelling as this is a requirement for the Level 2 Technical Certificate, although at some time during your career you may probably work on other 'domestic applications', such as flats, maisonettes, shops, offices, factories, etc.

Cold water systems

In simple terms, all we are trying to achieve here is to get a supply of cold water to the various appliances and components within a dwelling. This is achieved by connecting system pipework from

(a)

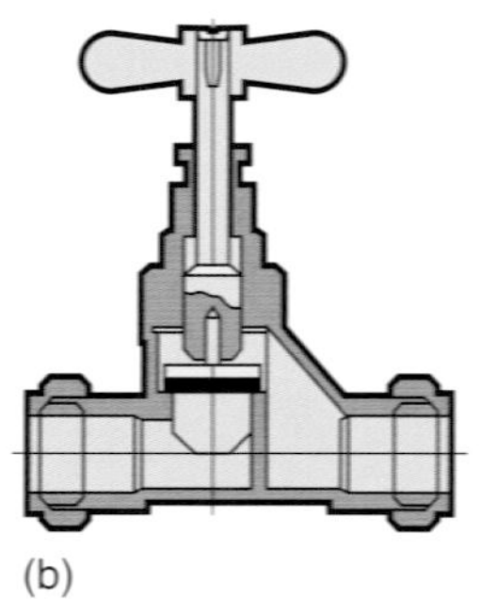
(b)

Figure 5.8 (a) and (b) Stop valves

the internal stop valve to the cold water storage cistern (CWSC), and depending upon the type of system to sanitary appliances, such as sinks and WCs.

Internal stop valve

The Water Regulations specify the requirements for internal stop valves as:

- Stop valves: a valve other than a servicing valve, used for shutting off the flow of water in a pipe.

There must be an adequate number of stop valves and drain taps fitted to the system as stated in Schedule 2, Paragraph 11 of the regulations; this reduces the amount of water discharge when fittings are to be replaced or maintained, and thus prevents the undue wastage of water.

Figure 5.8a shows a typical stop valve, and the figure to the right (Figure 5.8b) is a cross section through a valve showing the various components.

The stop valve should be manufactured to BS 1010 standards. Note the directional flow arrow on the valve body. The valve must be installed with the arrow pointing in the direction of the flow of water.

Try this

Why do you think it is important to be able to drain the system?

Figure 5.9 Combined stop valve and drain tap

Drain taps should be fitted at low points on installation pipework. Combined stop valve and drain tap is available, as shown in Figure 5.9. Drain taps allow the whole system or various sections to be drained down.

The stop valve and combined stop valve and drain tap images have been reproduced with the kind permission of Pegler Limited. Their contact details are included at the end of this unit.

Water mains pressure

Mains water pressure is measured in bars. Supply pressure is important because it needs to satisfy the demands of the consumer, as well as meeting appliance specifications that require a minimum water pressure to enable them to operate effectively, an example being instantaneous showers or water heaters. The operating pressures for such appliances should be included in the manufacturer's instructions.

The Water Supplier is responsible for the water mains delivery pressure to the building, so if it is poor they would need to be advised.

Water flow rate can be measured at the tap by using a flow meter device called a weir gauge. Alternatively, this can be carried out by using a container with a known capacity, e.g. a measuring beaker used for cooking, and with the tap fully open, recording the time it takes to fill the container and then calculating the flow rate per minute in litres per minute (refer to Figure 5.10).

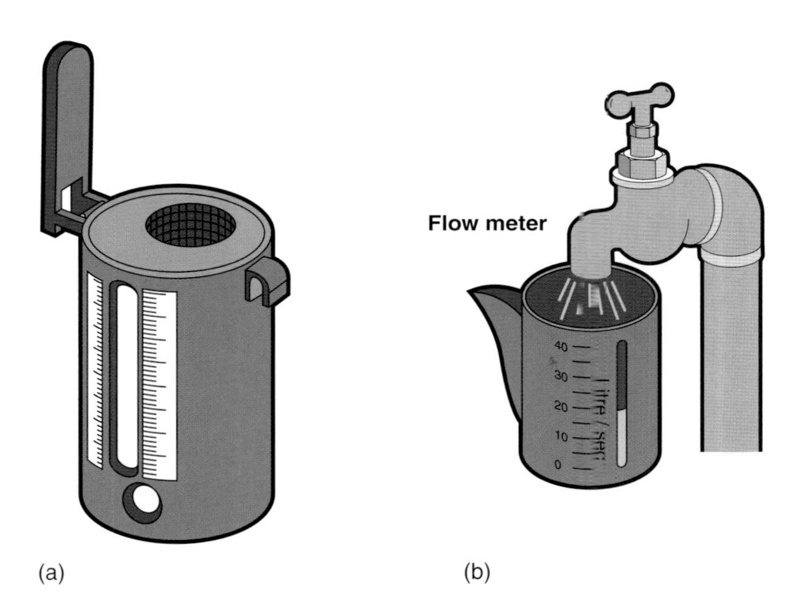

Figure 5.10 (a) Weir gauge (b) Flow meter

When water flows through the weir gauge, the flow rate is measured off the calibrated slot. The one illustrated is calibrated from 2 to 22 l per min. Water pressure gauges are also available.

> **Try this**
>
> What method is used in your college or centre for measuring water pressure? Have a look at manufacturers' catalogues to see what type of water pressure gauges are available.

Systems for treating water

You will remember that in the previous section on cold water supply and treatment, reference was made to water treatment from the perspective of the Water Supplier. When the water arrives to the consumer, usually no further treatment is necessary. This is not always the case in hard water areas.

Hard water supplies can cause problems and damage to a hot water system by the accumulation of limescale. This is an important factor to consider where a combination boiler may be installed, or in an area where the degree of water hardness is over 150 parts per million.

There are different methods for reducing the build up of limescale in domestic plumbing systems, including:

- Base exchange softeners
- Electrolytic scale inhibitors
- Magnetic scale inhibitors
- Digital electro magnetic conditioners.

> **Key point**
>
> The British Standards BS 6700 'Specification for design, installation, testing and maintenance of services supplying water for domestic use within buildings and their curtilages' makes recommendations for the installation of ion exchange water softeners.

Base exchange

The principle of a base exchange water softener is that the incoming cold water in hard areas is treated as it flows through a tank containing resin particles. The resin attracts and absorbs the salts' hardness, mainly calcium and magnesium, from the water. At the same time it replaces them with sodium from the resins.

This method is termed as ion exchange. After a while the resin becomes saturated with hardness salts and needs to be regenerated using salt solution to put sodium back into the resin. The hardness salts are released from the resin and washed down the drain.

Electrolytic scale inhibitors

The principle of design of these devices is to actually alter the way in which the scale behaves in a system, by hindering the formation of calcite crystals and keeping the beneficial mineral content in the water intact (see Figure 5.11).

When fitted on the incoming supply, electrolytic scale inhibitors offer whole-house protection.

Magnetic scale inhibitors

These work on the principle of magnetic force, see Figure 5.12. When water flows through the device, it crosses a magnetic field.

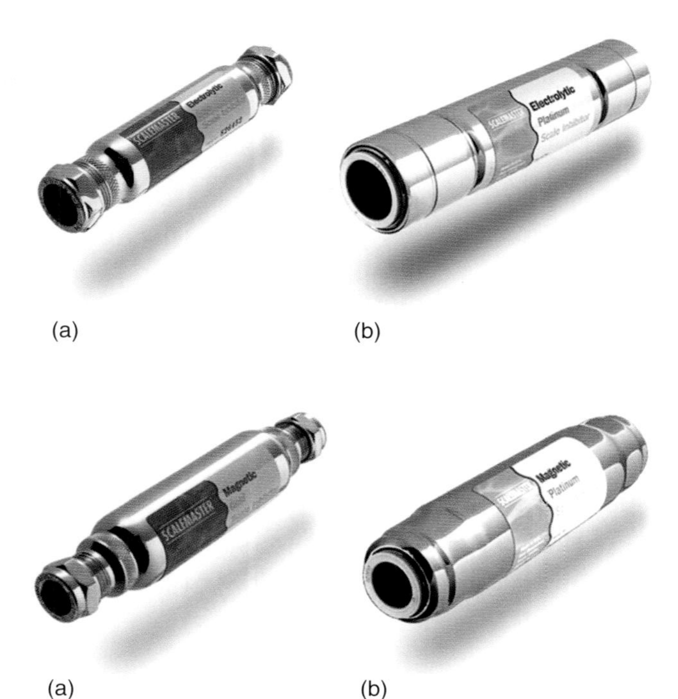

Figure 5.11 Electrolytic scale inhibitors (Reproduced with permission of Scalemaster)

Figure 5.12 Magnetic scale inhibitors (Reproduced with permission of Scalemaster)

This causes the tiny crystals that form limescale to become magnetised and adhered to each other; which tends to alter the size and shape of the crystals and makes them less likely to attach to the plumbing system and components.

The magnetic attraction is not permanent and wears off quickly, especially when water is stored for a longer period of time, so it is recommended that the magnetic scale inhibitors are installed on the supply to protect a specific appliance, e.g. combination boilers, water heaters, electric showers, dishwashers, washing machines, etc. Some types of these in-line inhibitors may require additional equipotential cross bonding.

Digital electromagnetic conditioners

These work on the principle of generating electromagnetic waves from a control box. The electromagnetic waves are then transmitted through the cold water system via a thin aerial wire wound on the pipework.

Numerous scientific reports have supported the theory that protection from limescale can be achieved by using the electromagnetic waves. With most installations, it is not necessary to cut the pipe, but this will depend on the model type and the manufacturer. In all the cases, they do require a permanent live electrical supply.

Try this

What do you think is the main difference between a direct and an indirect system of cold water?

Water system types

There are two types of cold water system:

- *Direct*
- *Indirect*

With a *direct* system, the cold water is supplied *directly* from the mains supply to all the draw-off points within the building, i.e. kitchen sink, bath, washbasin, WC, etc, as shown in Figure 5.13.

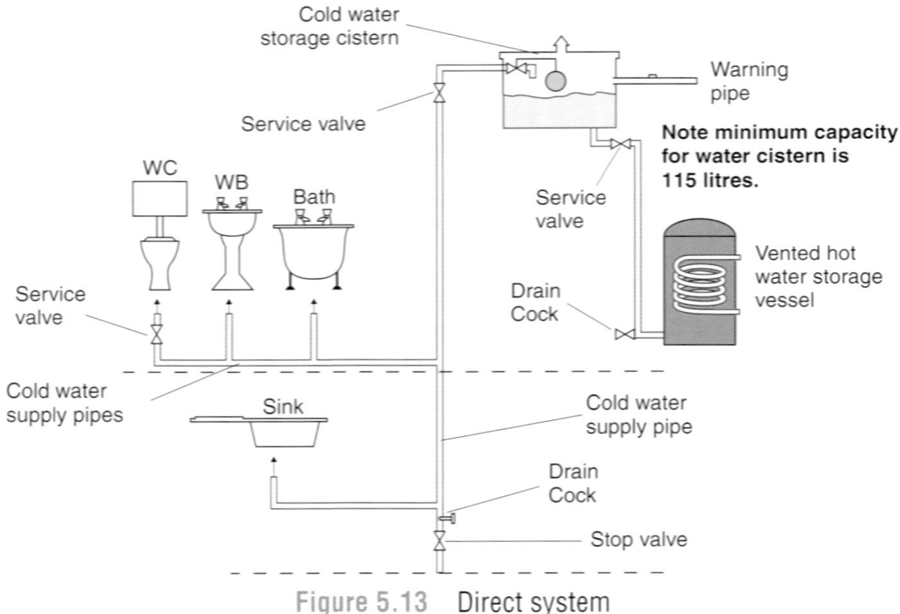

Figure 5.13 Direct system

With an *indirect* system, only one draw-off point, usually the kitchen sink tap is supplied directly, and this is used for drinking water purposes.

The supply then goes on to feed the cold water storage cistern (CWSC); which is usually found in the roof space of the building or dwelling. The CWSC is used to supply water to the remaining draw-off points *indirectly*.

Direct system

The installation of a direct water system offers a supply of water at mains (high) pressure to all draw-off points within the building. These installations are permitted by Water Suppliers in regions where the mains supply can provide adequate quantities of water at sufficient pressure.

Indirect system

The installation of an indirect water system offers a supply of low-pressure water from the storage cistern to the designated draw-off points within the dwelling or building. Provision should be made for one outlet to be directly fed from the mains, supplying wholesome water for drinking, cooking food and washing purposes (refer to Figure 5.14).

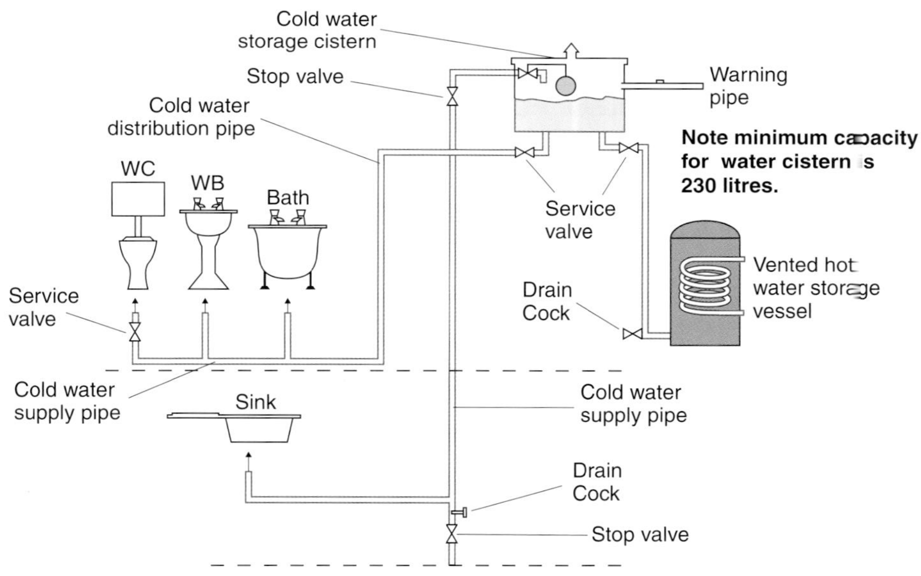

Figure 5.14 Indirect system

Activity 5.3

Take a close look at Figures 5.13 and 5.14 and read the text again. Think about what the advantages and disadvantages might be for both systems. Jot down your thoughts on separate sheets of paper and include them in your portfolio. A suggested answer is given at the end of this book.

Requirements for Cold Water Storage Cisterns (CWSC)

The CWSC is a very important part of a domestic cold water system, and as such is well covered by the Water Regulations. Figure 5.15 shows a CWSC that fully meets the Water Regulations requirements of Schedule 2, Paragraph 16(4) for CWSC installations.

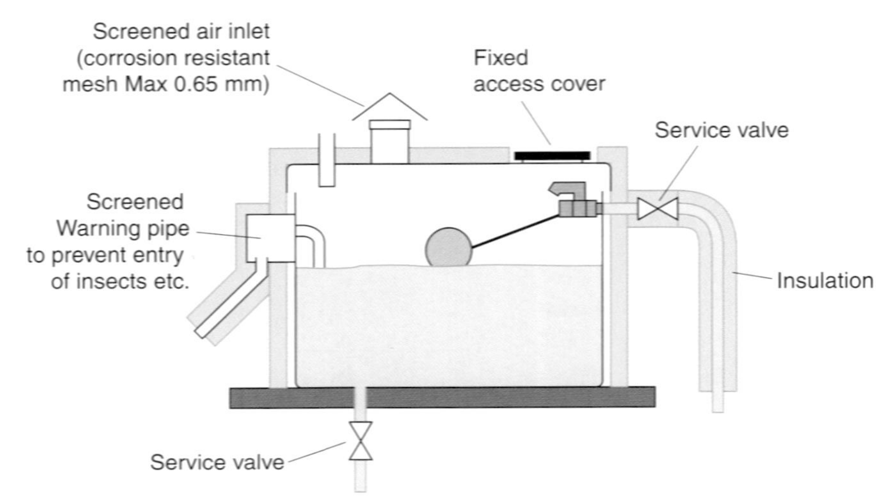

Figure 5.15 Requirements of CWSCs

Try this

When a storage cistern is located in the roof space of a building or dwelling and the space contains thermal insulation, the thermal insulation below the base of the cistern should be omitted. Why do you think this is so?

This additional measure will help to protect the cistern against freezing in severe weather conditions, and enables warmer air from the room below to circulate below the base of the cistern.

Cisterns storing water for domestic purposes should be:

- Fitted with a float-operated supply valve to maintain the correct water level in the cistern. The float valve must be fitted as high as possible within the cistern. It must comply with the requirements of BS 1212 Part 2 or 3, in order to maintain an air gap and prevent the possibility of back siphonage into the incoming supply.

This meets the requirements of the Water Regulations Schedule 2, Paragraph 16(1). Float valves complying with BS 1212 Part 1 may be used, provided they have an adjustable float lever connection.

- Fitted with service valves on the inlet and outlet pipes for maintenance or servicing purposes and for shutting off the flow of water in a pipe to a water fitting

  This meets the requirements of the Water Regulations Schedule 2, Paragraph 16(3)
- Covered with a securely fixed lid, that is not airtight, but excludes light or insects. The lid shall incorporate a screened air inlet, and where the vent pipe from the hot water cylinder passes through it, be appropriately sleeved.
- Fitted with a screened overflow/warning pipe to warn of overflow discharge. The overflow pipe must have a minimum internal diameter of 19 mm, and in all instances, be greater in size than the inlet pipe.
- Installed so as to minimise the risk of contamination of the stored water. The pipework connections must also be positioned to allow the water to circulate freely preventing the possibility of stagnant water.
- Insulated to protect against frost and freezing, and to prevent heat gain of the stored water. Keeping the water temperature below 20°C will restrict the possibility of microbiological growth, e.g. bacteria such as Legionella.
- Fully supported on the whole of its base area. This will prevent undue stress on pipe connections and distortion of the cistern when filled with water, which could result in leaks.
- Installed in an accessible position, to allow ease of maintenance and for cleaning purposes. There must be a minimum clearance height of 350 mm above the top of the cistern.

In order to reduce the risk of waterborne diseases such as Legionella in water storage systems, it is recommended that the outlet connection be positioned as near as possible to the bottom of the cistern. This allows any small particles that may be present to pass through the system, preventing an unhealthy build up of sediment at the base of the cistern.

Materials used for cisterns

Today, cisterns used for domestic installations are made from non-metallic materials conforming to the requirements of British Standard BS 4213 rather than galvanised ones that you may still come across on maintenance work. Typical materials are:

- Polyethylene
- Polypropylene
- Polyolefin
- Olefin copolymer
- Fibre glass
- Reinforced plastic.

CWSCs manufactured from these materials are:

- Light
- Hygienic
- Corrosion resistant
- Flexible.

Try this

What advantages do you think CWSCs made from nonmetallic materials might have over their metallic rivals?

Cisterns are available in a range of sizes and shapes including square, rectangular or circular. They are generally manufactured in black as this restricts the growth of algae (a type of green microorganism).

Holes cut out for pipe connections should be circular and made by using a hole-saw, and not burnt through with a piece of heated 22 mm copper tube! The joint between cistern wall and fitting should be made using plastic washers.

Where a jointing sealant is required, it must conform to BS 6956 Part 5, approved for use on water. Oil-based sealants are not acceptable for use on plastic cisterns, as they react against plastic materials, causing them to break down.

Plastic cisterns in particular should be fully supported, and not just rested on ceiling joists. This is normally done by fitting a flat surface between the ceiling joists and the base of the cistern using plywood, floorboards or blockboard.

Cistern connections and control valves

Inlet controls

The Water Regulations require that a pipe supplying water to a storage cistern must be fitted with an effective adjustable shut-off device. The device used on most domestic applications is called a float-operated valve.

When the water reaches the required level, the device will close, shutting off the incoming flow of water. This will be below the overflow level of the cistern. The valves must comply with the relevant British Standard, and are categorised as Parts 1, 2, 3 or 4.

There are several types of float-operated valves, as shown in Figures 5.16–5.18:

- Portsmouth type
- Diaphragm valve made of brass
- Diaphragm valve made of plastic.

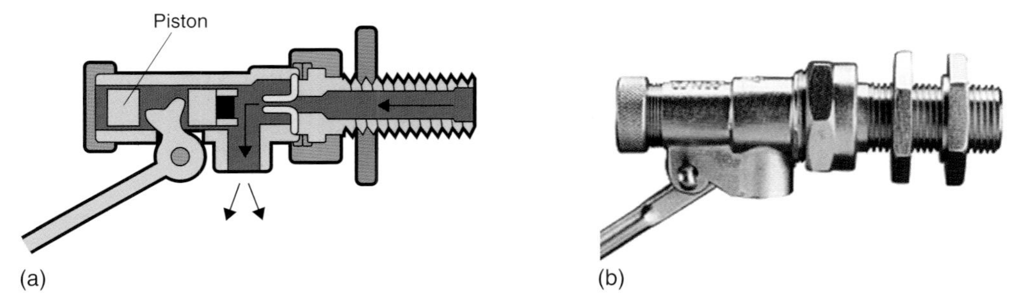

Figure 5.16 Portsmouth valve to BS 1212 Part 1 [(b) Reproduced with permission of Pegler Limited]

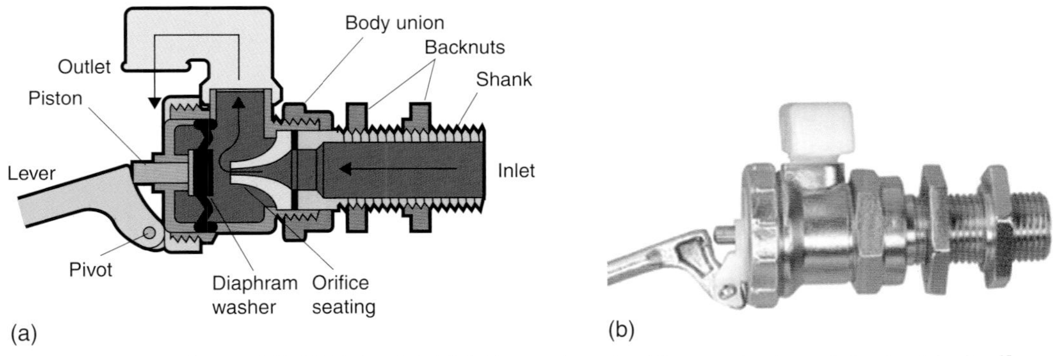

Figure 5.17 Diaphragm valve to BS 1212 Part 2 [(b) Reproduced with permission of Pegler Limited]

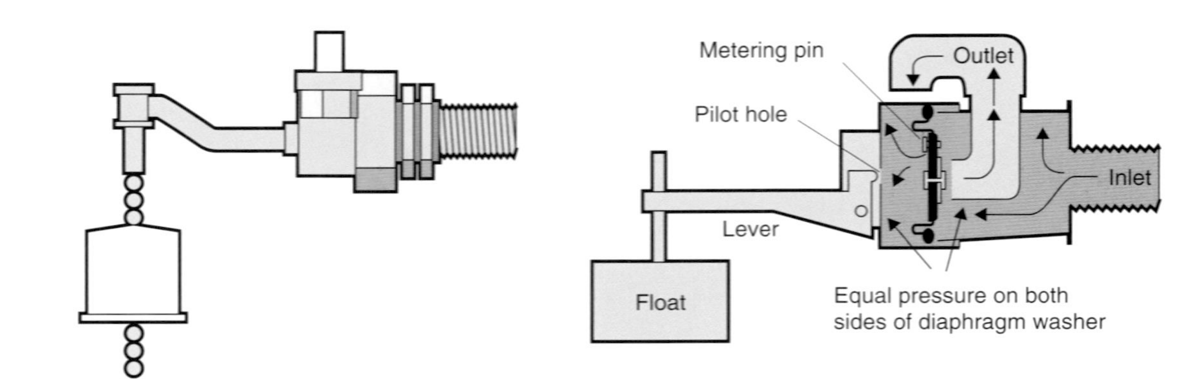

Figure 5.18 Diaphragm equilibrium float valves conforming to BS 1212 Part 4

The Portsmouth valve is not suitable for use in WC cisterns, unless there is a suitable backflow prevention device, e.g. a double check valve fitted before the valve. If used on a CWSC, the valve must have an adjustable float lever connection.

Brass diaphragm float valves with sliding float connection conform to BS 1212 Part 2 and can be used in any situation.

Plastic diaphragm float valves conform to BS 1212 Part 3. The required water level is achieved by means of a level adjustment screw, situated below the shut-off piston pin.

Diaphragm equilibrium float valves conforming to BS 1212 Part 4 are designed primarily for use in a WC flushing cistern.

Servicing valves

To conform to the requirements of the Water Regulations Schedule 2, Paragraph 16(2), inlet pipes to cisterns including WC cisterns must be fitted with a servicing valve immediately before connection to the cistern float valve. Servicing valves should also be fitted on outlet pipes, such as cold feed and distribution pipes.

Activity 5.4

What do you think is the main advantage of fitting servicing valves? Write your answer on a separate sheet and include it in your portfolio before checking against the suggested answer at the end of the book.

The valves should be installed in accessible positions, as close as reasonably practical, to the point of connection to the system (see Figure 5.19).

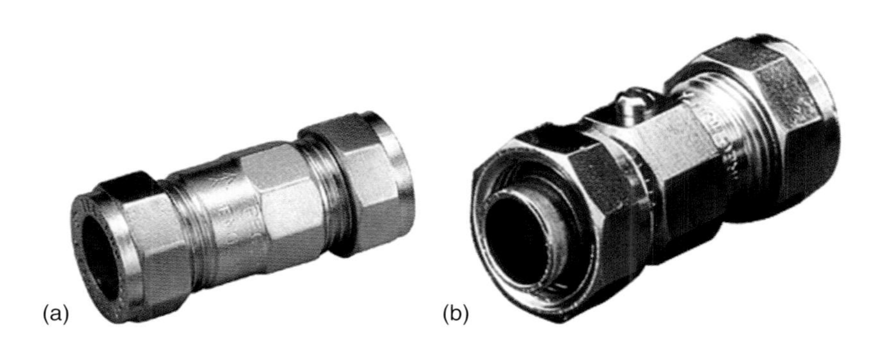

(a) (b)

(c)

Figure 5.19 Three types of servicing valves (Reproduced with permission of Pegler Limited)

The water can be isolated by turning the valve control with the aid of a screwdriver, 180° from vertical.

(a)

(b)

Figure 5.20 Single and double check valves (Reproduced with permission of Pegler Limited)

Use of check valves

Figure 5.20 shows a single check valve and a double check valve. The valves are used to prevent backflow and back siphonage and are installed to provide compliance with Water Regulations. The list below shows some typical applications where the Water Regulations require single and double check valves to be installed.

- **Single check valves**
  - Connected prior to water softener
  - Unvented heating systems
  - Supply to wet sprinkler system

– Downstream of meters and pressure reducing valves (no specific regulation but essential for correct operation of the equipment).

- **Double check valves**
  – Supply to hose taps
  – Supply to standpipes
  – Pipe connection to cisterns using Part 1 float valves
  – Supply to shower fitting.
- **Shower check valves**
  – Supply to shower spray head where shower hose pipe is unconstrained, e.g. air gap requirement cannot be guaranteed.

Try this

What do you think is the purpose of an overflow pipe and a warning pipe?

Warning and overflow pipes

Sometimes the water level in a cistern rises above its pre-set level; this is usually caused by a faulty float-operated valve. This excess volume of water is allowed to discharge through a pipe from the cistern, preferably outside the building to a position where it is likely to be seen. The regulations define each as:

A Warning pipe is a conspicuously placed pipe used to give a warning to the occupants of the building that a cistern is overflowing and needs attention.

An Overflow pipe is a pipe used to discharge larger volumes of water from larger cisterns where it will not cause damage to the building.

A small cistern, up to 1000 litres, must be fitted with a single warning pipe and does not require an overflow pipe. Larger cisterns, between 1000 and 5000 litres require an overflow pipe and a warning pipe. Figure 5.21 shows the location of the warning pipe in a cistern with a capacity less than 1000 litres.

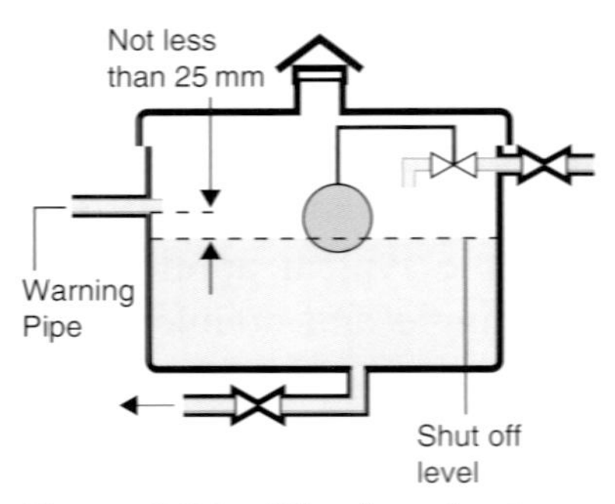

Figure 5.21 Warning pipe in a cistern with a capacity less than 1000 l

Case study 5.2

Leslie has been called out to a job to check that a CWSC installation conforms to the Water Regulations. If the float-operated valve becomes defective, the warning pipe must be capable of removing the excess water, preventing the valve from being submerged and the cistern from spilling over, so Leslie makes a note of the following points to check out:

- Warning pipes should have a minimum diameter of 19 mm, or be at least one size larger than the inlet pipe. They must be fitted with a screen or filter to prevent insects or vermin from getting in.
- Warning/overflow pipes must be constructed from rigid, corrosion-resistant material. Flexible hose connections are not allowed to form part of the pipe.
- Warning pipes from central heating feed and expansion cisterns must be kept separate from those from CWSCs
- Warning pipes must fall continuously from the cistern to the point of discharge
- Joining two or more warning/overflow pipes from similar types of cisterns to form a common warning pipe is acceptable, provided the discharge is visible and that one cistern does not discharge into the other.

Try this

If you have *safe* and *easy* access to your CWSC at home, check it out to see what material it's made from and note the other installation details.

Frost protection

Water is supplied to consumers at a reasonably low temperature, and in winter months, even a minor reduction in external temperature can cause freezing.

If copper pipework is exposed to sub-zero temperatures, the water in the pipe will freeze; remember, when water freezes it expands by approximately 10% and if there is not enough space for the ice to expand, pressure builds up on the wall of the pipe causing it to stretch and eventually split. Then, once the ice thaws you have what is termed as a burst pipe.

A burst pipe in a dwelling or building can cause immense water damage. Just think of the mess that is made if you spill a drink, and then imagine a burst pipe running at full pressure for two or three days.

The damage caused can prove to be costly; and remember, it also results in an undue waste of water which goes against the requirements of the Water Regulations, which state that all cold water fittings, including pipework located within the building, but outside the thermal envelope, or those outside the building, must be protected against damage by freezing.

Try this

One of the reasons why water pipes are laid underground to a depth of 750 mm is to protect the pipe from frost damage, and it is also a requirement of the Water Regulations under Schedule 2, Paragraph 7(4). Can you think of another reason why the pipe is laid to this depth?

Protection of underground pipework

In addition to frost protection, the other reason why the pipes are laid to 750 mm is that it gives them some protection from loads that may be placed above them, and to prevent someone putting a spade through them when digging the garden.

Where a supply pipe enters a building through a solid floor, it must be ducted; the pipe duct must be sealed at both ends to prevent gases, moisture or vermin getting into the building. If the pipe enters the building through a suspended timber floor, it should still be ducted. The pipe within the duct must be thermally insulated, and the ends sealed on completion; this is because the void between the suspended timber floor and the concrete over the site is exposed to external draughts.

Figure 5.22 shows the requirement for a suspended timber floor.

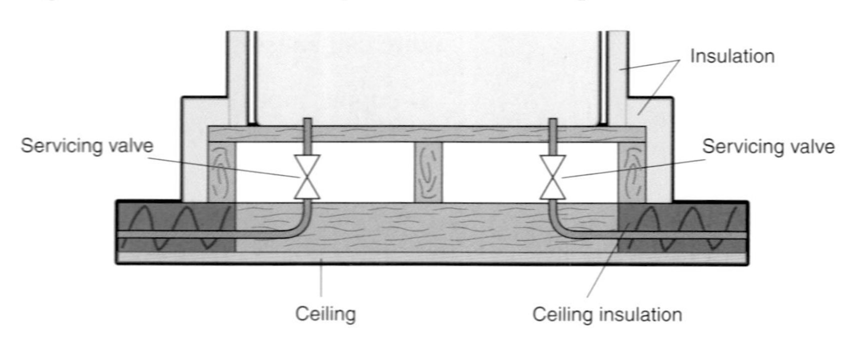

Figure 5.22 Requirements for suspended timber floors

Pipes and fittings used outside

This includes fittings such as:

- External taps used for garden hoses, car washing, etc
- Standpipes that are often used on building sites.

The above fittings must be protected against backflow, which can be achieved by fitting a double check valve to the installation, thus preventing potential contamination of the water supply, as shown in Figure 5.23.

Double check valves can also be susceptible to frost damage and therefore should be protected from freezing, preferably by installing them inside the building.

Activity 5.5

Figure 5.23 shows an external stand pipe installation. Can you identify the key design features? Write down your findings for your portfolio and check them out at the end of this book.

Key point

Location of pipework

When planning or installing piperuns you should try to avoid unheated spaces, and areas that are hard to keep warm, these are:

- Roof spaces, cellars, under floor spaces, garages or out-buildings
- External locations or 'cold surfaces' internally located, such as outside walls/external leafs, or ducts and chases in outside walls
- Areas where draught is evident, near windows, external doors, air bricks, ventilators and under ventilated suspended floors.

Protection of pipes and fittings

Where pipework is at risk from freezing, it must be protected using the approved pipe insulation material. Most people think that pipe insulation prevents the freezing of the water contained within the pipe by keeping out the cold; while, in fact, the insulation is designed to retain the heat energy in the water, thereby reducing the risk of freezing.

The efficiency of pipe insulation is dependent on the following factors:

- Its thickness
- Its thermal conductivity, in relation to pipe size.

So basically the thickness of insulation required will be dependent on:

- The pipe diameter
- The insulation type and its thermal conductivity
- The reason for the insulation, e.g. frost protection, heat loss/gain or to prevent condensation on the pipe
- Location of pipework, indoors/outdoors, in heated or unheated areas.

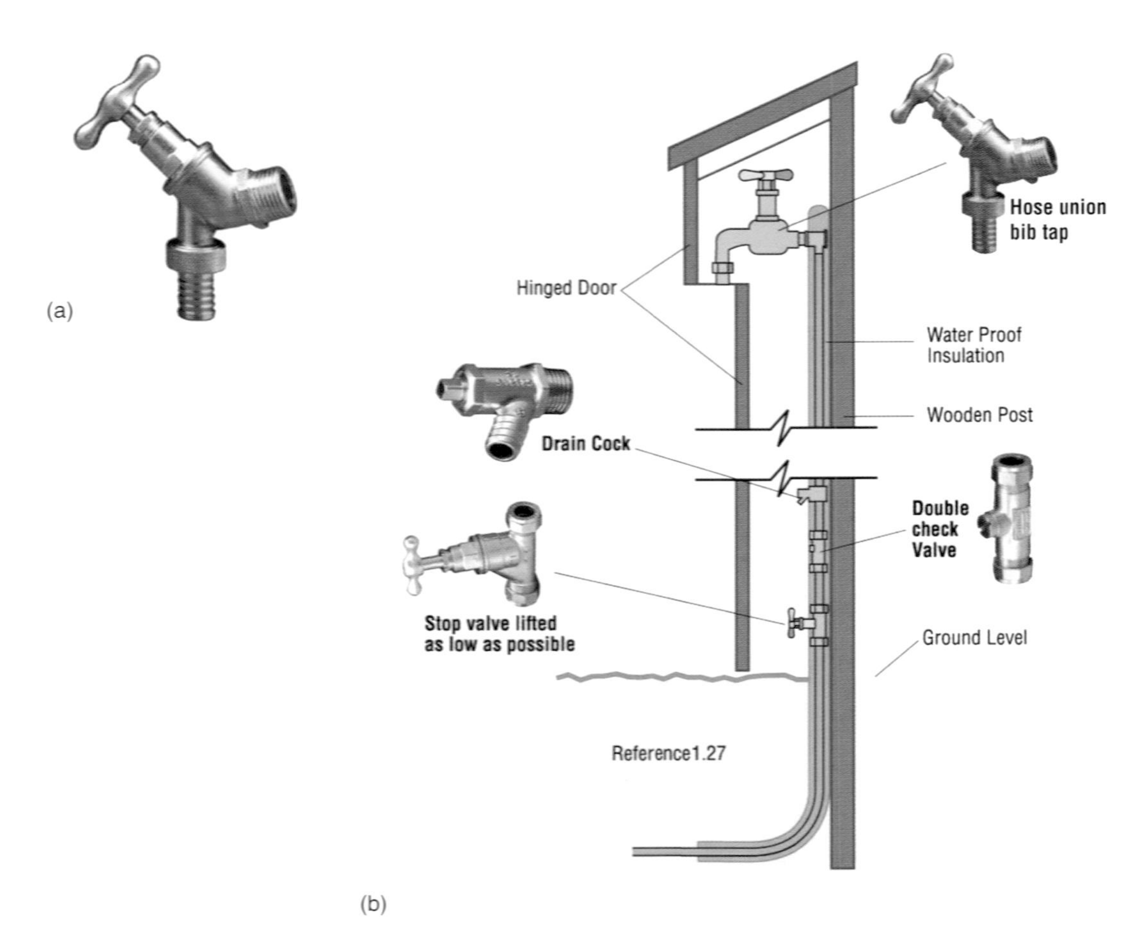

Figure 5.23 (a) Hose union bib tap (b) Requirements of external stand pipe

The minimum recommended thickness of insulation for 15, 22 and 28 mm water pipes must comply with the requirements of BS EN 1057.

Thermal insulation material for pipework should comply with BS 5422 (closed-cell-type) and be installed in accordance with BS 5970, and manufacturer's instructions.

Generally, thermal insulation materials must be resistant to, or protected from:

- Rain
- Moisture
- Subsoil water
- Mechanical damage
- Vermin.

Trace heating

This is a method of frost protection which is generally used on large industrial installations, but there are domestic products available. The installation involves attaching a low-temperature heating element to the outer wall of the pipe. This protects the pipe from freezing in severe weather conditions. The temperature control can be self-regulating or controlled thermostatically.

Hot tips

Obtain more information about trace heating by contacting manufacturers and obtaining their catalogue.

Pipes and cisterns in roof spaces

The current Building Regulations require roof spaces to be adequately ventilated to prevent condensation problems; the downside is that it increases the risk of pipe and cistern freezing where they are located in cold and draughty lofts especially through the winter months.

Figure 5.24 shows an insulated CWSC and associated pipework in a roof space.

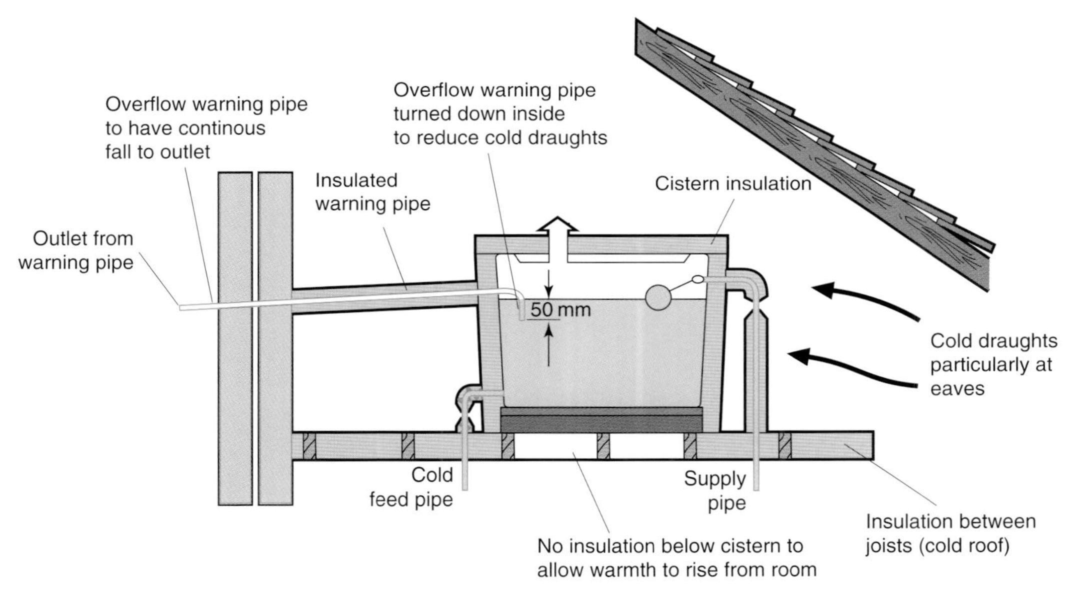

Figure 5.24 Details of insulated cold water storage cistern

The Water Regulations, Schedule 2, Paragraph 9 require that the water temperature in storage cisterns and cold water pipework should not exceed 25°C. However, the Building Regulations recommend a temperature of 20°C. Hence the use of insulation on pipework and components in high-temperature locations prevents rise in temperature from heat gain.

Test yourself 5.2

1. What are the minimum and maximum depth measurements for underground water service pipes?
2. List the two methods of treatment used for reducing limescale in hard water areas.
3. State two main purposes of stop valves and servicing valves on installation pipework.
4. Give two examples of where double check valves might be used.
5. The paragraph below describes the difference between a direct and indirect cold water system, fill in the missing words.

   All the pipes to the draw-off points (kitchen sink, bath, wash hand basin, WC, etc) in a ____________ system are taken ____________ from the ____________ and operate under ____________ pressure. With the ____________ system, one outlet usually the kitchen sink is fed directly from the rising main, before it continues to supply the ____________. The remaining draw-off points are fed from this source.
6. State two main forms of frost protection used on cold water installations.

Check your learning Unit 5

Time available to complete answering all questions: 30 minutes

Please tick the answer that you think is correct.

1. From which of the following would you find the legal requirements for the installation of cold water services?
   a. British Standards ☐
   b. Building Regulations ☐
   c. Construction Regulations ☐
   d. Water Regulations ☐
2. What is the minimum depth for an underground cold water service pipe?
   a. 350 mm ☐
   b. 550 mm ☐
   c. 750 mm ☐
   d. 1350 mm ☐
3. Which of the following items of equipment would be used to determine whether a cold water supply to a dwelling would be adequate to serve the plumbing appliances?
   a. Flow meter ☐
   b. Water meter ☐
   c. Manometer ☐
   d. Pressure gauge ☐
4. The most common use of type 'B' compression fittings is:
   a. On internal cold water supply pipework ☐
   b. To isolate float operated valves ☐
   c. On underground service pipework ☐
   d. To drain down the CWSC ☐
5. Drain taps on internal service pipework should be located:
   a. Immediately above the stop valve ☐
   b. Immediately below the stop valve ☐
   c. A minimum of 300 mm above the stop valve ☐
   d. A drain tap should not be fitted to the internal service pipe ☐
6. The minimum capacity of an indirect CWSC in a domestic dwelling should be:
   a. 75 litres ☐
   b. 115 litres ☐
   c. 230 litres ☐
   d. 415 litres ☐
7. The minimum capacity of an direct CWSC in a domestic dwelling should be:
   a. 75 litres ☐
   b. 115 litres ☐
   c. 230 litres ☐
   d. 415 litres ☐
8. A weir gauge would be used to measure the:
   a. Water pressure at a tap ☐
   b. Water flow rate at a tap ☐
   c. Depth of the CWSC ☐
   d. Temperature of the water in the CWSC ☐
9. On an indirect cold water system, the cold feed outlet from the CWSC to the domestic hot water cylinder should be:
   a. Level with the cold distribution outlet ☐
   b. Lower than the cold distribution outlet ☐
   c. Located at the base of the cistern ☐
   d. Higher than the cold distribution outlet ☐

(Continued)

Check your learning Unit 5 (Continued)

10. On a cold water system conforming to the Water Regulations, which of the following should be turned off to isolate the WC float valve?
    a. Supply stop valve ☐
    b. Appliance service valve ☐
    c. Water supplier's stop valve ☐
    d. Distributing service valve ☐
11. In areas where the water supply is poor during periods of peak demand, what type of system would usually be specified for a domestic dwelling?
    a. Direct Cold Water System ☐
    b. Boosted Cold Water System ☐
    c. Indirect Cold Water System ☐
    d. Secondary Cold Water System ☐
12. An overflow pipe as well as a warning pipe are required on cisterns with a capacity in excess of:
    a. 115 litres ☐
    b. 230 litres ☐
    c. 1000 litres ☐
    d. 1250 litres ☐
13. On a CWSC with a capacity of 115 litres, what is the required distance between the water level and the outlet of the warning pipe?
    a. 25 mm ☐
    b. 35 mm ☐
    c. 40 mm ☐
    d. 45 mm ☐
14. In a direct system of cold water supply, the purpose of the cold feed pipe is to:
    a. Replace the water to the hot water storage cylinder as it is drawn off. ☐
    b. Provide a means of filling the CWSC from the mains. ☐
    c. Supply the water to the various cold water outlets within the system ☐
    d. Prevent the cistern from overflowing should the float-operated valve fail ☐
15. On a new cold water system conforming to the Water Regulations, which of the following should be turned off to isolate the float valve to the CWSC?
    a. Appliance service valve ☐
    b. Supply pipe service valve ☐
    c. Water supplier's stop valve ☐
    d. Distributing service valve ☐
16. The Water Regulations require that the water temperature in CWSCs and cold water pipework should not exceed:
    a. 20°C ☐
    b. 21°C ☐
    c. 23°C ☐
    d. 25°C ☐
17. Where a CWSC is located in the roof space of a dwelling and the space beneath it contains thermal installation, what action should a plumber take?
    a. Increase the depth of insulation by 50 mm ☐
    b. Increase the depth of insulation by 100 mm ☐
    c. Decrease the depth of insulation by 50 mm ☐
    d. Remove the insulation from beneath the CWSC ☐
18. The minimum unobstructed space for a cistern of 1000 litres or less is:
    a. 250 mm ☐
    b. 350 mm ☐
    c. 400 mm ☐
    d. 450 mm ☐

Check your learning Unit 5 (Continued)

19. The purpose of pipework insulation material is to:
    a. Prevent the water in the pipe from freezing in cold weather ☐
    b. Protect the pipework and fittings from accidental damage ☐
    c. Retain the heat energy in the pipe to reduce the risk of freezing ☐
    d. Insulate the pipework should it come into contact with a live cable ☐

Sources of Information

We have included references to information sources at the relevant points in the text; here are some additional contacts that may be helpful.

- Pegler limited
  St Catherine's Avenue
  Doncaster
  South Yorkshire
  DN4 8DF
  Tel: 01302 560560
  Website: www.pegler.co.uk
- Yorkshire Fittings
  Head Office
  PO Box 66
  Leeds
  LS10 1NA
  Tel: 0113 270 1104
  E-mail: info@yorkshirefittings.co.uk
  Website: www.yorkshirefittings.co.uk
- Scalemaster
  Emerald Way
  Stone Business Park
  Stone, Staffordshire
  ST 15 OSR
  Tel: 01785 811636
  Fax: 01785 811511
  Website: www.scalemaster.co.uk

UNIT 6

HOT WATER SUPPLY

Summary

In the cold water unit, you looked at where water comes from and how it gets to a dwelling, and then at the various systems that supply to the consumer. This unit takes a look at what happens with the domestic hot water, and as there are some commonalities between hot and cold water installations such as provision for access to pipework, we have included that here.

A supply of hot water is an essential requirement for any habitable domestic dwelling, and it is also required in most working environments. It is mostly used for personal washing and cleaning purposes.

Hot water can be supplied by a number of methods, which can be categorised generally as centralised and localised.

This unit will cover:

- Differences between hot water storage and instantaneous supply
- Hot water storage systems
  - Types of storage systems – direct and indirect
  - Components of a hot water system
  - Various types of storage vessels
  - Thermal insulation requirements.

- Instantaneous hot water systems
  - Types of instantaneous heater
  - Single point and multi-point
  - Thermal stores
  - Instantaneous showers
  - Thermostatically controlled showers.
- Maintenance requirements of hot and cold water systems
- System testing
- Unvented systems

Whilst unvented systems are not a Level 2 requirement, it is included here as an introduction to the subject.

Hot water supply

The supply of hot water in domestic dwellings was first introduced into the majority of households around 1920–1930. From these early, relatively simple applications, emerging technology has vastly improved the design principles and use of components and appliances in today's modern domestic hot water systems.

Try to imagine what it would be like for you now if you did not have a readily available supply of hot water for bathing and cleaning purposes.

As with the cold water systems covered in the previous unit, the design and installation of hot water systems is covered under the current Water Regulations and British Standards BS 6700.

The type of building where a hot water system is installed will usually influence its design. As a plumber you are likely to work in buildings other than domestic dwellings. This could include washroom facilities in a factory canteen, where the installation of an instantaneous type water heater may be the more suitable and economical choice compared to that of a hot water storage system.

Typical hot water systems in domestic dwellings today can include atmospheric vented storage systems, unvented storage systems, directly fed thermal stores and combination boilers, often referred to in the trade as combi boilers.

Properties of hot water in practice

To produce hot water some form of heat is required; heat is a form of energy created by chemical change or friction. In the case of domestic hot water systems, they are heated by burning fuels, such as oil, gas or solid fuels or using electricity. Solar water heating panels collect heat directly from sunshine.

The intensity of heat is expressed as temperature, which can be measured using equipment such as a thermometer or a thermostat. The temperature in a domestic hot water system rarely exceeds 85°C.

In an atmospheric-vented domestic hot water system (systems open to the atmosphere), the water in the storage vessel should be stored at a temperature of not less than 60°C. This enables an approximate distribution temperature of 55°C, restricting the possibility of the growth of Legionella bacteria. This distribution

Key point

Whilst the current trend in new build domestic dwellings is towards the use of combination boilers or hot water and heating systems incorporating unvented units, plumbers are also likely to work on atmospheric vented systems on a regular basis.

Key point

Water, when heated between 4 and 100°C, expands by 24 times; so in open-vented systems, provision needs to be made in the design to allow for the hot water to expand. This is done using a vent pipe from the hot water drawn off from the storage vessel.

temperature may not always be achieved where hot water is provided by a combination boiler or instantaneous water heater.

It is strongly recommended that the water storage temperature does not exceed 65°C. Temperatures above this can result in the formation and accumulation of limescale, especially in hard water regions.

Water temperature on modern storage vessels is usually controlled by a cylinder thermostat like the one shown in Figure 6.1.

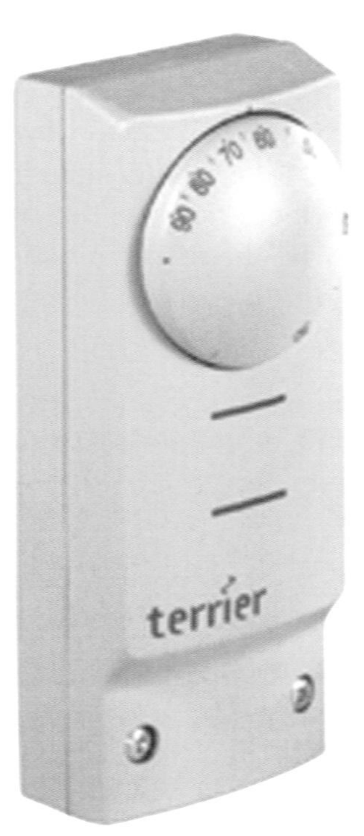

Figure 6.1 Typical cylinder thermostat (Reproduced with permission of Pegler Limited)

Key point

Where water is distributed at a temperature higher than 50°C, there is a potential risk of scalding to the user, the young, elderly and infirm being particularly at risk.

Types of hot water systems

There are a number of factors that will influence the design and type of hot water system. Much will depend on the type of building that it is going to be installed into, availability of fuel types and of course the requirements of the consumer. The following factors should be taken into consideration prior to system selection and design.

- Fuel cost and overall efficiency of the system
- Hot water demand requirements
- Storage and distribution temperature
- Installation and maintenance costs
- Safety of the user
- Waste of water and energy.

System selection

Hot water systems can range from a simple single point outlet, to the more complex centralised boiler systems supplying hot water to a number of outlets. BS 6700 sets out several ways of supplying hot water, and generally they are divided into two types, centralised and localised. Figure 6.2 illustrates the various options.

Let us take a look at these in a little more detail.

Centralised system

With this type of system, the heated water can be stored, usually centrally within the building, supplying a system of pipework to various draw-off points. Stored means water held in a vessel until required, with the water temperature usually controlled by a thermostat.

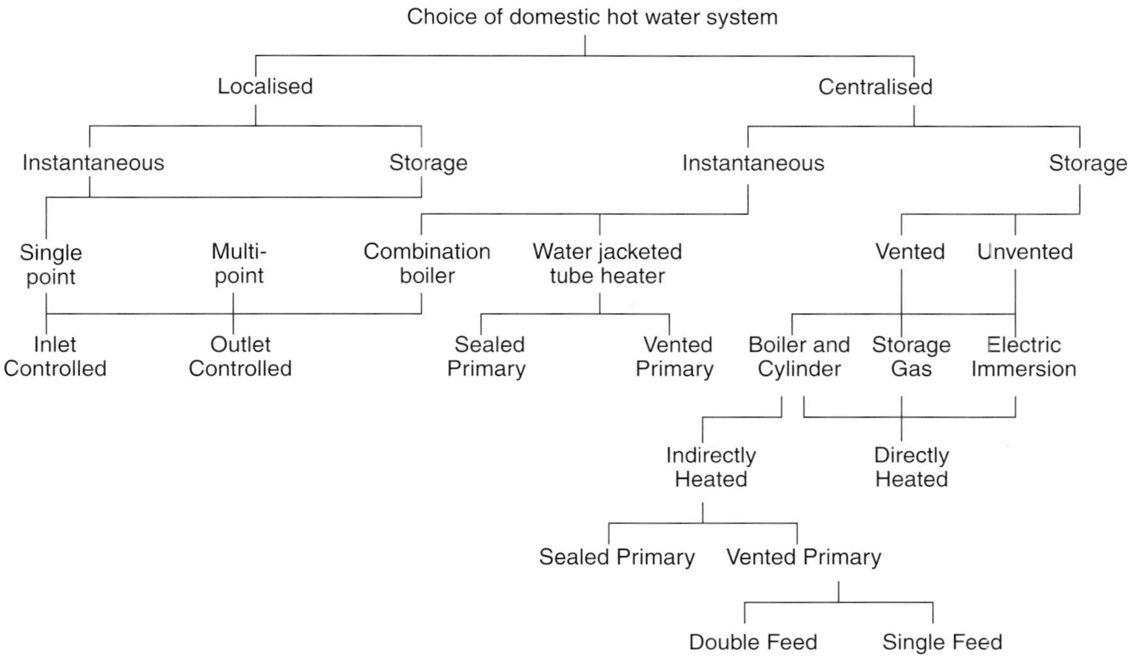

Figure 6.2 System selection flow chart

Localised system

With this type of system, the water is heated locally to meet the requirements of the consumer, a typical example being a single point instantaneous water heater sited over a sink. These are often used in situations where a long distribution piperun from a stored system would lead to an unnecessary wait for hot water to be drawn off. Not only does this save energy and reduce the wastage of water, it also helps prevent the risk of growth of microorganisms such as Legionella in the system.

Instantaneous supply

Instantaneous, as the name implies, means that there is a readily available (instant) supply of hot water at the outlet, and only the amount of water drawn off is heated.

That completes the first section on hot water, which is really an introduction to hot water systems. In the next session you will look at hot water systems in more detail.

Test yourself 6.1

1. What is the difference between the quantity of heat and intensity of heat?
2. In order to comply with the requirements of the Water Regulations, what must be fitted on supply pipework feeding an instantaneous water heater?
3. In an atmospheric-vented hot water system, the temperature of hot water should be stored at not less than:
   - 40°C
   - 50°C
   - 55°C
   - 60°C
4. Water when heated between temperatures 4 and 100°C expands by approximately:
   - 1/10
   - 1/20
   - 1/24
   - 1/50
5. Write a short paragraph to explain what is meant by a centralised system and a localised system?

Hot water storage systems

Your work as a plumber will bring you into contact with a wide range of domestic hot water systems, so in this unit we will expand on the various system types in detail, starting with hot water storage systems. For all the systems shown in this unit, make a mental note of how the hot water systems are connected to the input services (i.e. the cold water supply).

Storage systems, in particular the indirect types, are still fairly common in domestic dwellings, although the use of unvented hot water systems are becoming more popular.

Hot water systems can be categorised as being:

- Direct
- Indirect
- Sealed/unvented
- Mains thermal store
- Combination.

Direct system (vented)

Take a good look at Figure 6.3. It will give you a general idea of what a direct hot water system using gravity circulation looks like, and how it works.

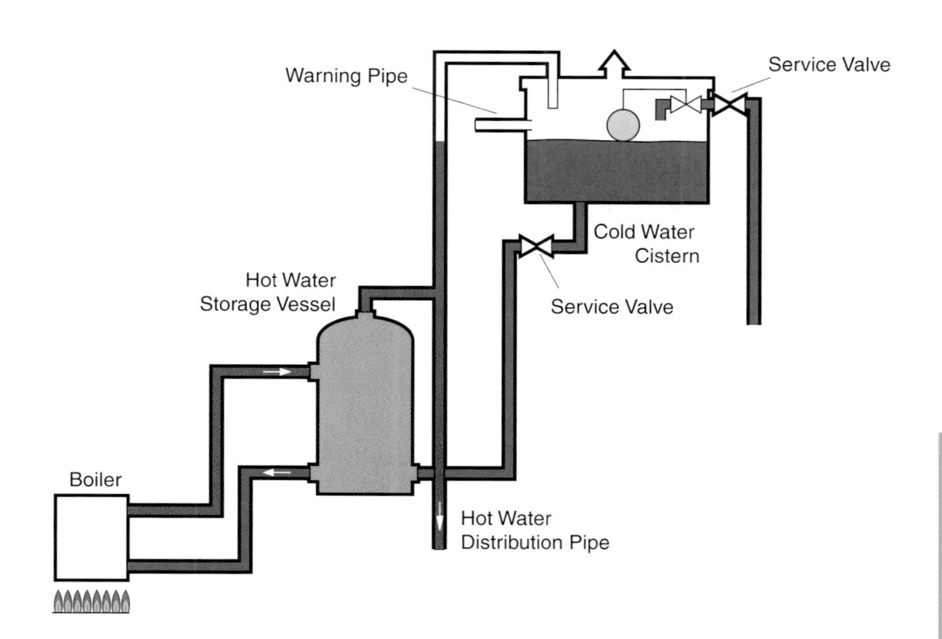

Figure 6.3 Vented direct system

Systems of this design are not commonly used today, and you will probably come across them only on *maintenance and repair work*.

Try this

Can you recall what is meant by the term 'principle of convection' and how that is relevant to a hot water system?

The system works on the principle of convection that you encountered in the unit on key plumbing principles. Water is heated from a gas circulator or a solid fuel boiler, or in some instances, hot water can be provided solely using an electric immersion heater, controlled by a thermostat.

The immersion heater is inserted directly into the water through a boss fitting at the top of the storage vessel, or in some cases two small immersion heaters are used at the top and the base of the storage vessel thus ensuring that the whole of the vessel is heated during peak demand. The temperature is controlled by a thermostat.

Here are some important factors about direct gravity systems:

- The primary pipe circuits to storage vessels should not be less than 22 mm diameter for gas systems and 25 mm for solid fuel systems

- The open vent pipe from the storage vessel should not be less than 22 mm diameter, and must not be valved
- Primary circulation pipes should rise from the boiler; this assists the water circulation and prevents airlocks forming in the system and aid in system draining
- The hot water draw-off pipe, directly off the top of the storage vessel, should incorporate a minimum 450 mm horizontal run between the vessel and the vertical open vent pipe. This prevents one pipe circulation (convection currents rising up the pipe) preventing the heat loss of the water, reducing the fuel consumption.
- Liquid corrosion inhibitors must not be introduced into the system, as the water in the boiler and storage vessel is not separated, and directly supplies the draw-off outlets.
- The cold feed pipe from the cistern to the storage vessel must not have another connection made into it. The pipe should be sized in accordance with the recommendations of BS 6700 Paragraph 2.5.
- The heating appliance (boiler/circulator) must have a non-ferrous heat exchanger
- Direct systems are not recommended in hard water regions.

Activity 6.1

Before moving on, think about the last bullet point. What do you think the reason is that direct systems are not recommended in hard water regions? Make a note for your portfolio and then check with the suggested answer at the end of this book.

Key point

You may find gravity primaries on older existing systems, but Building Regulations (Part L1 A and B 2006) require that all new systems, or system upgrades, have to have fully pumped primaries. The requirements of the Building Regulations (Part L1 A and B 2006) is covered in greater detail in the central heating unit.

Indirect systems (vented)

Take a good look at Figure 6.4, it will give you a general idea of what an indirect hot water system using gravity circulation looks like and how it works.

Please note that the illustration is not to scale and the size of the feed and expansion cistern would be smaller than the cold water storage cistern (CWSC).

In this system, water is heated by a boiler, which is usually adequately sized to allow for the provision of a central heating circuit.

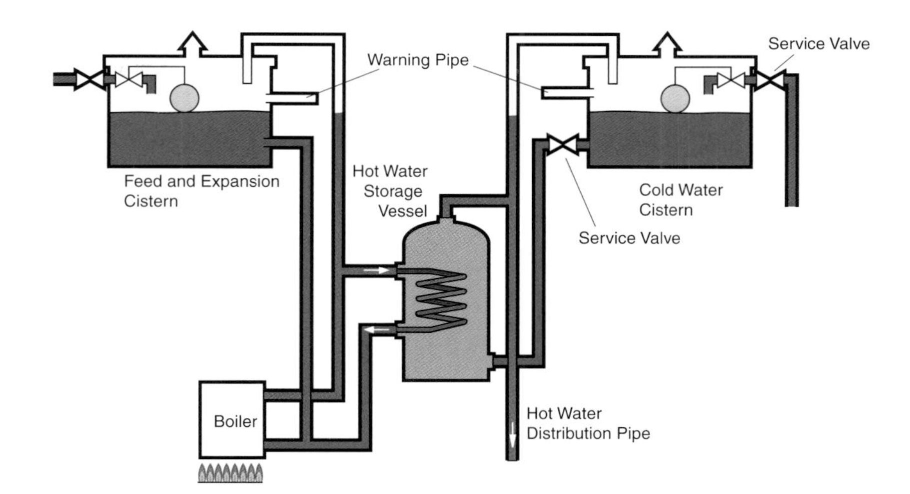

Figure 6.4 Vented indirect system

Key point

The system is termed as indirect because the water within the storage vessel is heated indirectly, via a heat exchanger.

The hot water from the boiler is conveyed to the storage vessel via primary flow and return pipes.

The storage vessel is fitted with an internal heat exchanger, usually in the form of a coil or an annulus. The heated water passes through the heat exchanger, heating the stored water within the vessel, and then returning to the boiler.

Figure 6.5 shows an example of an indirect cylinder with an annulus heat exchanger, and one with a coil.

This water does not mix with the stored water within the vessel, and requires an independent feed and expansion cistern to supply it.

Indirect copper storage vessels are categorised as double feed indirect, manufactured to BS 1566: 2002.

Here are a few important factors about double feed indirect hot water systems.

- The system requires a separate feed and expansion cistern
- The system requires an open vent and cold feed pipe connecting into the primary pipe circuit, or can be separately fed into the boiler
- The heating appliance (boiler) can have a cast iron heat exchanger
- If the open vent pipe is not connected to the highest point in the primary circuit, then an air release valve must be fitted, preferably an automatic one
- No type of shut-off valve should be fitted to the open vent pipe.

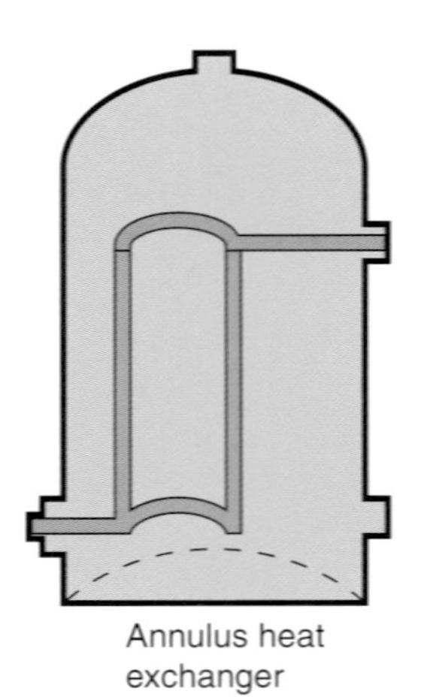

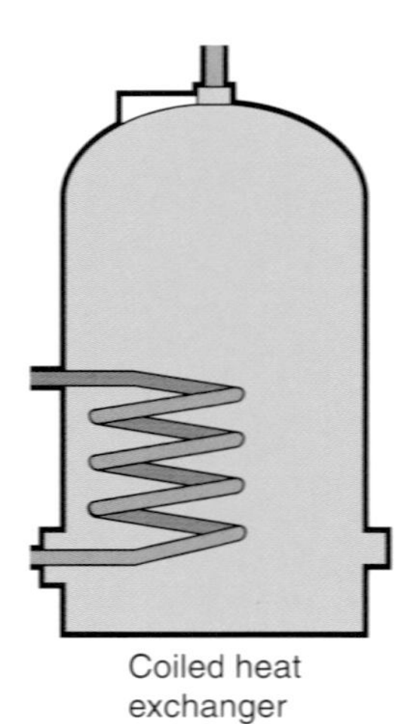

Figure 6.5 Indirect cylinder with annulus heat exchanger, and with coil

Use of gate valves

Gate valves are usually located in low-pressure pipelines such as the cold feed from the CWSC to the hot water storage cylinder, and the cold water supply on an indirect system. You might also find them used on supplies to thermostatic shower valves. Figure 6.6 shows the component parts of a gate valve.

Gate valves are sometimes referred to as a full way gate valve because when it is open, there is no restriction through the valve. The gate valve is fitted with a hand wheel or wheel head attached to the spindle. When the head is turned clockwise the threaded part of the spindle is screwed into the wedge-shaped gate raising it towards the head.

> **Activity 6.2**
>
> Figure 6.6 shows a gate valve that has been stripped down. See if you can identify the component parts, and then check your findings with the answer given at the end of this book.

Figure 6.6 Component parts of a gate valve

The main problem with gate valves is leaking packing glands. These can be repaired without turning off the water by using PTFE tape. In some cases the wedge or gate can stick. For the hot or indirect cold water supply, this will mean isolating the water to the CWSC and draining the system through the appropriate taps. The bonnet should be removed from the valve body, and the wedge or gate and stem should be cleaned and lubricated using petroleum jelly.

Single feed vessels

Single feed storage vessels are no longer used on new installations, but are still available for replacement purposes. The vessels are referred to as single feed 'self-priming' cylinders, often called primatics. When installed, the system looks very similar to a direct system (see Figure 6.7).

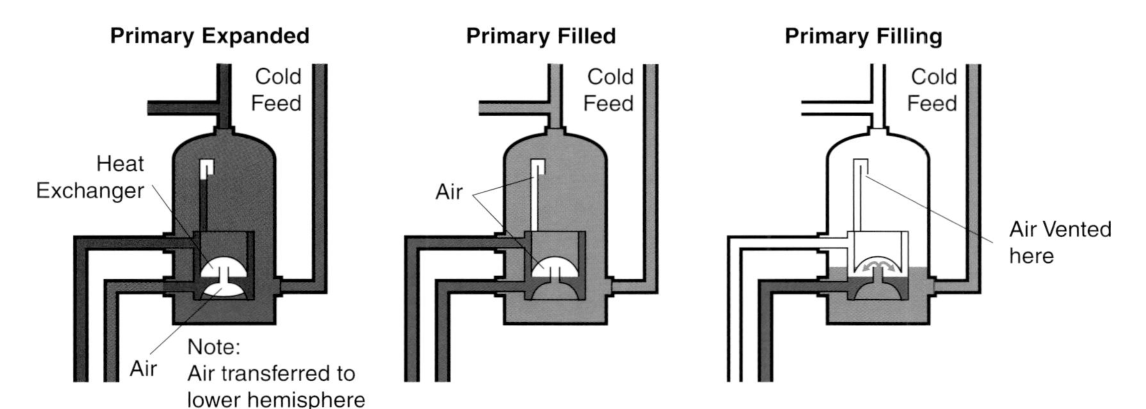

Figure 6.7 Single feed vessels

Key point

Circulating pumps should not be fitted to the primary circuits, as this could result in the loss of the air seal. The loss of the air seal would in effect convert the system into a 'direct system', resulting in the mixing of primary and secondary waters, and it is for this reason that corrosion inhibitors or other types of additives must not be added into the primary circuits.

Storage vessels should conform to BS 1566 Part 2, 2002 installed as per manufacturer's instructions, to vented systems only.

Activity 6.3

Using college or centre resources or obtaining manufacturer's instructions write a short description of how the single feed system works. Include it in your portfolio and then check your answer against the one at the end of this book.

Secondary circulation, dead legs and dead ends

The Water Regulations provide recommendations for the maximum lengths of uninsulated hot water pipes and are given in Table 6.1.

Table 6.1

Outer diameter of the pipe (mm)	Maximum length (m)
12	20
Over 12 and up to and including 22	12
Over 22 and up to and including 28	8
Over 28	3

Key point

Secondary circulation systems are used to prevent system dead legs. A dead leg on a hot water system, in this context, is an outlet (tap) at the end of a long run of pipework. When the outlet is closed, the water in the pipe cools down. When the outlet is opened, because of the length of the piperun, it takes a long time for the cool water to run out before the hot water appears. This wastes water and energy and on larger systems can lead to Legionella risks.

Secondary circulation systems, in simple terms, consist of a ring main from the hot water draw off, which is returned about 25% down the cylinder. The circulation is pumped, and branches are taken off to the various hot water appliances. Another method of tackling the problem of dead legs is with the use of trace heating.

Dead ends usually occur on systems where an appliance has been removed, but the pipework has not been cut back to the original pipework branch.

Leaving dead ends is a bad practice, as this will lead to the stagnation of the water in the pipe, creating conditions for the build up of silt and sludge. On larger systems, these conditions are ideal for supporting the growth of Legionella bacteria.

Storage vessels

Grades of storage vessels

Storage vessels are graded by manufacturers and are available in various sizes and capacities. The grading indicates fitness for installation in open-vented conditions and relates to the working head which is the distance from the base of the cylinder or vessel to the top of the water level in the cold storage cistern.

The grades are:

Grade 1	25 m maximum
Grade 2	15 m maximum
Grade 3	10 m maximum

Capacities and sizes

Copper storage vessels are available in a wide range of sizes, having different storage capacities. A typical manufacturer's product ranges from 600 mm in height × 400 mm in diameter having a storage capacity of 62 l, through to 1800 mm in height × 600 mm in diameter, having a storage capacity of 440 l.

Most of the domestic dwellings are fitted with a 900 mm × 450 mm vessel, having a storage capacity of 117 l or 1050 mm × 400 mm with a storage capacity of 140 l. The figures are based on double feed indirect vessels.

Corrosion resistance

The new British Standard now defines the specification requirements for corrosion resistant (copper) cylinders, and manufacturers

Figure 6.8 High-performance storage vessel – outside (Reproduced with permission of Albion Water Heaters)

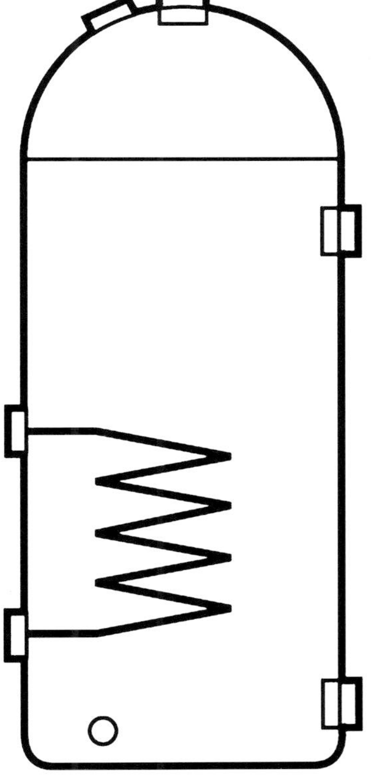

Figure 6.9 High-performance storage vessel – inside

give a five-year guarantee against corrosion. Sacrificial anodes are now no longer fitted to cylinders, unless requested to do so.

High-performance storage vessels

Vessels of this type offer energy efficiency and lower fuel running costs. They give ultrafast recovery times for hot water, and this improved efficiency and performance allows for a reduction in both the size of the vessel and its storage capacity (see Figures 6.8 and 6.9).

The vessel is designed for use on *indirect fully pumped* hot water systems, incorporating a motorised zone valve and a thermostat temperature control (this will be covered in greater detail in the central heating sections).

Unlike conventional indirect copper cylinders, which utilise a single internal 28 mm diameter coil heat exchanger, high performance vessels are fitted with a multi-coiled heat exchanger, generally consisting of four coiled tubes. This method ensures that more water is in contact with the outer wall of the heat exchanger, vastly improving the heat transfer between the primary and secondary water.

A traditional indirect hot water system has a heat recovery time of approximately 40–50 min, whereas a high-performance system offers a heat recovery time of 6–10 min at a temperature of 60°C. These figures will be dependent on the boiler output size and storage capacity of the vessel.

The vessels are available in various sizes having storage capacities of 45, 80, 120, 140, 160, 180 and 210 l. The storage capacity for an average three-bedroom domestic dwelling with one bathroom and thermostatic gravity shower would be 45 l, resulting in dimensions of 600 mm height × 350 mm diameter.

Thermal insulation

Thermal insulation was covered in some detail in the previous unit on the cold water supply, where you looked at the requirements for CWSCs, and the insulation of associated pipework. The main point to remember with hot water systems is that the storage vessel must be insulated. Today, new cylinders and storage vessels are prelagged during the manufacturing process, using expanded choroflu-orocarbon (CFC)-free foam insulation. The insulated cylinder must comply with the requirements of Part L of the Building Regulations.

Key point

When the water temperature in the storage vessel drops below 60°C, the 'heat recovery time' is the period of time it takes for the stored water in the vessel to be re-heated to a temperature of 60°C.

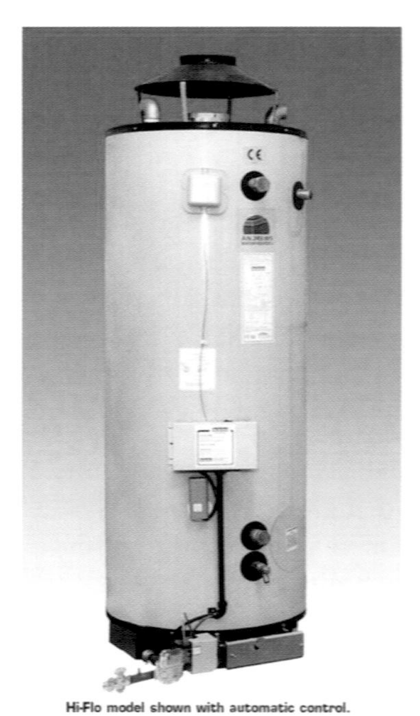

Hi-Flo model shown with automatic control.

Figure 6.10 Storage heater

Part L1 of the Building Regulations 2006 requires that all pipes that are connected to a storage vessel should be insulated for up to a metre in length, or up to the point to where they become concealed. This requirement includes the primary flow and return and vent pipe connections, reducing the heat loss from the storage vessel. The thermal insulation of hot pipework can be achieved using spilt sectional closed-cell-type pipe insulation.

Combination storage units

These types of units incorporate a CWSC and hot water storage vessel and when installed, should be positioned so that the base of the vessel is not lower than the level of the highest hot water draw-off outlet. The units must be positioned at a height in order to give adequate pressure of flow at the taps.

Storage units are available in direct or indirect patterns. Direct units can be heated solely by an immersion heater; indirect units can be heated by a boiler/water circulator and immersion heater.

Combination storage units are available in various sizes and capacities, and are manufactured from copper to conform to BS EN 1653 and BS 3198: 1981.

Storage heaters

Figure 6.10 shows a typical gas storage water heater; electrically heated types are also available.

These are not commonly found in domestic dwellings, but you may come across them in commercial/industrial buildings on maintenance work.

Activity 6.4

What type of situation do you think a combination storage unit would most likely to be installed, and what potential disadvantages could you see with this type of combination storage system? Record your answers for your portfolio and check them out with the answer given at the end of this book. Find out more about combination storage units by doing your own research. You might try Albion (www.albion-online.co.uk)

The gas storage heater is basically a self-contained unit, complete with gas burner and storage vessel. The gas burner is usually in the form of a large gas ring, located directly below the stored water. The heater is usually designed to be fed from either a CWSC (indirect) or (direct) from the cold mains supply. Heaters of this type are classified as being outlet-controlled (flow of water at taps) and can supply multiple outlets.

Test yourself 6.2

1. What type of hot water storage vessel should be used in hard water areas?
2. On a gas-fired direct hot water system, what is the minimum diameter in millimetres for the primary pipe circuits?
   a. 15 ☐
   b. 22 ☐
   c. 25 ☐
   d. 28 ☐
3. Why should corrosion inhibitors not be added to the primary circuit of a single feed system?
4. What type of storage vessel would be most suitable for a ground floor flat?
5. What safety devices prevent the temperature in a hot water storage vessel from reaching the boiling point?
6. Which storage system would you recommend if a fast heat recovery time is important to the customer?
7. Briefly describe the insulation requirements of Part L of the Building Regulations.

Instantaneous hot water systems

The heat source for instantaneous systems can be either gas or electric, or in the case of a water jacketed tube heater (thermal store), it is oil.

The speed of heating the water is limited, so the flow rate of the water needs to be controlled to a rate where it can be heated properly. Due to the reduced flow rate it is not possible to supply a number of outlet points all at once, so they would not be installed in situations where there is a high demand. For example, you would possibly find a multi-point in a one-person flat or a single point in an office kitchen area or sink in a WC compartment.

Water-jacketed tube heaters (thermal store) are available in a range of specifications and are capable of supplying properties with two large or three standard baths and two en-suite showers.

Key point

Instantaneous systems work on the basis of passing cold water from the service pipe through a heat source, which heats the water by the time it comes out at the outlet.

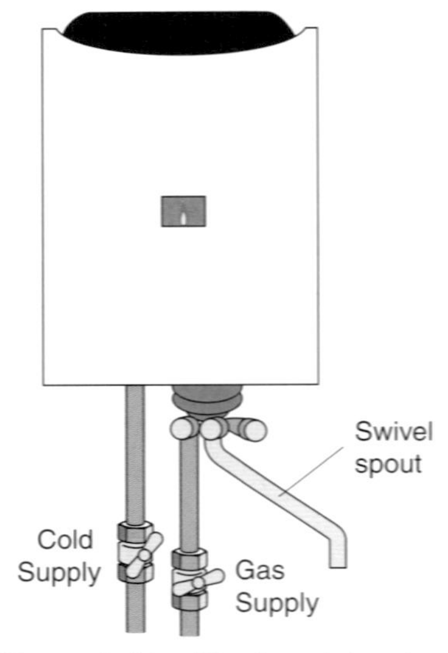

Figure 6.11 Single point water heater

Types of instantaneous heaters include:

- Single point
- Water-jacketed tube heater or thermal store
- Multi-point.

Single point

As the name suggests, this type of heater is designed to provide water at a single point, normally over a sink, and is heated by gas or electricity (see Figure 6.11). Hot water is provided through either a swivel spout outlet or by a single hot tap connection, using a special tap which allows for the expansion of the heated water. It is an ideal installation for an office, kitchen or single WC.

To obtain a constant flow rate temperature at the outlet, the heater is fitted with a water governor, which controls the flow rate of the incoming water as it passes through the heat exchanger.

To maintain constant temperature at the outlet, the water flow rate and pressure must also be constant, as any fluctuations will affect its performance.

Water-jacketed tube heater or thermal store

Thermal stores are supplied directly from the incoming cold water main and unlike unvented storage systems, they are exempt from the Building Regulations Approved Document G3, and notification to the building inspectorate is not required.

Thermal stores are suitable for new and existing installations. They are available in various sizes and capacities, manufactured from copper and can be of either direct or indirect type.

Direct type

Direct thermal stores are fitted with a single integral coil. Mains cold water enters the coil via the lower tapping. The primary water contained within the vessel is heated directly by the boiler, and this water heats the water in the coil as it passes through it. Hot water is distributed to the draw-off points from the top tapping of the coil, passing through the blending valve, which ensures that the distribution temperature does not exceed 55°C.

Direct thermal stores are designed for open-vented primary systems and are not suitable for primary sealed systems (see Figure 6.12).

Activity 6.5

- Can an indirect thermal store unit be installed in a sealed system?
- What is one of its main advantages?

Write your answer in your portfolio and check your understanding against the answer given at the end of the book.

Indirect type

Indirect thermal stores are fitted with two internal coils; the water contained within the vessel is heated indirectly by the boiler via the lower multicoil heat exchanger. The stored water within the unit is static, and does not mix with the primary water, as shown in Figure 6.12.

The upper coil provides hot water to the draw-off outlets, working on the same principle as a direct system, with the addition of an expansion vessel fitted before the blending valve.

Multi-point

This type of heater, as shown in Figure 6.13, is designed to provide water to more than one point, typically kitchen sink, bath and wash hand basin. It would be an ideal installation for a single bedroom flat.

On this appliance, the gas burner is located beneath the heat exchanger. Hot water is provided when a tap is opened, allowing water to pass through the heater. This causes the gas valve to open due to the drop in pressure at the differential valve. This pressure drop is caused by water passing through the venturi, which creates a negative pressure as it sucks the water from the valve.

The diaphragm is connected to a push rod and as it lifts, it opens the gas line, allowing gas to flow through the burner, which is ignited by the pilot light. When the hot water tap is turned off, the pressure within the differential valve is equalised, and this causes the diaphragm to close, shutting off the supply of gas to the burner.

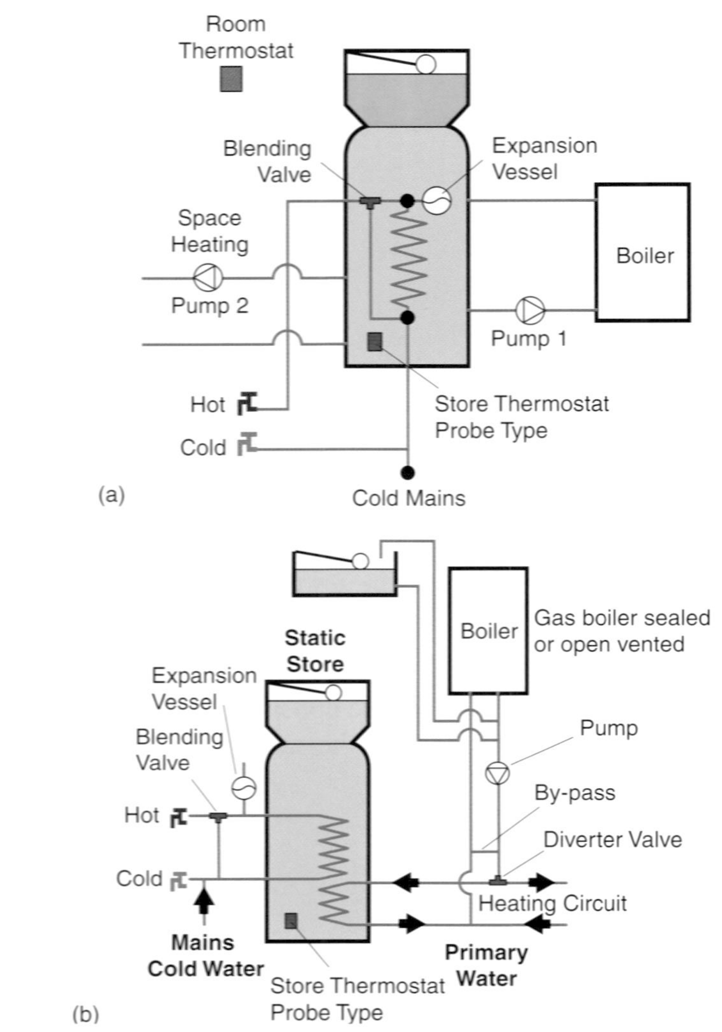

Figure 6.12 (a) Direct and (b) Indirect thermal stores

Other types of appliances that provide instantaneous hot water are combination gas central heating boilers. These provide a supply of hot water on demand, at a constant temperature.

The electric multi-point

The electric multi-point is basically a small tank of water with an electric heating element inside. Due to the low volume of water, it quickly heats up as it is drawn through the heater. The temperature at the outlet will be related to the water flow rate and the kilowatt rating of the heater.

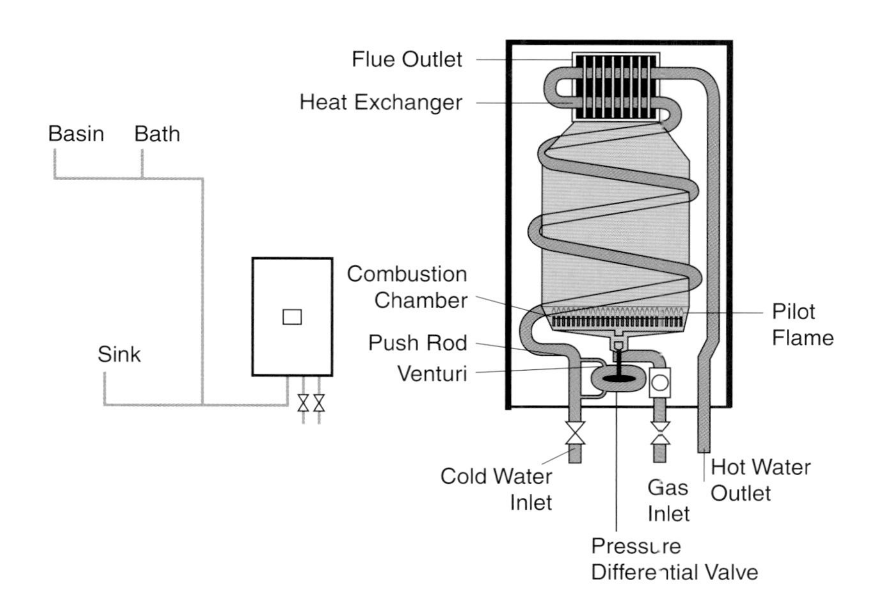

Figure 6.13 Gas multi-point heater

Electric instantaneous showers

The electric heater, as shown in Figure 6.14, is designed for water mains connection, although some can be fed indirectly. The electrical rating can be up to 9.6 kW depending on the type of shower, so it is important that the supply is adequate, and wired directly from the mains distribution unit (MDU).

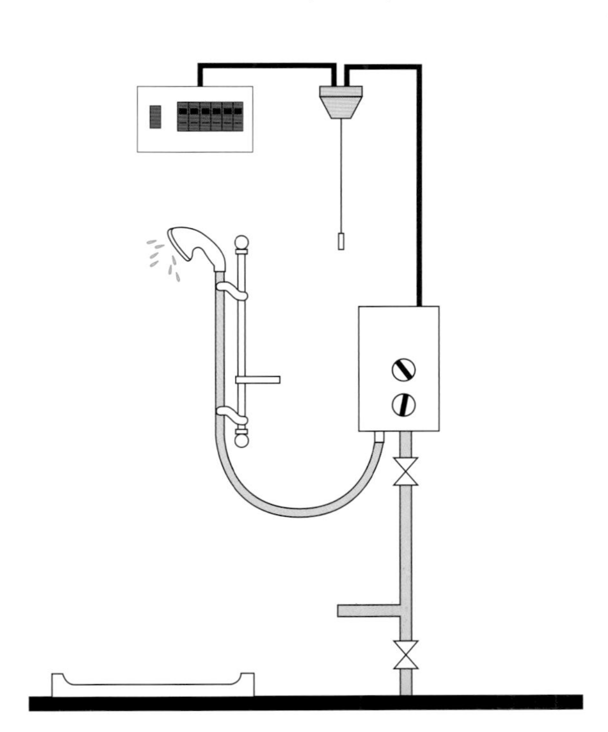

Figure 6.14 Electric instantaneous shower mains fed (Reproduced with the kind permission of Mira Showers)

For a rating of 9.6 kW, the fuse requirements would be 45 A and the cable 10 mm^2. The shower should also be isolated with a switch, which should be located outside the shower room with easy access or operated by a pull cord.

Mains fed

The Water Regulations require that provision is made to prevent backflow, and this can be done either by installing a double check valve, or by using a rigid connection to the shower head. Mains connections can only be used on thermostatic mixing valves using manufacturer's valves designed specifically for this purpose.

Any pressure variations in the cold water supply to the shower will be handled by the flow governor. Most electrical instantaneous showers are fitted with a flexible hose outlet.

Gravity fed instantaneous shower

Figure 6.15 shows a gravity fed instantaneous shower and is reproduced with the kind permission of Mira Showers. Contact details are provided at the end of the unit.

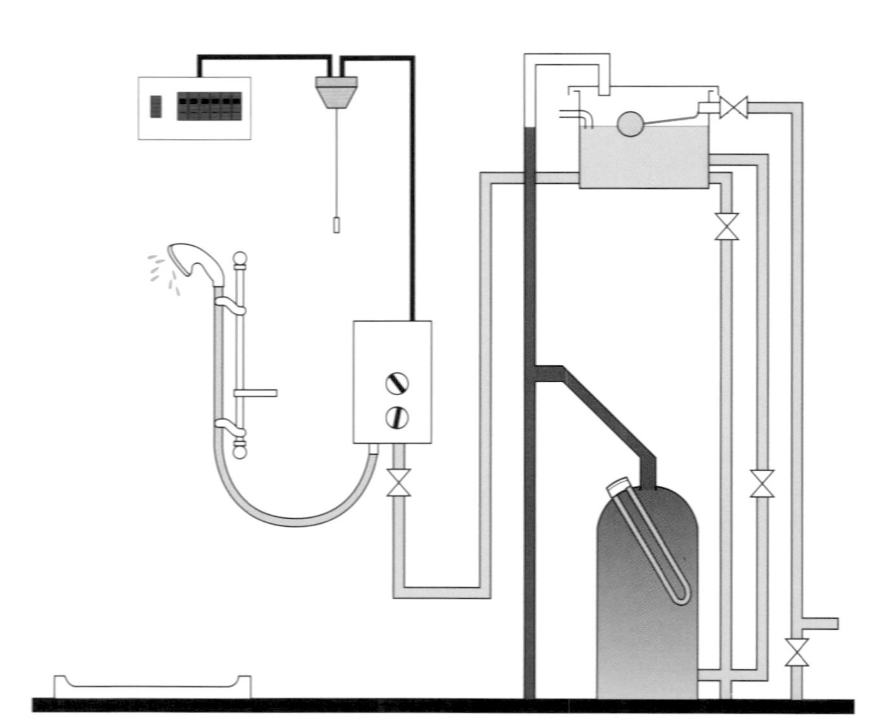

Figure 6.15 Electric instantaneous gravity fed shower (Reproduced with the kind permission of Mira Showers)

Try this

Carry out some research into the various manufacturers' products available for the main types of instantaneous hot water heaters and showers. This may be done through the internet or manufacturers' catalogues, or if you get time, visit your local plumber's merchants. Use this information to improve your understanding of how these appliances work and include whatever work you do in your portfolio. Include on your list:

- Single and multi-point appliances
- Electric instantaneous showers

Test yourself 6.3

1. List the main types of instantaneous water heaters:
   -
   -
   -
2. From your answers above:
   - Select and sketch a typical appliance
   - Label the component parts
   - Describe briefly how each type works.

Requirements common to hot and cold water systems

Taps and valves used in cold and hot water plumbing systems

Taps and fittings for both cold and hot water should conform to BS 1010 Parts 1 and 2, BS 1552 and BS 5433, they are usually made of brass pressings or castings, and are chrome-plated to enhance appearance and improve ease of cleaning.

Plastic taps and valves are also available. These are manufactured from thermosetting plastic called acetal.

Taps and valves must be:

- Sufficiently strong to resist normal and surge pressure
- Easily accessible to renew seals and washers
- Made of corrosion-resistant materials
- Capable of working at appropriate temperatures
- Suitable for their purpose.

Taps used for supplying sanitary appliances (sinks, baths, wash basins and bidets)

Standard pattern taps

Figure 6.16 shows two examples of sink taps. There is a vast range of taps used for sanitary appliances, and with the exception of the ceramic disc-type taps, all work on the same principle.

Maintenance requirements of taps

Since there is a vast range of tap designs, it would be impossible to show every type and their maintenance requirements here; examine Figure 6.17 as an example of the maintenance requirements.

Figure 6.17(b) shows the seating (left), and the washer assembly (right). The seating is where the washer 'seats' on the body of the tap when turned into the off position. This stops any water from passing through. Sometimes, the seating deteriorates due to fatigue and wear. A re-seating tool can be used to grind the seating smooth again and provide a good seal between the body of the tap and the washer.

The servicing requirements of both cold and hot taps are really the same thing, it is just the isolation of and turning on the supply that is slightly different.

> **Try this**
>
> Find out more about a re-seating tool, what it looks like and how it works.

Figure 6.16 (a) Standard high neck sink pillar taps (b) Standard sink mixer taps (Reproduced with permission of Pegler Limited)

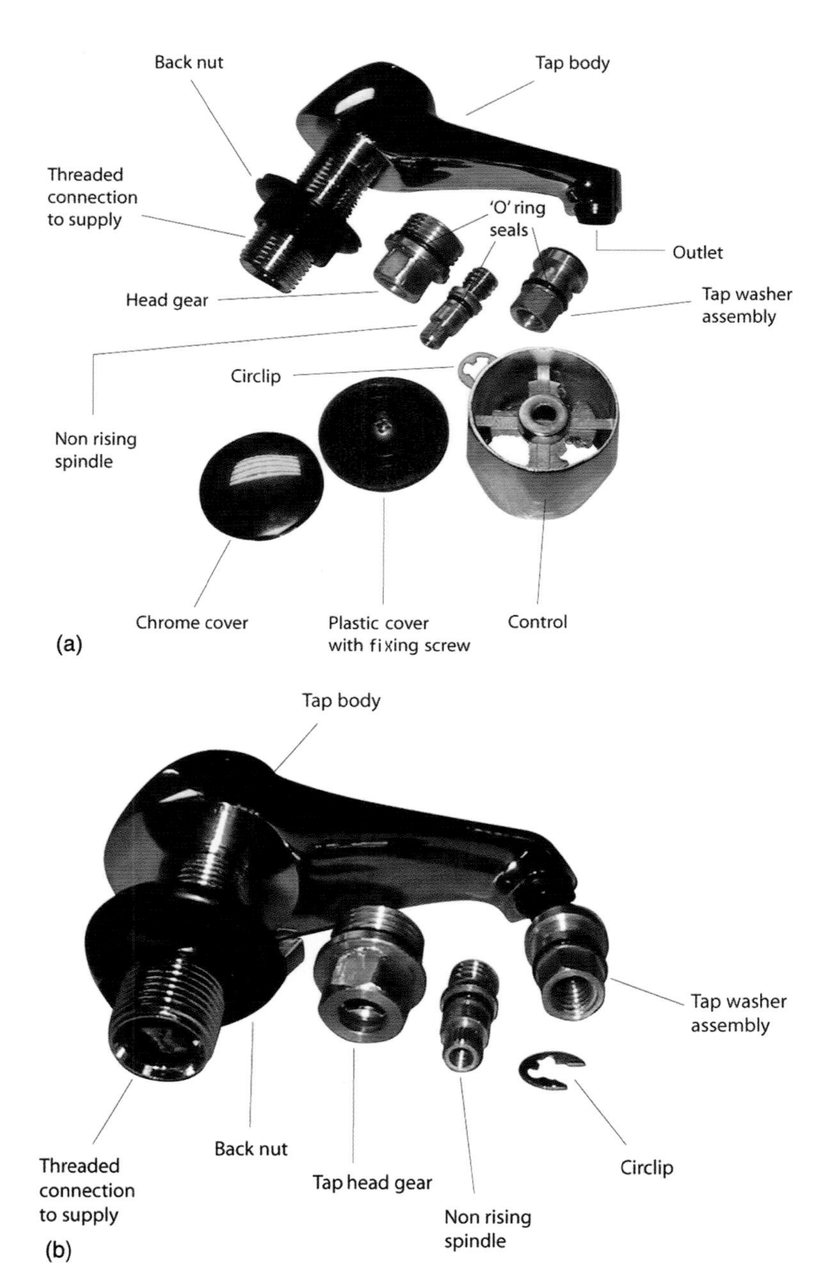

Figure 6.17 (a) Pillar tap components (b) Pillar tap seating and washer

Common faults

Depending on the type of taps fitted (and there are a lot on the market) faults tend to be dripping taps, leaking taps (glands), tap heads difficult to turn and noise when the cold tap is turned on.

Maintenance checklist:

- Isolate the supply
  In some properties, it is possible that servicing valves have been fitted to the cold and hot water supply under the sink. This is a good practice, but not a requirement of the Water Regulations currently. If they are fitted, turn them off.
  Assuming they are not fitted, on a direct cold water system, turn off the cold water supply to the system and drain through the sink tap
  What if it is an indirect system, how would you isolate the supply? Try to think back to the cold water unit about what to do before moving on.
  - If it is an indirect cold water system, turn off the service valve to the cold water distribution pipe and drain through the sink tap
  - For a hot water tap, turn off the gate valve to the domestic hot water cylinder
  - Drain down the hot water system pipework through the sink tap
- Repair or replace the defective part (refer to the stripped down image)
  Again, the procedure will depend upon the type of taps fitted. Most taps fitted to kitchen sinks are chrome-plated brass. Some may be plastic.
  - The first job is to remove the cover from the tap head assembly; these are usually screwed through the top of the tap cover, and concealed behind a chrome or plastic cap. The cap is lifted by a small flat-headed screwdriver, and the cover screw (which is usually cross-headed) can then be removed.
  - You are then left with a head gear similar to the stop tap dealt with earlier
  - The process for re-washering the tap is the same
  - You may come across sink taps using ceramic discs rather than more traditional washer and seating methods of control. The ceramic discs do wear and allow leakage. These discs can be replaced but you may have difficulties in locating the parts. It may be more cost effective to replace the head gear completely.
- Reassemble the component
  - It is the same procedure for the stop tap, but you will also need to replace the cover. If the cover is plastic, make sure

that you do not overtighten the fixing screw, or it could split the plastic.

- Turn on and test
  - Make sure the cold and hot taps to the sink are turned off
  - Turn on the stop tap if it is a direct system
  - Turn on the service valve on the cold water distribution pipe if it is an indirect system
  - Turn on the gate valve to the hot water storage cylinder
  - Return to the sink taps and make sure the supply of pressure to the cold water tap is satisfactory (direct) and that the cold water (indirect) and hot water supply is flowing smoothly
  - Test the operation of the taps.

Ceramic disc-type tap

Ceramic disc taps are designed to last a lifetime. Unfortunately, in hard water areas they do not and it is as well to know how they operate and how to maintain and service them. It is very rare for the ceramic discs to wear out and the need to replace them is even rarer. It is most often just a case of dismantling the moving parts and servicing them.

Should you be unlucky enough to have a tap where the parts do actually fail, then it is unlikely you will be able to replace the discs and will probably need to replace the whole cartridge within the unit.

The disc tap should be virtually indestructible because:

- The ceramics do not wear out
- The distance moved by any part of the tap in any one go is only a quarter of a turn.

Try this

Find out as much as you can about taps and valves used on hot and cold water systems. You can do this by

- Sending off for manufacturers' catalogues
- Looking on manufacturers' websites
- Making notes similar to our maintenance checklist in your portfolio about any maintenance work carried out on taps and valves both at work or in your college or centre.

Cold and hot water installation details

Noise in systems

Noise in systems is usually caused by vibration. Not only is vibration a source of annoyance to the occupier of the building, but in severe cases it can also cause damage to pipework and fittings, eventually causing leaks. Noise in systems can be categorised as:

- Expansion noise
- Water hammer
- Flow noise.

Expansion noise

Mainly found in hot water pipework systems. As the pipe gets hotter and expands, it causes creaking sounds which are often repeated as it cools and contracts. The use of relevant pipe clips, brackets or pads between pipes, fittings and pipework surfaces should help to deal with expansion and contraction.

Water hammer

This is probably the most common cause of complaint from customers. When a valve is closed suddenly, shock waves are transmitted along the pipework making a loud hammering noise. The problem is made worse where cold water supply pipework is not adequately clipped.

Key point

The shower head needs to be at least a metre below the bottom of the CWSC to ensure that adequate pressure from the shower head is achieved.

Defective float valves and tap washer on cold taps help to cause water hammer, but can be cured through regular maintenance of the valve or taps. The velocity of the cold water supply will further affect the problem, so reducing the flow rate by turning down the stop valve will help to reduce the chance of water hammer.

Flow noise

Pipework noise becomes significant at velocities over 3 m/s, so the system should be designed to operate below 3 m/s even if it means increasing the diameter of the pipe.

Air locks

One of the most common causes of problems in hot water systems is that of air locks. Air locks are usually caused when pipework from CWSCs, hot water storage vessels and boilers (primary gravity circulation) are not installed correctly. Two installation details that highlight this problem are shown in Figure 6.18.

Pipework running horizontally should be level or to an appropriate fall allowing air to escape from the system. Horizontal runs of gravity primary circulation pipework should also be installed to the appropriate falls.

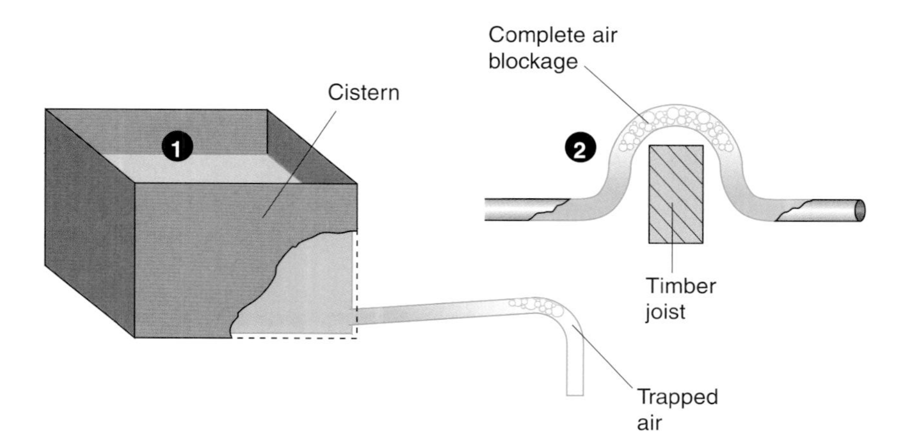

Figure 6.18 Common causes of air locks

Showers and bidets

On the gravity fed shower, the cold and hot supplies need to be of an equal pressure, and the shower head a minimum of 1 m below the bottom of the CWSC, as shown in Figure 6.19.

Maintenance of shower mixer valves

There are a number of shower mixer valves in the market and they come with full installation and maintenance instructions. Figure 6.20 is an example of a magnified view of a 'Victorian' pattern valve.

When you attend a college or centre to do the practical element of your technical certificate, you will be given practical instruction on the service and maintenance of thermostatic showers. With regard to your practical training and assessment, you will have to carry out the maintenance of a shower mixer valve. To try to give detailed maintenance requirements here would prove difficult given the various types of showers available. Try to read some of the manufacturers' catalogues. Contact details are at the end of the unit.

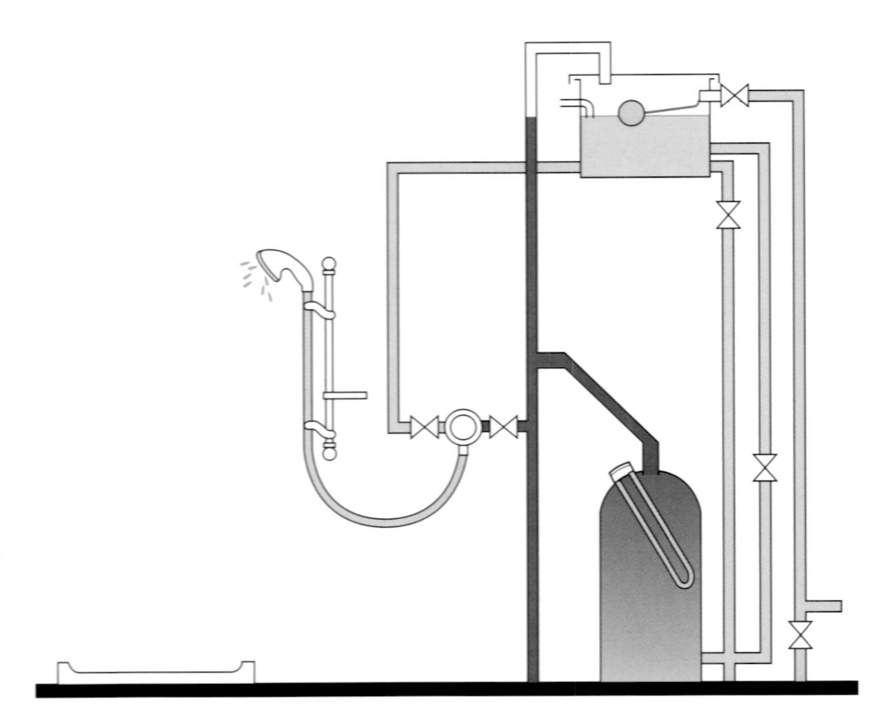

Figure 6.19 Installation details for a gravity fed shower with mixing valve (Reproduced with the kind permission of Mira Showers)

Common problems with thermostatic showers

One of the main problems is a constantly dripping shower head. This is caused by the deterioration of seals within the main body of the shower on the outlet side of the shower head.

Key point

Manually controlled showers rely on the cold and hot water supplies being at the same pressure. If one becomes partially blocked (sometimes pieces of insulation find their way into systems), flattened or lifted above the other, then it could affect the temperature.

Another problem is erratic temperature control, particularly on thermostatically controlled showers due to failure of the thermostatic control mechanism. The shower can also fail to work properly due to the entry of dirt or grit.

It is essential to make sure that shower heads are inspected, dismantled and cleaned at least on a quarterly basis, particularly if they are located in residential-type accommodation. The reason for this is to reduce the risk of Legionnaire's disease, which thrives in limescale or dirty conditions and is transmitted in shower aerosols.

If a customer complains of poor pressure from the spray handset or outlet, check the installation to make sure that there is sufficient head to provide adequate pressure. Remember, this should be a minimum of 1 m from the base of the CWSC to the point at which the water exits the shower head. If you find that there is insufficient head, then one possibility, other than raising the height of the CWSC, is to install a shower pump.

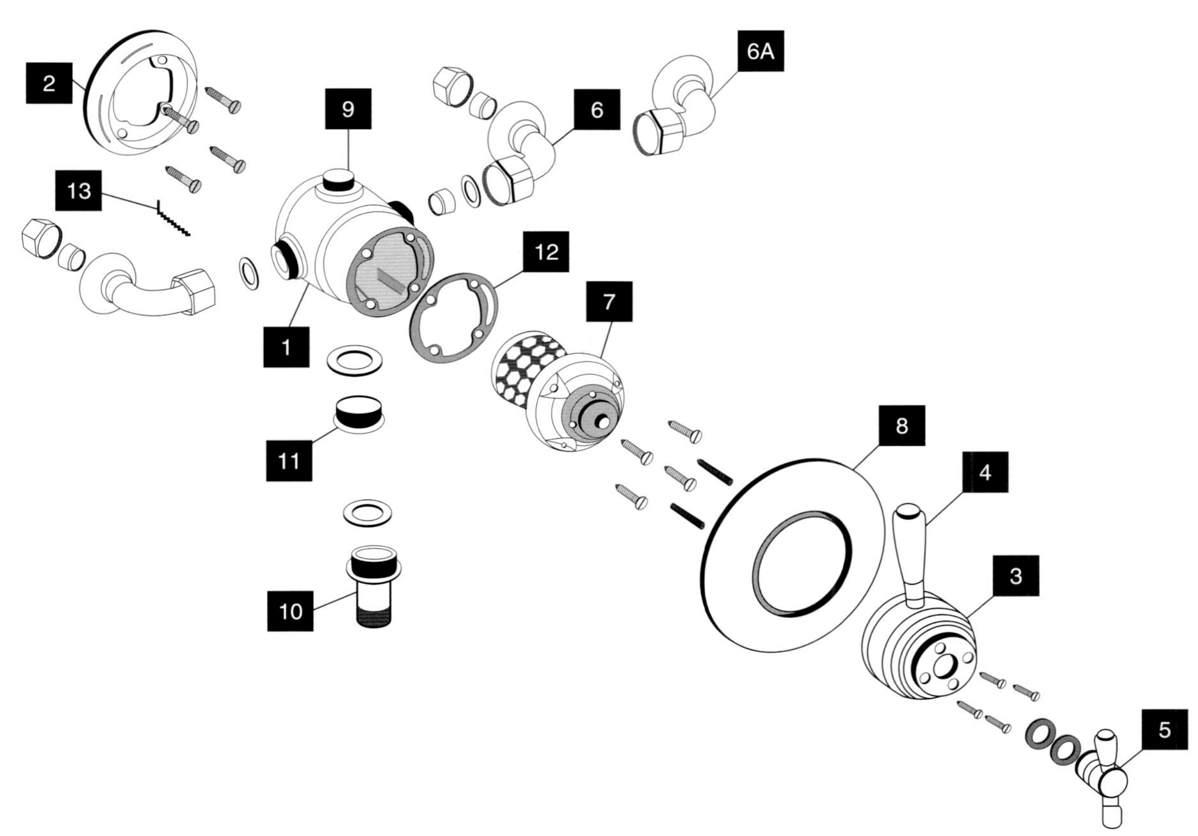

Figure 6.20 Exploded view of thermostatic shower mixer valve parts. List of components: 1. Shower body, 2. Wall mounting plate, 3. Temperature control lever, 4. Ceramic temperature control handle, 5. On/off control, 6. $\frac{3}{4}''$ BSP elbow, 7. Thermostatic cartridge, 8. Wall plate, 9. $\frac{3}{4}''$ outlet connector, 10. Hose connector, 11. Blanking plug, 12. Gasket with filter, 13. Allen key with grub screw

Shower maintenance

The cold and hot water supply to the shower from the CWSC and hot water storage vessel should be isolated using service valves. Otherwise, it will mean draining down at least 225 l of water. Some shower manufacturers include isolating check valves in the actual shower unit which improves the ease of servicing. Once the supply is turned off, maintenance can begin!

To strip the component means gaining access to either the thermostatic or manual cartridge. This is usually done by removing the on/off control cap. Behind this should be screws holding the temperature control housing.

Repair or replace components by replacing a set of 'O' ring washers (leaks) or replacing a thermostat assembly (erratic temperature) control. It could be as simple as cleaning out the internal ports.

Once the mixer has been re-assembled, the supply is turned on and the mixer shower tested.

If servicing valves had not been previously fitted, then they should be installed during the maintenance work.

For nonthermostatic shower mixers, the cold water connection to the storage vessel should be above the cold water supply to the shower. This is to prevent scalding; should the water supply to the CWSC be accidentally turned off, and the contents of the cistern allowed to drain down.

Shower pumps

Pumped or boosted showers are used to overcome the problems of restricted head.

The shower pump is suitable for boosting an individual mixer shower or supplying a tank-fed instantaneous electric shower, by increasing the static pressure and flow rate to the shower outlet (see Figure 6.21). They are not designed for use on mains-fed

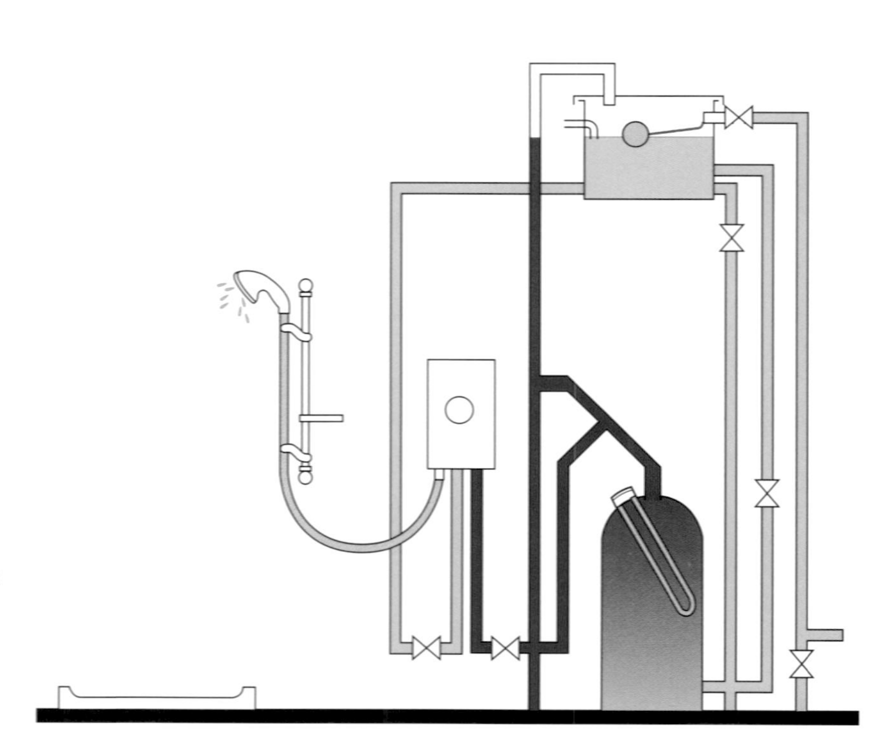

Figure 6.21 Installation details for a gravity fed shower using a shower pump (Reproduced with the kind permission of Mira Showers)

installations. They can be used on installations with as little as 300 mm head. Manufacturers provide detailed installation instructions, including the electrical connection requirements, and these should be strictly followed.

Connecting bidets to cold and hot water supplies

There are two types of bidets:

- Over the rim fitments, as shown in Figure 6.22
- Ascending sprays which discharge water from a nozzle located at the base of the appliances.

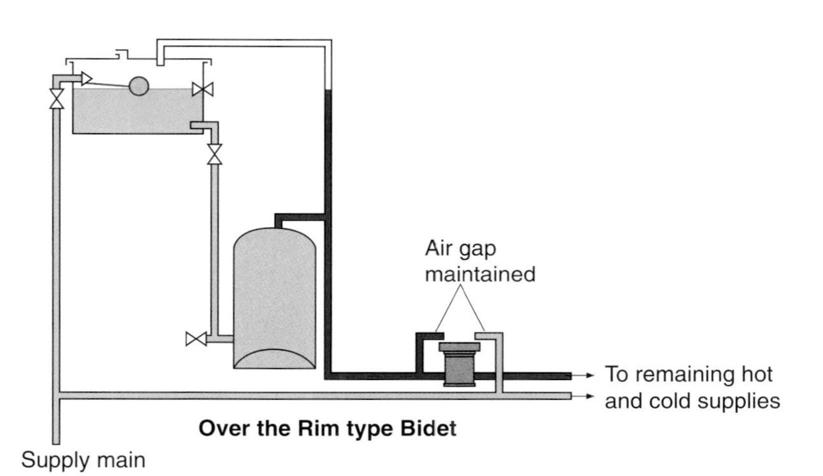

Figure 6.22 Over the rim fitment and installation requirements

Key point

It is important that others are advised before the testing commences, this could be the householder where you are working, the other trades or supervisor on a new building site, or a customer/customer's representative in other premises.

The danger with the ascending spray is that it could be possible for the nozzle to become submerged in the bowl of the bidet, and thus constitute a possible risk of contamination of the water supply. For this reason, a separate dedicated cold water supply should be taken from the CWSC, and the hot water supply should run via separate distribution pipes separate from other draw offs.

Over the rim fitment

Bidets with over the rim supply fitments are popular because they are easier to install and do not require the same amount of pipework and fittings. Figure 6.22 shows a typical pipework arrangement for the over the rim fitment.

De-commissioning of cold and hot water systems

De-commissioning domestic systems basically means turning off the supply to the system, removing system pipework and components, and making sure that the hot or cold supplies are sealed or left so they cannot be turned back on.

Activity 6.6

Can you think of a typical de-commissioning situation? Make a note for your portfolio and check out the answer at the end of this book.

Test yourself questions are found on page 254

Soundness testing on cold and hot water systems

It is essential to carry out soundness testing on completed hot water and cold water installations if you are to leave the installation leak free. Soundness testing of cold and hot water systems usually includes

- Visual inspections
- Testing for leaks
- Pressure testing
- Final checks.

Procedures for carrying out soundness testing

BS 6700 provides the standard for soundness testing on cold and hot water systems.

Visual inspection

This includes making sure that all pipework and fittings are thoroughly inspected to ensure

- They are fully supported, including cisterns and hot water cylinders
- They are free from jointing compound and flux
- That all connections are tight
- That terminal valves (sink taps, etc) are closed

- That inline valves are closed to allow stage filling
- The storage cistern is clean and free from swarf.

It is useful at this stage to advise the customer or other site workers that soundness testing is about to commence. It would also help if notices were placed on sinks, etc advising that the pressure testing was in progress.

Testing for leaks

When testing for leaks, you should follow this checklist:

- Slowly turn on the stop tap to the rising main
- Slowly fill in stages to the various service valves, and usually inspect for leaks on each section of pipework, including fittings
- Open service valves to appliances, fill the appliance and again visually test for leaks
- Make sure cistern levels are correct
- Make sure the system is vented to remove any air pockets prior to pressure testing.

Pressure testing

Pressure testing of installations within buildings is done using hydraulic pressure testing equipment. BS 6700 has separate procedures for testing rigid pipes and plastic pipes.

Activity 6.7

Obtain a copy of BS 6700 and make notes for your portfolio of the BS specification for testing rigid and plastic pipes then check it out with the answer given at the end of this book. Find out from manufacturers' catalogues or websites the type of hydraulic equipment used to carry out the test.

Final system checks

After the system tests have been completed, carry out a final visual check for leaks. Advise the customer and/or other site workers that testing is complete.

Make sure the system is thoroughly flushed out before it is commissioned.

Test yourself 6.6
(Note: This test yourself section covers the last three sections)

1. The Water Regulations state five requirements for the use of water fittings on plumbing systems, state three of them.
2. Where would a gate valve be used?
3. When installing pipework from a CWSC to a hot water storage vessel, what precautions should be taken to avoid air locks?
4. What is meant by the term dead leg and how is this different from a dead end?
5. List the three main causes of noise in systems.
6. What four stages are normally carried out while carrying out soundness testing?
7. Which British Standard covers soundness testing?

Unvented storage systems

Unvented storage systems are a requirement for the Level 3 Technical Certificate and S/NVQ but not for Level 2. However, there is a short section here to provide a basic understanding of the concept.

This information in itself is not enough to allow you to work on unvented systems. In order to work on them, a plumber must have been on an approved unvented course and be registered with an accredited certification body. You will be able to obtain details by contacting BPEC Certification Limited on 024 7642 0970. Their website is www.bpec.org.uk

Key point

In a **package system** the vessel has all the safety devices fitted at the factory, but all the operating devices are supplied by the manufacturer in a 'kit form' for on-site assembly, as shown in Figure 6.23.

Figure 6.23 Package system (Reproduced with permission of Albion Cylinders)

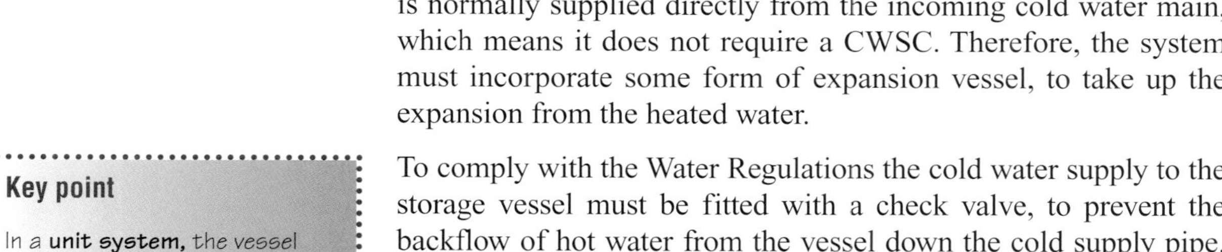

The storage vessel on an unvented hot water system does not require an open vent pipe for the expansion of heated water and it is normally supplied directly from the incoming cold water main, which means it does not require a CWSC. Therefore, the system must incorporate some form of expansion vessel, to take up the expansion from the heated water.

Key point

In a **unit system,** the vessel has all the safety and operating devices fitted by the manufacturer at the factory, ready for site installation.

To comply with the Water Regulations the cold water supply to the storage vessel must be fitted with a check valve, to prevent the backflow of hot water from the vessel down the cold supply pipe. A pressure-reducing valve must also be fitted to the cold water supply to prevent damage to the system from excessive water pressure.

Unvented hot water systems rely on mechanical devices for the safe control of water temperature and hot water expansion. Systems that are in excess of 15 l are covered by the Building Regulations Approved Document G3. This requires that the installation of a system must be carried out by an approved registered installer. The system should be purchased as a package or unit, being complete with all the factory set safety devices (see Figures 6.24 and 6.25).

Figure 6.24 Unit system (Reproduced with permission of Albion Cylinders)

The regulations state that at no point of time must the temperature of the hot water within the vessel reach 100°C. This is ensured by the use of three safety devices:

- Water temperature thermostat, set at 60°C
- High temperature cut-off device, cuts out if the water temperature exceeds 90°C
- Temperature relief valve opens at a temperature above 95°C.

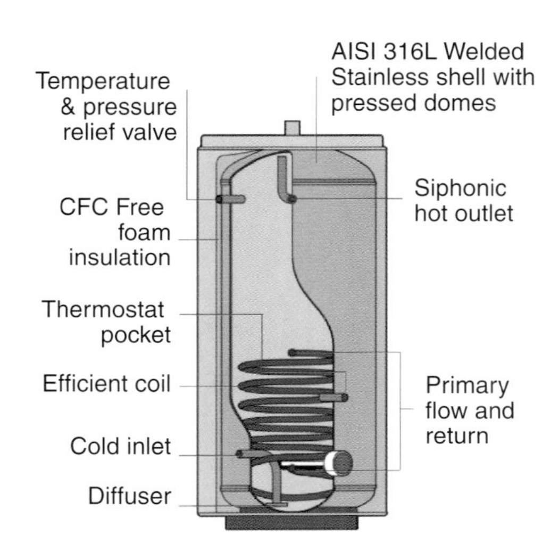

Figure 6.25 Indirect storage vessel (Reproduced with permission of Albion)

UNIT 6

The system must also be fitted with a pressure relief valve, to safeguard against failure of the expansion vessel, removing the expanding water safely from the system. The valve on a domestic installation is normally factory set to release at a pressure above 7 bar.

Unvented hot water storage vessels are manufactured from materials, such as copper, copper-and glass-lined, and stainless steel. They are available in various sizes and capacities, and can be of either direct or indirect type, and should be labelled as complying with BS 7206. The storage capacity for an average domestic dwelling with one bathroom and thermostatic shower would be 125 l.

The advantages of this system are:

- Higher pressure at the draw-off points
- Quicker installation
- Less pipework
- Omits CWSC and thermal insulation, reduces the risk of frost damage in a dry loft space
- Stainless steel units offer longer life span (25-year guarantee)
- Flexible siting locations.

The disadvantages of this system are:

- Incoming cold water supply pressure must be adequate to supply both hot and cold outlets (approximate minimum requirements 1.8 bar)
- Regular inspection and servicing of the unit is essential
- In hard water regions, limescale build up could prove to be a problem
- No reserve supply of water if the mains supply is interrupted
- Installation components are more costly.

Figure 6.25 shows an unvented indirect storage vessel.

There are no test yourself questions for this section.

Check your learning Unit 6

Time available to complete answering all questions: 30 minutes

1. Where could a plumber source the statutory requirements for the installation of hot water distribution pipework?
   a. Water Regulations ☐
   b. British Standard 6700 ☐
   c. Building Regulations ☐
   d. Construction Regulations ☐
2. From which of the following publications could a plumber find specific guidance on system selection?
   a. British Standard 6700 ☐
   b. Water Regulations ☐
   c. British Standard 1192 ☐
   d. Building Regulations ☐
3. The temperature of water in a vented storage vessel should be:
   a. 50°C ☐
   b. 60°C ☐
   c. 85°C ☐
   d. 100°C ☐
4. What is the specific heat capacity of water?
   a. 1.2 kJ ☐
   b. 2.4 kJ ☐
   c. 3.4 kJ ☐
   d. 4.2 kJ ☐
5. The main principle of gravity circulation of hot water is known as
   a. radiation ☐
   b. convection ☐
   c. induction ☐
   d. conduction ☐
6. The two main types of storage hot water systems are:
   a. Vented and instantaneous ☐
   b. Instantaneous and unvented ☐
   c. Vented and unvented ☐
   d. Multi-point and single point ☐
7. Where should a service valve be located in order to isolate the hot water supply from the hot water storage vessel to the taps?
   a. Primary return ☐
   b. Primary flow ☐
   c. Cold feed ☐
   d. Open vent ☐
8. What type of electrical device is often fitted to a hot water storage vessel to supplement the hot water supplied from a boiler?
   a. Immersion heater ☐
   b. Diverter valve ☐
   c. Heat pump ☐
   d. Gas circulator ☐
9. The minimum diameter for the primary flow and return pipework to a direct cylinder is:
   a. 15 mm ☐
   b. 22 mm ☐
   c. 25 mm ☐
   d. 28 mm ☐
10. Which of the following hot water storage system installations is likely to require the most pipework and fittings?
    a. Single feed indirect ☐
    b. Single feed direct ☐
    c. Water jacketed tube heater ☐
    d. Double feed indirect ☐
11. A vessel which creates an air lock to prevent primary water mixing with the hot supply is called a
    a. Single feed indirect cylinder ☐
    b. Single feed direct cylinder ☐

(Continued)

Check your learning Unit 6 (Continued)

c. Water jacketed tube heater ☐
d. Double feed indirect cylinder ☐

12. In order to permit an indirect hot water cylinder to be emptied, a draining valve should be installed
 a. At the lowest point on the hot water distributing pipework system ☐
 b. Combined with a draining tee where the cold feed enters the cylinder ☐
 c. Combined with a draining tee where the primary return enters the boiler ☐
 d. Combined with the full way gate valve on the cold feed pipework ☐

13. A secondary circulation supply to hot water draw-off taps is used to
 a. Improve the pressure at the tap outlets ☐
 b. Rreduce the pressure at the tap outlets ☐
 c. Prevent the possibility of dead legs ☐
 d. Avoid fitting a pump to the supply ☐

14. According to the Water Regulations, the maximum permissible length of dead leg, without insulation, from the top of a hot water cylinder to the outlet tap for a 22 mm pipe is
 a. 3 m ☐
 b. 6 m ☐
 c. 8 m ☐
 d. 12 m ☐

15. A service valve should be installed on the pipework to an instantaneous electric water heater
 a. As close as possible to the water heater ☐
 b. Immediately after the supply stop valve ☐
 c. On the outlet side of the water heater ☐
 d. A minimum of 300 mm from the water heater ☐

16. An instantaneous water heater designed to supply water to a bath, sink and wash hand basin is usually described as:
 a. Single point ☐
 b. Multi-point ☐
 c. Indirect ☐
 d. Direct ☐

17. A high-performance storage vessel is designed to:
 a. Reduce the amount of pipework compared to an indirect system ☐
 b. Provide ultrafast recovery times for the hot water supply ☐
 c. Reduce the amount of pipework compared to a direct system ☐
 d. Reduce the recovery time of the hot water supply to conserve energy ☐

18. Most pre-lagged storage vessels use expanded foam insulation:
 a. Excluding CFC ☐
 b. Containing CFC ☐
 c. Containing CO ☐
 d. Containing LPG ☐

19. The statement: 'all pipes connected to a storage vessel should be insulated for up to a meter in length' is a requirement of Building Regulations Part:
 a. P ☐
 b. G ☐
 c. L ☐
 d. J ☐

Sources of Information

Here are some additional contacts that may be helpful. This list is not exhaustive; you might wish to find additional ones to these.

Fittings

- Pegler Limited
  St Catherine's Avenue
  Doncaster
  South Yorkshire
  DN4 8DF
  Tel: 01302 560560
  Website: www.pegler.co.uk

Cylinders and storage vessels

- Albion
  Shelah Road
  Halesowen
  West Midlands
  B63 3PG
  Tel: 0121 585 5151
  e-mail: customer.care@albion-online.co.uk
  Website: www.albion-online.co.uk
- Elsy & Gibbons Limited
  (Combination units only)
  Simonside
  South Shields
  Tyne and Wear
  NE34 9PE
  Tel: 0191 4270777
  Fax: 0191 427 0888
- Gledhill Water Storage Limited
  Sycamore Estate
  Squires Gate
  Blackpool
  FY4 3RL
  Tel: 01253 401494
  Fax: 01253 349657
- Range Cylinders
  Tadman Street
  Wakefield
  West Yorkshire
  WF1 5QU
  Tel: 01924 376026
  Fax: 01924 385015

Water heaters including showers

- Applied Energy
  Tel: 08709 000430
  Website: www.applied-energy.com
- Kohler Mira Ltd
  Cromwell road
  Cheltenham
  Gloucestershire
  GL52 5EP
  Tel: 0870 241 0888
  e-mail: technical@mirashowers.com
  Website:www.mirashowers.com
- Baxi Potterton (gas water heaters)
  Main address
  Baxi Potterton
  Brownedge Road
  Bamber Bridge
  Preston
  PR5 6UP
  e-mail: enquiries@potterton.co.uk
  Tel: 08706 060780 (Main switchboard-automated)
  Website: www.potterton.co.uk

General

- Copper Development Association
  5 Grovelands Business Centre
  Boundary Way
  Hemel Hempstead
  Herts HP2 7TE
  United Kingdom
  Tel: 01442 275700
  Technical enquiries e-mail
  helpline@copperdev.co.uk
  Website: www.cda.org.uk

  CDA provide some excellent general information on hot and cold water systems

UNIT 7

Above Ground Discharge Systems

Summary

Above ground discharge systems (AGDS) are essential to ensure that we are able to keep outselves and our household environment clean and hygienic. In the unit on AGDS we will look at:

- Sanitary systems
- Sanitary appliances
- Sanitary pipework and fittings, and the various appliances on the market, and how to install them
- How to test and maintain systems
- Rain water drainage – that is guttering and fall pipe systems
- De-commissioning systems.

Sanitary systems

Introduction

The majority of systems in new domestic dwelling installations are known as primary ventilated stacks. You may still hear this referred to as 'single stack on site'.

Plumbers are involved in maintenance work and some of the earlier above ground discharge systems you may still come across were loosely classified as 'two pipe'. This is where the pipework from the WC was separate from that of the washbasin, bath and sink. The waste water only joined forces once it entered the drain.

Today, all pipework systems used on domestic dwellings are based on an one-pipe system with four variations on the theme.

The design and installation of systems must comply with Building Regulations Part H1 2002, which supersedes the 1992 edition. As well as the guidance given in H1, BS 8000 - 13; Workmanship on building sites Part 13: Code of practice for above ground drainage and sanitary appliances provides additional information on good practice related to AGDS.

BS EN 12056 provides additional guidance on such a pipework system design including waste pipe connections, trap seal and ventilation of system pipework.

H1 2002 can be downloaded from the website called the Planning Portal, www.planningportal.gov.uk. If you do not have internet access, their address is:

Communities and Local Government
Eland House
Bressenden Place
London
SW1E 5DU
Tel: 020 7944 4400

BS 8000 Part 13 and BS EN 12056 can be obtained from BSI on 020 8996 9001, but there will be a cost, alternatively you can try the library, but your college or centre should have a copy.

Types of systems

There are four types of systems in use today and these are:

- Primary ventilated stack systems
- Stub stack and systems using air-admittance valves

Key point

H1 lays down a number of 'rules' that you need to remember which relate to any system design.
The H1 rules are based around:

- The size of the discharge pipes
- The maximum length of the discharge pipes
- The gradient of the discharge pipes
- The location of branch connections
- The connection of the soil pipe to the drain.

- Ventilated discharge branch systems
- Secondary modified ventilated stack systems.

All these systems can be installed either inside or outside the building.

Primary ventilated stack system

You are most likely to install this type of system in the majority of domestic dwelling situations.

Figure 7.1 is an example of a primary ventilated stack, which complies with H1.

There are limitations to the minimum pipe sizes, maximum lengths of the branch connections and their gradients.

These can be summarised as shown in Table 7.1

In this system you should have the appliances grouped closely together. There is some flexibility, however. For example, if you did install a shower with a 50 mm waste it could be located up to 4 m away from the stack as opposed to 3 m if using 40 mm pipe. The size of the branch pipes should always be at least the same diameter as the trap.

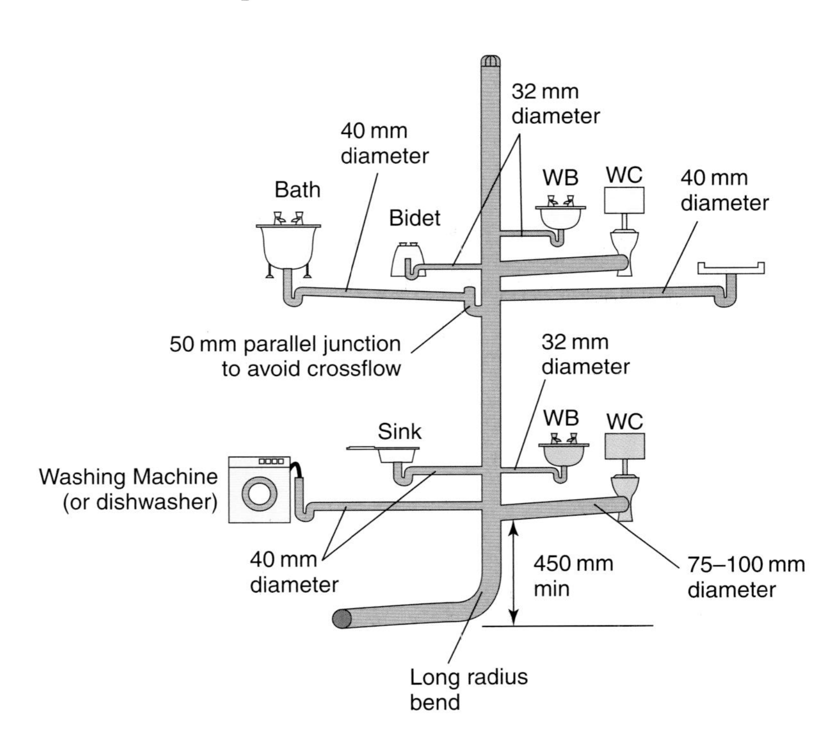

Figure 7.1 Primary ventilated stack

Table 7.1

	Pipe size (mm)	Max Length (m)	Slope
Basin	32	1.7	18–80 mm fall per m run
Bath	40	3.0	19–90
*Shower	50	4.0	18–90
WC	100	6.0	18 h mm/m min

Branch connections

The location of a branch pipe in a stack should not cause cross flow into another branch pipe (cross flow happens when two branches are located opposite to each other). Cross flow can be prevented by working on the following details, as shown in Figure 7.2.

You might find on some installations that it is easier to run the kitchen sink waste pipe into a gulley rather than pipe to the stack. This is allowed as long as the pipe end finishes between the grating or sealing plate and the top of the water seal. *These branch connection principles also apply to the ventilated discharge branch and secondary ventilated stack systems mentioned a little later.*

Key point

The primary ventilated stack system is popular because it costs less in materials and installation time, as it does not need a separate ventilating pipe like other one-pipe systems. This is good for the plumber and good for the environment (less material production).

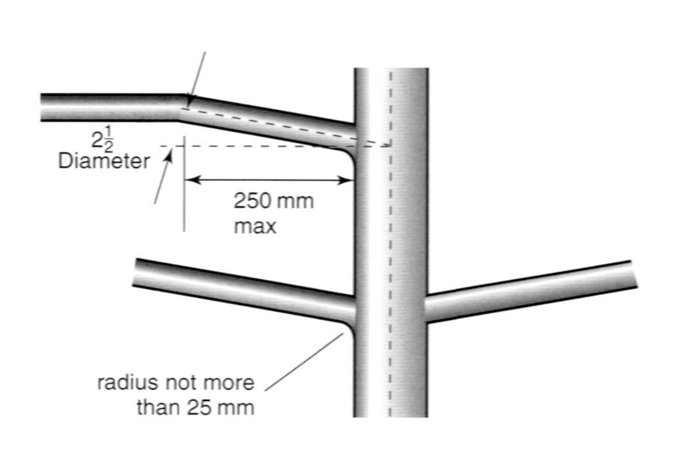

(a)

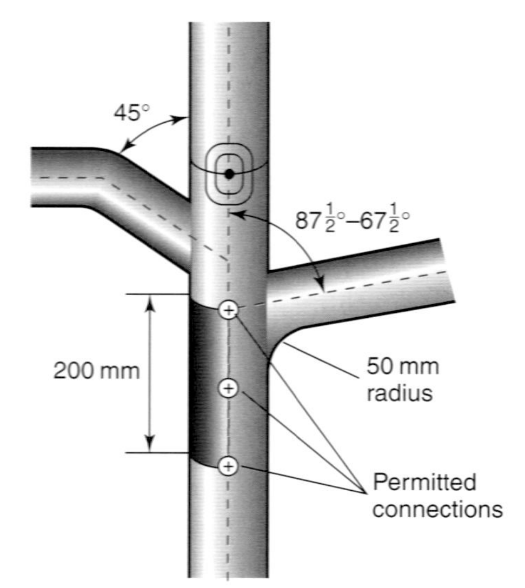

(b)

Figure 7.2 Branch pipe locations

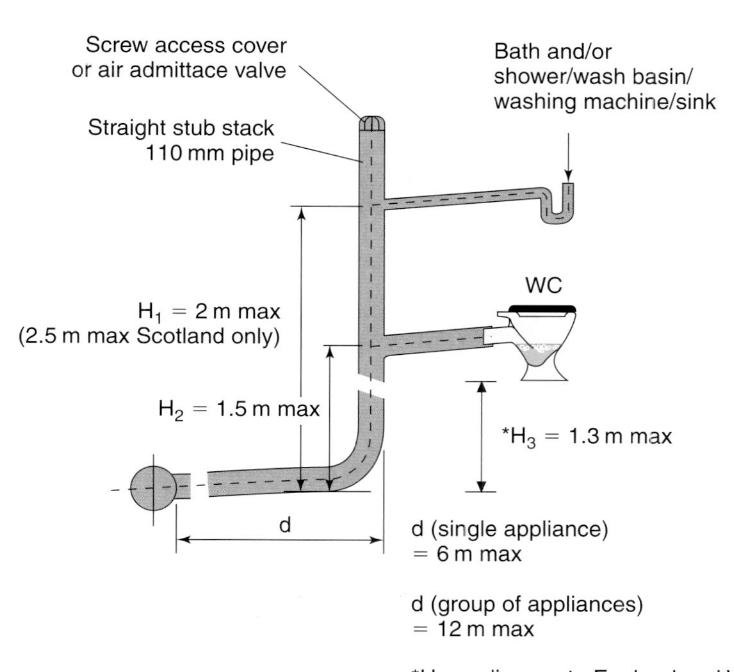

Figure 7.3 Stub stack

Connection to the drain

In single dwellings up to three storeys, a branch discharge pipe should not be connected lower than 450 mm above the invert of the tail of the bend at the foot of the stack.

Stub stacks

Figure 7.3 shows, when a group of appliances or a WC on a ground floor is connected directly to an underground drain, a stub stack of 110 mm diameter pipe can be used. Ventilation is necessary if the distance from the highest appliance connection to the stack to the invert of the drain is in excess of 2 m, or if the distance from the crown of the WC connection to the invert of the drain is in excess of 1.3 m.

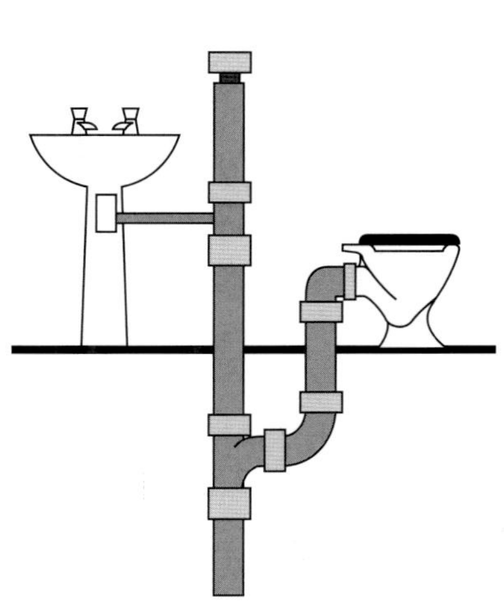

Figure 7.4 Use of air-admittance valves

Use of air-admittance valves

An air-admittance valve as shown in Figure 7.4 is a means of adding ventilation to the drainage system, preventing the loss of water seals in traps and consequent release of foul air into the building.

Figure 7.5 shows a cross section through an air-admittance valve. Figure 7.5 is reproduced with the kind permission of Hepworth Plumbing Products.

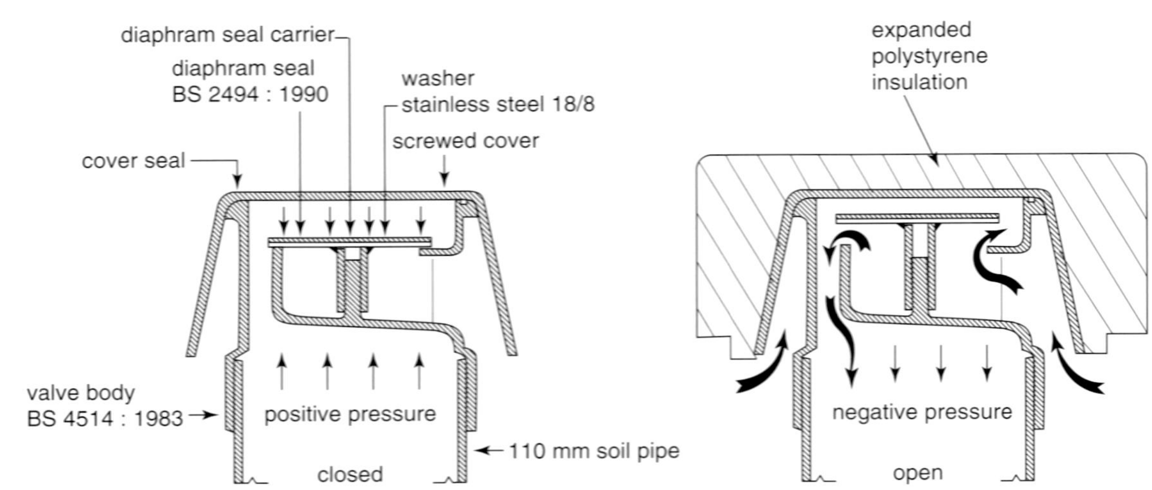

Figure 7.5 Cross section through air-admittance valve (Reproduced with permission of Hepworth Plumbing)

The valve, which operates on negative pressure (valve open lets air in) and positive pressure (valve closed keeps smells out), can be located in the roof space or the pipe boxing arrangement to the WC.

Air-admittance valves are packaged in formed polystyrene boxes, the tops of which should be fitted to the valve top after installation, providing insulation for the valve in use. The valve should be situated in a non-habitable area of the building, e.g. the roof space, where it will be easily accessible and there is reduced risk of freezing.

Key point

Air-admittance valves should not be located outside of buildings due to risk of freezing.

The underground drain or branch drain which serves a stack or stacks to which air-admittance valves are fitted may require additional ventilation at a position further upstream from the stack connection. This minimises the effects of excessive back pressure if a blockage should occur in a drain. In determining the requirement for additional ventilation to the underground drainage system, the following can be used for general guidance.

- Up to and including four domestic dwellings up to three-storeys high, additional drain ventilation is not necessary
- Where an underground drain serves more than four dwellings which have soil systems fitted with an air-admittance valve, the drain must be vented as follows:
  - In the case of five to ten dwellings, additional conventional ventilation must be provided at the head of the underground drainage system

– In the case of eleven to twenty dwellings, additional conventional ventilation must be provided at the head and midpoint of the underground drainage system.

All multi-storey dwellings require additional venting of their underground drainage system if more than one building, equipped with air-admittance valves, is connected to a common drain not ventilated by conventional means.

Activity 7.1

What advantage do you think the air-admittance valve might offer over the traditional venting method? Check out your answer at the end of this book.

Other ventilated stack systems

Where the requirements described earlier in the primary ventilated stack system can be met, separate ventilation will not be needed. On some larger domestic properties this is not always possible (although the use of resealing taps is an option – you will study traps shortly). The alternative is to install separate ventilation pipework. This can be done in two ways:

- Ventilating each appliance into a second stack – the ventilated discharge branch system
- Directly ventilating the waste stack – secondary ventilated stack system.

Figure 7.6 shows what the systems look like.

The ventilated discharge branch system

Activity 7.2

In this system, a ventilating pipe is extended to connect to each of the individual branch pipes throughout the system. Can you think of where this system might be used? Make a note in your portfolio, then check it out against the suggested one at the end of this book.

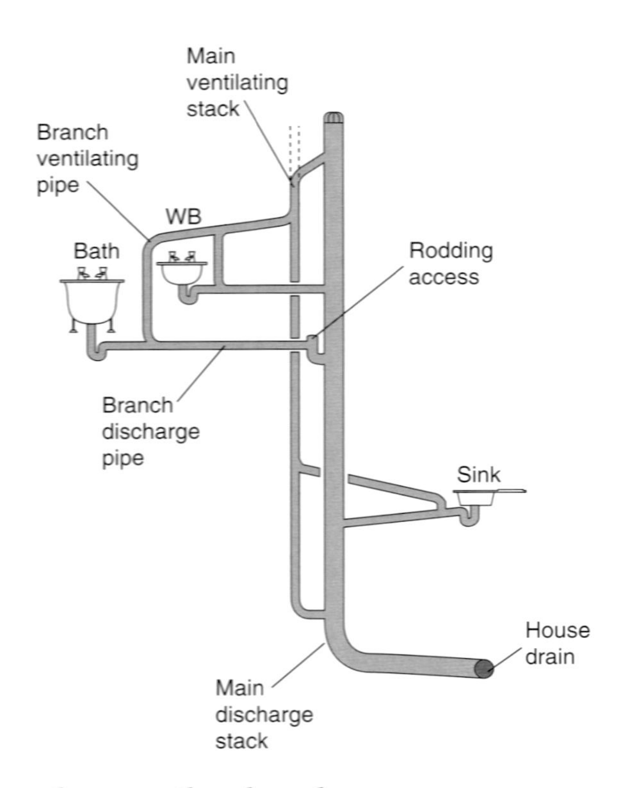

Figure 7.6 The ventilated discharge branch system

The secondary ventilated stack system

As you can see from Figure 7.7, only the main discharge stack is ventilated. This arrangement will prevent any pressure fluctuations, either negative or positive.

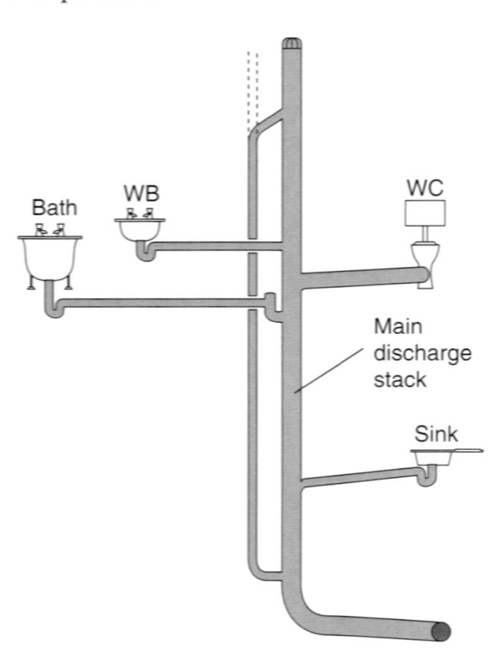

Figure 7.7 The secondary ventilated stack system

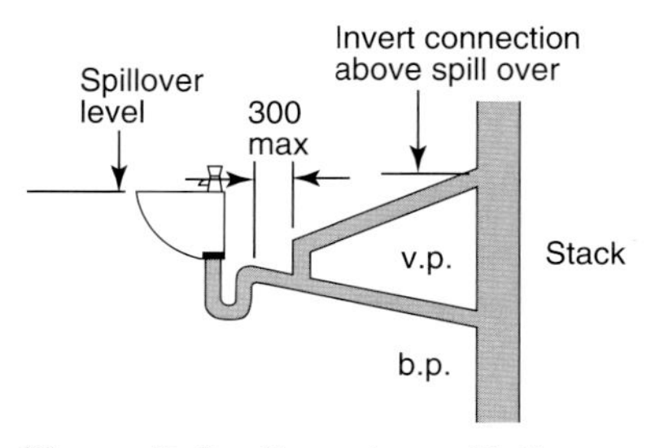

Figure 7.8 Branch ventilation pipes 'rules'

Branch ventilation pipes 'rules'

- The branch vent pipe must not be connected to the discharge stack below the spillover level of the highest fitting served (see Figure 7.8)
- The minimum size of a vent pipe to a single appliance should be 25 mm. If it is longer than a 15 m run or serves more than one appliance, then it is 32 mm minimum.
- The main venting stack should be at least 75 mm. This also applies to the 'dry part' of the primary vented stack system.

General discharge stack requirements

An external stack would be terminated, as shown in Figure 7.9 and a terminal should be fitted to prevent the possibility of 'bird nesting'.

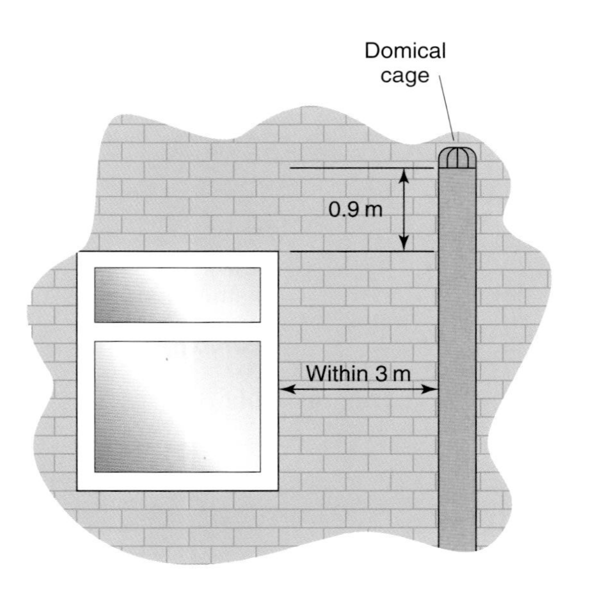

Figure 7.9 Stack termination

A vent cowl could be fitted where the stack is sited in exposed windy conditions.

Access

All stacks should have access for cleaning, and clearing blockages. Rodding points and access fittings should be placed to give access to any length of pipe which cannot be reached from any other part of the system. All systems pipework should be easy to get to in case of repair.

Activity 7.3

Why do you think it is necessary to install a trap on an AGDS? Check out your answer against the suggested one at the end of this book.

Traps

Key point

BS EN 12056 Part 2 says about traps:

'Any trap connected to a discharge pipe of 50 mm or less discharging into a main stack should have a seal of 75 mm'.

Traps are mainly manufactured in plastic (polypropylene to BS 3943), although they are also available in copper or brass/chrome-plated brass for use on copper pipework, where a more robust installation is required. Most trap fitting connections are either push-fit or compression-type.

Trap specifications

Where a trap diameter is 50 mm or above, a trap seal of 50 mm is required. This is because the size of the pipe means it is unlikely to discharge at full bore which is one of the two causes for the loss of trap seal.

If the discharge pipe from the trap runs into a gulley or hopper head, a seal of 38 mm is allowed; the gap between the gulley and the pipe provides an air break should the trap loosen its seal, thus no smell can enter the building.

A summary of the depth of trap seal is given in Table 7.2.

Table 7.2

Appliance	Diameter of trap (mm)	Depth of seal (mm of water)
Wash basin	32	75
Bidet		
Bath shower	40	50
Food waste disposal sink	40	75
Washing machine	40	75
Dish washer		
WC pan outlet		
Below 80 mm	75	50
Above 80 mm	100	50

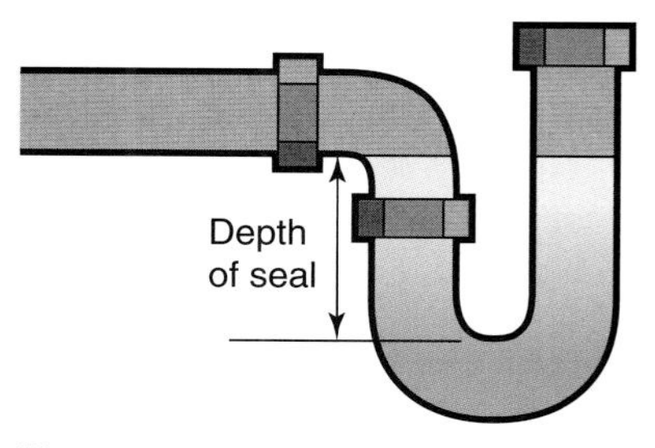

Figure 7.10 How to measure trap seal

The depth of the trap seal is measured as shown in Figure 7.10. The size of the trap is also governed by the size of the waste pipe it is connected to. Table 7.3 gives the minimum size of the waste fitting and trap.

Table 7.3

Type of appliance	Waste fitting size (in)	Discharge pipe and trap size (mm)
Sinks, showers, baths, washing machines and urinals	1½	40
Washbasins, bidets, drinking fountains and trough urinals	1¼	32
Stall and slab urinals	2½	65

Trap designs and location
'Trap gallery'.

Key point

Apart from specialist traps such as washing machine traps, etc there is a trap missing. Here is a clue: 'This uses a valve located on the top of the trap. If the pressure drops inside the pipework, the valve is activated, allowing air to enter the system, and equalising the pressure'. Do you know what it is? Find out in the text shortly.

Activity 7.4

Figure 7.11 shows the montage of just about all of the trap types that are available (one has been left out on purpose). Try to name each type of trap by matching a number against the name.

- 'P' Trap
- 'S' Trap
- Tubular swivel trap
- Bottle trap
- Straight through or wash basin trap
- Hepworth discharge pipe valve
- Straight through or wash basin trap
- Low-level bath trap
- Shower trap
- Running trap
- Re-sealing trap.

The answers are at the end of the book.

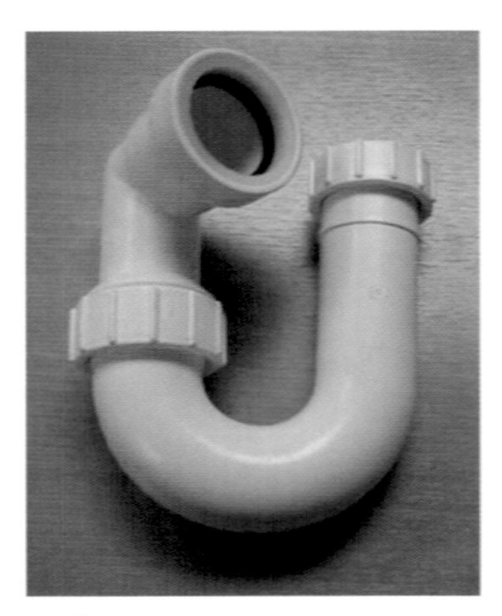
(a) Swivel trap

(b) Shower trap

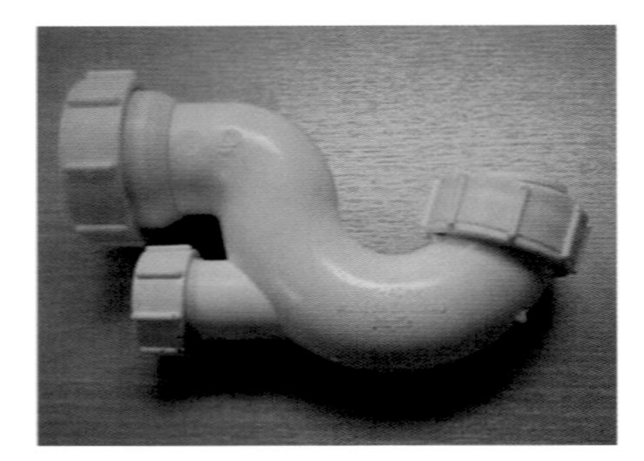
(c) Low-level bath trap

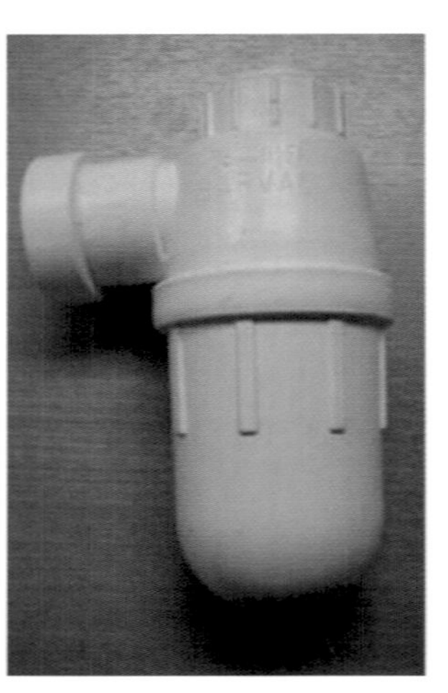
(d) Bottle trap

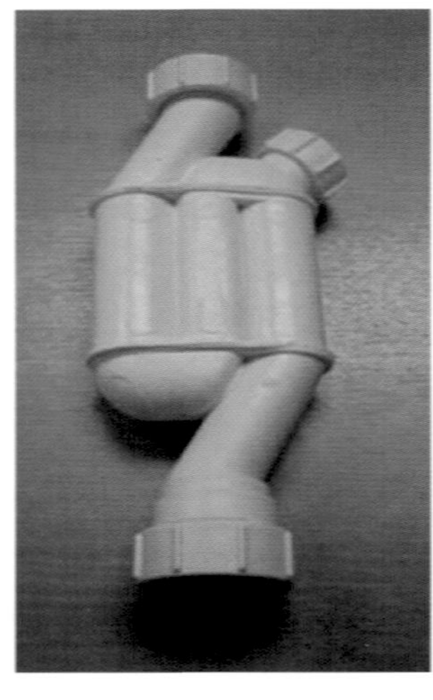
(e) Straight through trap

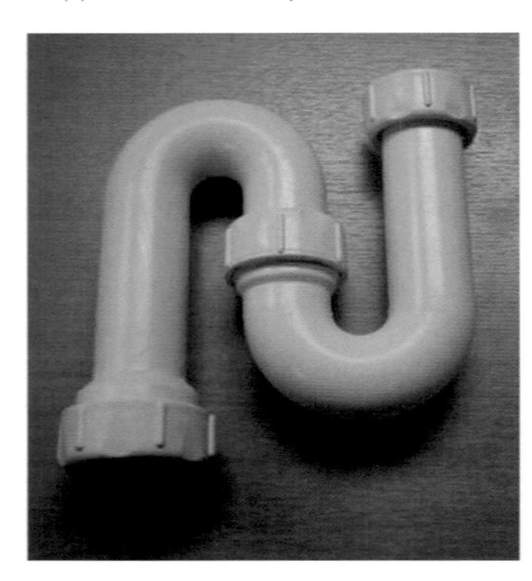
(f) 'S' trap

Figure 7.11 Trap gallery

Tubular swivel traps

On new jobs, tubular swivel traps are often used on sinks with multiple bowls because of their multi-positions, which provide a number of options when connecting to pipework. They are also particularly useful on appliance replacement jobs as they give more options when connecting to an existing waste pipe without using extra fittings or altering the pipework.

Bottle traps

Bottle traps are often used because of their neat appearance. They are easier to install in small areas such as behind a wash basin. They should be avoided on sink as they are prone to cause food blockage.

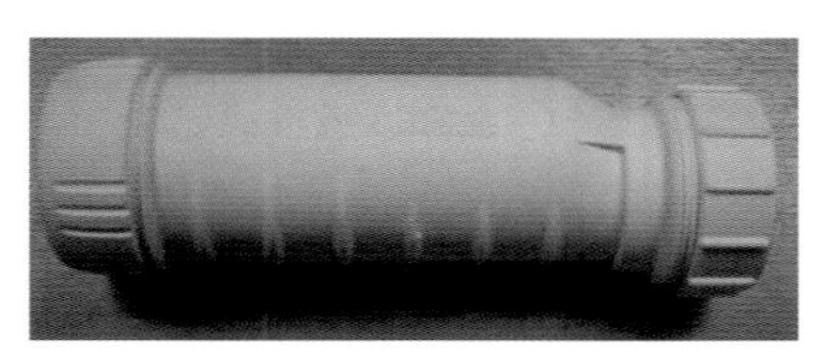
(g) Hepworth valve

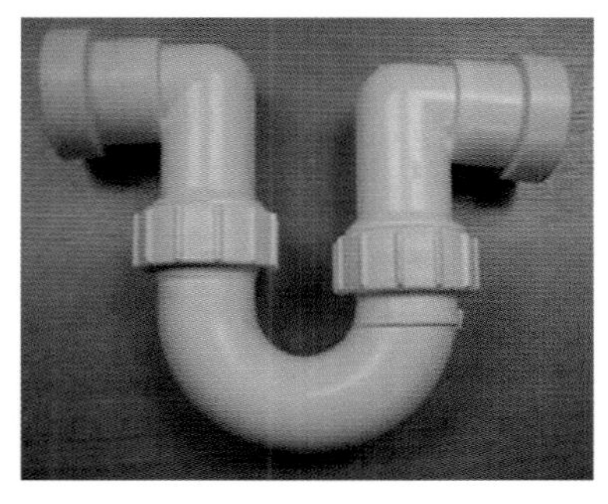
(h) Running trap

(i) P trap

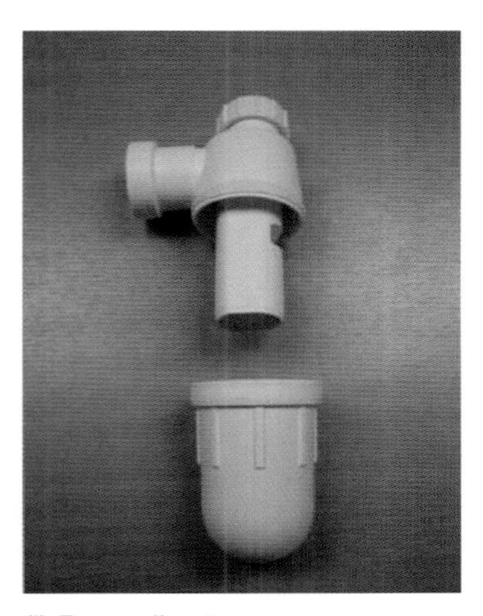
(j) Re-sealing trap

Figure 7.11 Continued

'P' and 'S' traps

'P' or 'S' traps are so named due to their shape, they are available in tubular design, or with a joint connection (like the ones in the figures), which have a few additional features useful when fixing them to pipework and fittings.

'P' traps are often used where the waste pipe is installed directly through a wall from the appliance and into a drain or directly into a stack. The 'S' trap would be used where the pipe has to go vertically from the trap through a floor or into another horizontal waste pipe from another appliance. 'P' traps (and bottle traps) can be converted to 'S' traps using swivel elbows.

Hepworth discharge pipe valve

The Hepworth valve works on the simple principle of using an internal plastic membrane. The membrane allows water to flow through it when the water is released, then closes to prevent foul air from entering the building.

The valve can be used on systems meeting BS EN 12056 Part 2. It is ideal for fitting behind pedestals and under baths and showers, and is supplied with a range of adaptors, so it can be used in various situations.

Low-level bath traps, bath traps and shower

These are designed so they can fit in tight spaces under baths and shower trays. Remember, a bath or shower trap is 40 mm wide and the minimum depth of a 40 mm trap connected directly to a primary ventilated stack is 50 mm. Some of these traps are shallow traps, i.e. they only have a 38 mm seal, which means they would need to have a separate ventilating pipe in practice, though a plumber would fit an anti-siphon version of the shallow trap.

Straight through or wash basin trap

Wash basin traps are used as an alternative to an 'S' trap where space is limited. They are also easier to hide behind pedestal basins. The main problem with this design is the two tight bends which slow down the flow of water.

Running traps

You might see these used in public toilets or schools where one running trap is used for a range of untrapped wash basins. On domestic installations it could be used in cases where a 'P' or 'S' strap arrangement is not possible or is difficult to achieve due to limitation of space or an obstruction. Running traps are sometimes used with a washing machine waste outlet or dishwashers, although specialist traps are also available for these appliances.

Resealing and anti-siphon traps

If an above ground discharge system is designed and installed correctly, the loss of trap seal should be prevented. We will look at some of the other reasons for the loss of trap seals next, but these traps are designed to prevent seal loss due to the effects of siphonage. These types of traps could be specified or fitted in situations where normal installation requirements cannot be met.

You probably think this looks like a bottle trap, but its internal construction is somewhat different. This type has a bypass within the body of the bottle. A dip pipe allows air to enter the trap via the bypass arrangement. As the seal is lost due to siphonage, air is allowed into the trap, thereby breaking the siphonic effect.

Anti-vac trap

This uses an anti-vacuum valve located on the top of the trap. If the pressure drops inside the pipework, the valve is activated, allowing air to enter the system, and equalising the pressure.

Access

Whatever type of trap is fitted, it is important that you can get to the trap for cleaning. As you have seen, some traps have cleaning eyes; others can be split at their swivel joints to enable a section of the trap to be removed.

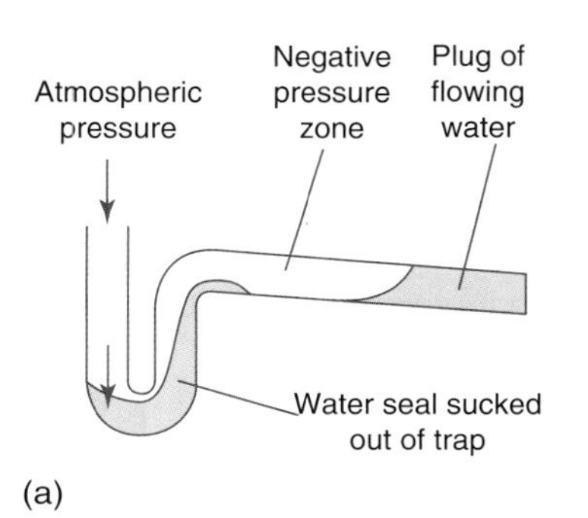

(a)

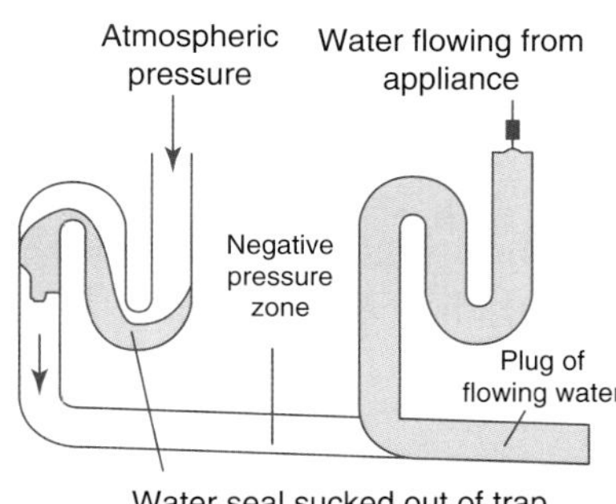

(b)

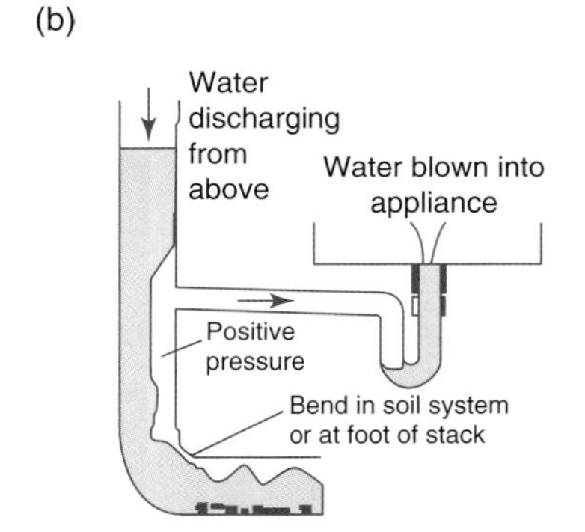

(c)

Figure 7.12 (a) Self-siphonage (b) Induced siphonage (c) Compression (Reproduced with permission of Hepworth Plumbing)

Trap failure

The reasons that a trap fails are usually down to bad design which can lead to self or induced siphonage, but it may be down to natural causes such as evaporation or foaming. Below are three of the main causes of trap failure.

- Self-siphonage
- Induced siphonage
- Compression.

Illustrations of Figures 7.12((a)–(c)) have been reproduced with the kind permission of Hepworth Plumbing Products.

Self-siphonage

As the water discharges, a plug of water is formed. This creates a partial vacuum (negative pressure) in the pipe between the water plug and the basin, which is enough to siphon the water out of the trap. Self-siphonage can be prevented by ensuring that the length of waste pipe is within the regulations for single stack installations, or that the waste pipe is ventilated. The use of resealing traps is another option, see Figure 7.12(a).

Self-siphonage is most common in wash basins as its shape allows water to escape quickly.

Induced siphonage

This is caused by the discharge of water from an appliance which is connected to the same waste pipe as other appliances. As the water plug flows past the joints of the second appliance, a negative pressure is created between the pipe and appliance which siphons the water out of the trap. This arrangement is not acceptable on a primary ventilated stack, unless the final branch pipe is a size bigger than the largest diameter waste pipe from the appliance, e.g. basin into bath waste, the bath waste would need to be 50 mm. In the above diagram, fitting a branch ventilation pipe between the two traps would solve the problem (or fitting a resealing trap), see Figure 7.12(b).

Compression

In Figure 7.12(c), you can see that as the water is discharged from an appliance into the main stack (usually WC at first floor level), it compresses at the base of the stack, causing back pressure. The back pressure can be enough to force the water out of the trap, thus loosening the seal.

The use of large radius bends and the minimum of 450 mm length between the vent of the drain and lowest branch pipe are in the regulations in order to prevent this.

Other trap-related problems:

- *Evaporation*
  This is the most common form of natural seal loss, which happens in very warm and dry weather, but is unlikely to occur in traps with a 75 mm seal.
- *Wavering out*
  This is caused by the effects of wind pressure across the top of the soil and vent pipe, particularly in exposed locations. This causes the water in the trap to produce a wave movement and wash over the weir of the trap. It does not happen often and can be avoided by fitting a 90° bend or a cowl to the top of the vent pipe.
- *Foaming*
  Foaming occurs when excessive detergent has been discharged into the sanitary system. The build up of foam in the waste pipe can cause depletion of the water seal in the trap.
- *Momentum*
  This is blowing the seal of a trap with the force of water and can happen if a bucket of water is poured forcefully into a sink or toilet.
- *Capillary action*
  This only occurs in ‘S’ trap arrangements, and does not happen often. If a thread of material becomes lodged as shown in

the drawing, water could be drawn from the trap by the effect of capillary attraction (remember your science sessions!)

- *Leakage*
  Leakage is pretty self-explanatory. It usually occurs if the trap has been damaged or split prior to installation and the damage cannot be detected visually (e.g. hair-line crack), and the water gradually seeps from the trap. Water can also leak from the cleaning eye, as shown in the figure.

Test yourself 7.1

1. State the specific regulation related to the design and installation requirements of sanitary pipework.
2. State three of the five main requirements of the regulation.
3. Which of the following British Standards covers workmanship for above ground drainage and sanitary appliances?
   a. BS 8000 ☐
   b. BS 8080 ☐
   c. BS 8808 ☐
   d. BS 8088 ☐
4. Of the four main types of the above ground discharge systems, which two are missing from the list below?
   a. Ventilated discharge branch system ☐
   b. Secondary modified ventilated stack system ☐
   c. ☐
   d. ☐
5. Given the size and maximum length of discharge pipes, which other three 'rules' are referred to in the regulations?
6. What is meant by the term 'cross flow'?
7. When an air-admittance device on a vent pipe is subject to negative pressure would it be open or closed?
8. State five main causes for the loss of trap seal.

Pipework system installation

Introduction

The fact that the Technical Certificate contains a practical element means that you will practise the installation of above ground drainage and sanitary systems, while being assessed. That you have also been given the design requirements of the pipework lay-outs means that you should know where everything goes. Here we will provide a background to the system installation requirements.

Workmanship relevant to the installation of above ground drainage and sanitary systems is also covered in BS 8000 Part 13.

Preparing to install above ground drainage system (AGDS) pipework

On new work such as multi-dwelling housing developments, the soil and vent position will be set by the position of the drain connection, but you will need to ensure that any builders' work necessary for the installation of the pipework is or will be complete prior to commencing your work, so there will not be any delays to the ongoing drainage and installation work.

It is an idea to agree the sequencing of the work with other site contractors where applicable. If the preparation is being carried out on your behalf, make sure there is adequate clearance around the pipes and fittings to be installed.

Types of material

Plastic is probably the most commonly used material for domestic AGDS pipework installations. There is a wide range of fittings used for soil and waste installations and these can be summarised as:

- Push-fit
- Solvent-welded
- Compression.

For push fit waste systems, polypropylene to BS 5254 is used. For solvent welded systems PVC, MuPVC (modified uPVC) and ABS plastics are used. These meet the requirements of BS 5255. Compression fittings are available which are suitable for both types of pipe, but these tend to be restricted for use on waste pipes and, in particular, traps.

The most common plastic for soil and vent pipes and fittings is uPVC to BS 4514 using ring-sealed push-fit joints. Pipe diameters for domestic use include 82 mm and 110 mm. This is the outside diameter of the pipe and they are available in lengths of 2.5, 3 and 4 m, either as pipe ends or single-socket pipes.

The waste pipe is available in sizes of 32, 40 and 50 mm diameters, in 3 m lengths. The choice between push-fit or solvent-welded installations will probably be made for you on larger contracts such as multidwelling housing developments. On one-off installations, the choice is often personal in terms of what you prefer to work with.

Activity 7.5

Can you think of the advantages of both methods? Record your thoughts for your portfolio, and check them out at the end of this book.

Range of fittings

Soil and waste systems

There is a wide range of soil waste pipe fittings, too many to include here.

Try this

Obtain a manufacturer's catalogue or use their website to get an idea of what is available for soil and waste pipe and fittings. Try designing a single stack system for the appliance layout in your home, and then try estimating what fittings you would need from the manufacturer's information.

Installation requirements

Access fittings

Access to enable the pipework to be cleaned and clearing any blockages is a requirement of the regulations. This can be done by inserting access plugs into waste and soil tee junctions or by installing purpose-made fittings. All traps are also provided with a cleaning eye, or in the case of a bottle trap, the bottom portion of the trap can be removed.

Fixing details

Plastic waste pipes should be clipped at the spaces shown in Table 7.4.

Soil and vent pipes are normally clipped vertically every two metres of length.

Table 7.4

	Horizontal (m)	Vertical (m)
32 mm	0.8	1.5–1.7
40 mm	0.9	1.8
50 mm	1.0	2.1

Drill types and plugs were discussed in common plumbing processes.

Soil stack connections to drains

A soil stack can be connected at ground level to an underground drainage system of uPVC, cast iron or earthenware material. A range of adaptors and couplers is available for these connections. A typical connection detail is shown in Figure 7.13.

Provision would be made above the coupler for access to the soil pipe for cleaning/rodding purposes.

On a new job, say on a large building site, the location of the connection point to the drain will be on the drawings and in the specification. But when you go into the dwelling, it is usually pretty obvious because all the outlets to the drain will be in place for the soil and vent pipe or downstairs WC.

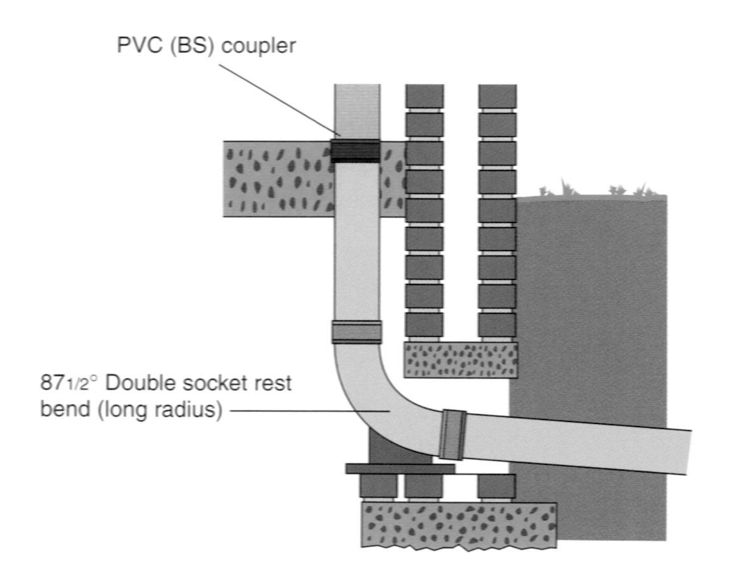

Figure 7.13 Soil stack connection to drain

On a replacement job, the connection will be the same as the existing soil and vent pipe, so you will use this. The soil and vent pipe is likely to be connected to an earthenware drainage pipe. Fittings are made to connect the two. Some drain pipe terminations, particularly on older systems, are finished with a collar. Drain connectors are inserted into the collar and a joint is made using a sealing compound. This is finished off with a sand and cement joint of a one to four mix. Remember about access? Check the need for including an access pipe fitting.

Drainage systems

Because the Technical Certificate syllabus requires you to know how to connect to a drain, you will need to have some idea of the systems you might be connecting to.

Overview of drainage design

There are three main types of drainage systems:

- The separate system
- The partially separate system
- The combined system.

The separate system

Here foul water runs into a separate sewer, and the rainwater likewise into one for surface water, as shown in Figure 7.14.

Partially separate system

This system still uses two pipes, but some of the surface water is discharged into a water course, soakaway or a drainage ditch, as shown in Figure 7.15.

The combined system

In this system the water from the sanitary appliances (foul water) and the rainwater all go into the sewer (see Figure 7.16).

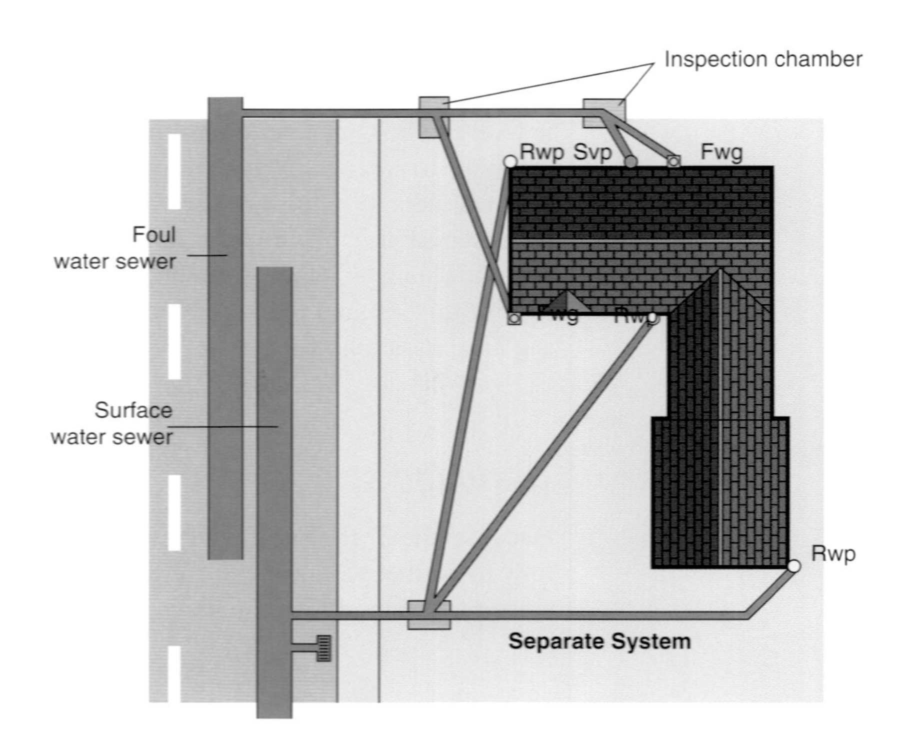

Figure 7.14 Separate System

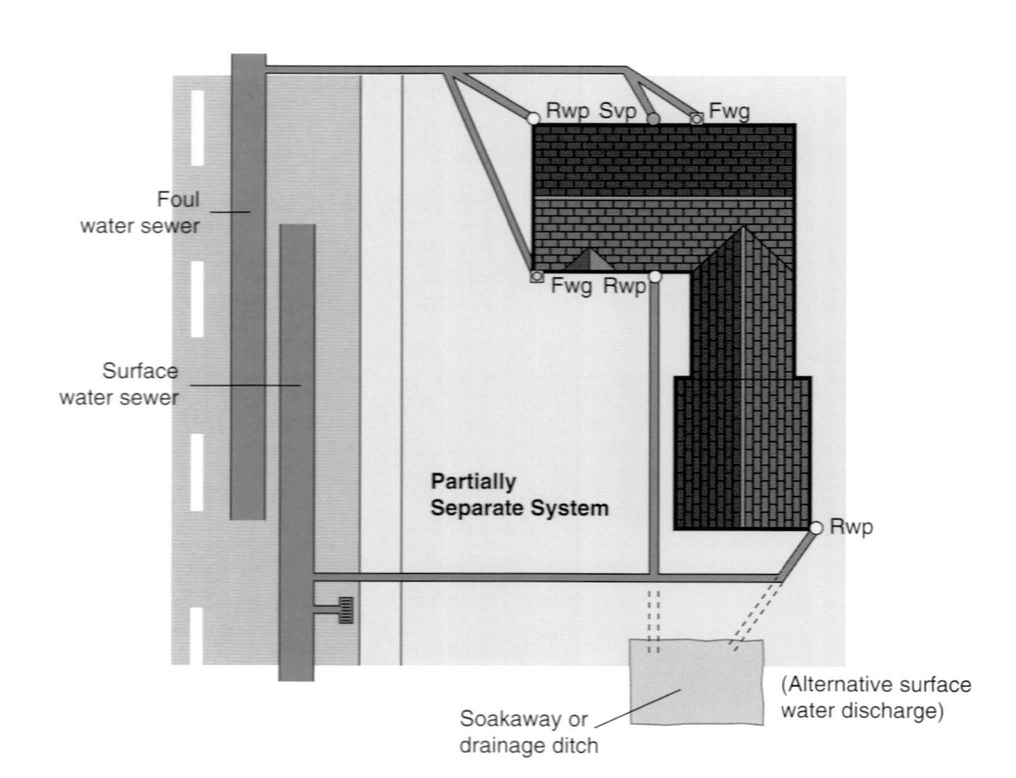

Figure 7.15 Partially separate system

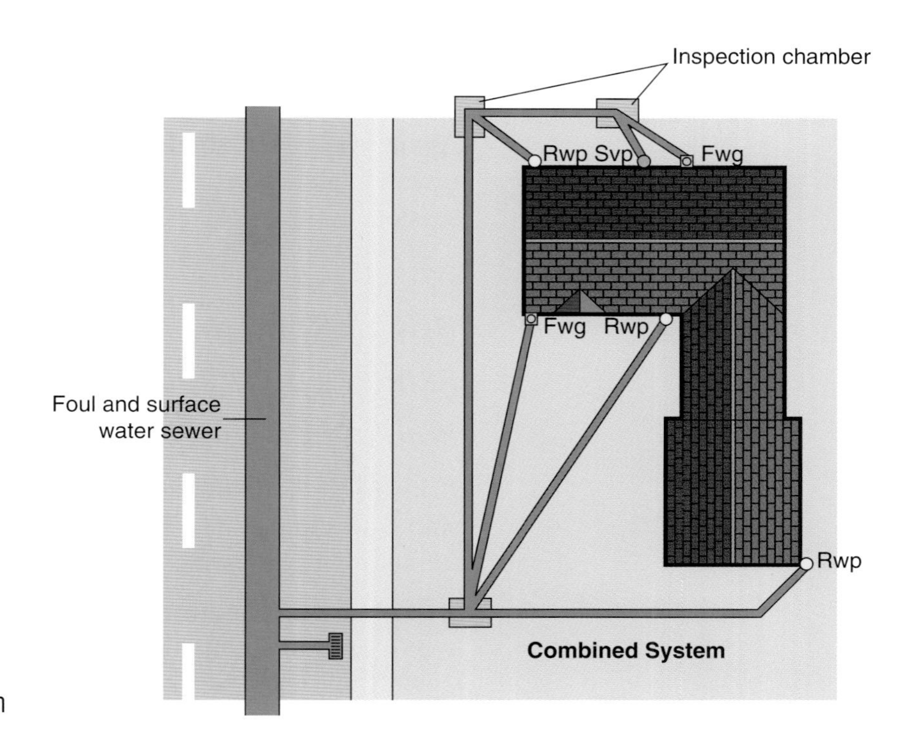

Figure 7.16 Combined System

Test yourself 7.2

1. Which are the three main types of fittings used to join soil and waste installations?
2. What are the main types of plastic used for waste and soil pipe fittings?
3. What is the spacing requirement for clips on a horizontal run of 32 mm pipe?
4. Explain what is meant by the term 'adequate access' to AGD system pipework.
5. What are the three main types of underground drainage systems?

Sanitary appliances

Introduction

Sanitary appliances play a vital role in all our lives. They mean we can keep ourselves and our eating utensils clean, and we can use the toilet in private and in comfort.

In this section, we will take a quick look at the range of appliances on the market; there is quite a lot, so we will only show a few examples.

The Level 2 Technical Certificate is mostly about domestic installations, but here we will extend coverage to urinals and flushing cisterns, as you may come across when working in public toilets or larger buildings.

Again, we place some responsibility on you to do your own research, and obtain as many catalogues as you can to support this session.

British Standards and regulations for sanitary appliances

British Standards

British Standard 8000 is the Code of Practice covering workmanship on building sites. Part 13 covers above ground drainage and sanitary appliances including:

- Material handling
- Site storage of components
- The preparation of work, materials and components
- The installation of sanitary appliances
- Inspection and testing.

Water Regulations

The Water Regulations do not refer specifically to the actual sanitary appliances, but they deal with how they perform in terms of the water fitments used, undue consumption, water wastage and protection against backflow. Here are a few examples:

WC cisterns and urinals are limited to the capacity of water for flushing, as are automatic flushing cisterns. (New WC cistern capacity is now 6 l.)

Backflow prevention has a bearing on sanitary appliances, which are categorised by the Water Regulations in terms of risk. This covers WC suites, wash basins, bidets, baths, shower trays and sinks. Backflow prevention is achieved either through mechanical (use of valves) or non-mechanical methods (use of air gaps between outlet and appliance water levels) to prevent the possibility of water being drawn back into the supply pipework and into the main. This could occur if an open tap outlet was allowed to become submerged in a sink or bath, the main outside stop valve was turned off and water drained from the system. As the water is drained down from the system, water from the sink or bath would be drawn into the pipe work due to siphonic action.

Types of appliances and materials

Hot tips

It is not possible to show examples of every variation of each appliance, so we strongly recommend that you obtain manufacturers' catalogues. The images we have shown here have been reproduced with the kind permission of Armitage Shanks and Ideal Standard. Refer to their excellent publication entitled 'The blue book'; this can be accessed on its website www.thebluebook.co.uk.

They can also be contacted on:
Armitage Shanks
Armitage
Rugely
Staffordshire
WS15 4BT
Tel: 01543 490 253

Ideal-Standard
The Bathroom Works
National Avenue
Kingston upon Hull
HU5 4HS
01482 346 461

Activity 7.6

What items do you think are covered by the term 'sanitary appliances'? Jot down your thoughts for your portfolio, and then check your answer against the suggested one at the end of this book.

Materials used in sanitary appliances

Materials used in sanitary appliances can be the following:

- Vitreous china
- Fireclay
- Cast and formed acrylic
- Enamelled steel
- Stainless steel
- Enamelled cast iron.

Vitreous china

Vitreous china conforms to BS 3402: 1069 (high-grade ceramic ware used for sanitary appliances). It is made from a mixture of white burning clays and finely ground materials, which after firing at high temperature, and when unglazed, does not have a mean water absorption greater than 0.5% of the dry weight.

This 99.5% vitreosity means that, even if unglazed, it cannot be contaminated by bacteria and will remain hygienic. However, it is still coated on all exposed surfaces with an impervious non-crazing vitreous glaze giving either a white or coloured finish. It is hollow cast, which means careful handling when transporting the appliances and when installing. This material is ideal for use with sanitary appliances, such as wash basins, WC suites, bidets and urinal bowls.

Fireclay

Fireclay is solid cast as opposed to hollow cast vitreous china. This makes it ideal when very heavy duty ware is required. The clay body is of sufficient strength that it can be used to manufacture very large sanitary pieces as well as offering heavy duty WCs, basins and shower trays. It is also used to make modern sinks, like Belfast sinks, which are now very popular in domestic kitchens.

Cast and formed acrylic

Cast acrylic is used mostly for the production of baths. Cast acrylic is a composite of resins and minerals, which is cast onto the reverse of a vacuum acrylic shell producing a 'factory made' product usually with a minimum wall thickness of 8 mm. Cast acrylic is very rigid, in fact it is the most rigid of all synthetic bath materials including cast polymer baths.

Formed acrylic baths are manufactured from acrylic sheet. The sheet is vacuum formed to produce the required shape and then fully reinforced. The advantage of acrylic sheet is its low thermal conductivity, light weight, and high gloss which is easy to maintain.

Other synthetic materials

Materials using a blend of quartz silica and acrylic resins, or quartz rock particles and acrylic resins are moulded to produce sinks for modern kitchen applications.

Injection moulding is used to produce WC seats and some cisterns are manufactured using high-impact plastic.

Enamelled steel

Sheet steel is pressed into the desired shape and then coated with layers of vitreous enamel which is then fired at a high temperature. The enamel and steel bond during firing to give the enamel a glasslike finish which is both practical and durable. Enamelled steel is used for the production of baths and sinks.

Stainless steel

Stainless steel is used extensively in the production of sanitary ware, ranging from sinks to special applications such as WC bowls for use in public toilets.

Sinks

Kitchen sinks

Figure 7.17 shows the photograph of a stainless steel sink, this particular model is a right-hand drainer, with a single hole for use

Figure 7.17 Typical sink design and waste fitting

with a monoblock tap. They can be single drainer left or right hand, double drainer, single and basket (suitable for connection to small waste disposal unit) and double sink with single drainer. They are designed to take a 1½″ waste pipe. This can be slotted for use with a sink overflow (like the one shown), or plain. Pillar and mixer taps can be used, in which case the sink top would have two holes.

Sinks are also produced in stainless steel, plastic-coated pressed steel, fireclay, and plastic in the form of acetyl.

Belfast sinks

Figure 7.18 Belfast sink

Originally, Belfast sinks, as shown in Figure 7.18 were used in older properties, as well as in applications, such as schools and residential accommodation. The hot and cold water was usually supplied via a pair of bib taps, and were ideal for filling and emptying mop buckets, and washing out or cleaning utensils.

Belfast sinks today, however, have also found a use in the domestic market in designer kitchens, where other tap options, such as pillar and monoblock taps can be used. They are designed to take a 1½″ slotted waste, which takes away any water that finds its way through the overflow.

London sinks are similar but without the overflow. Both those types of sink are made of heavy duty fireclay which is manufactured in a similar way to vitreous china. They are available in white only and are designed to take a 1½″ threaded waste.

Next is a Gloucester sink, designed on the same principle as the Belfast and London. Figure 7.19a shows the modern application of this appliance. Figure 7.19b shows an under counter fireclay sink.

(a)

(b)

Figure 7.19 Gloucester sink (Reproduced with permission of Ideal Standard Armitage Shanks)

Water closets (WCs) and cisterns

WCs and cisterns are categorised as:

- Close coupled wash down or siphonic WC pans and cisterns
- Wall-mounted WC pans
- Back to the wall WC pans
- Low-level WC cisterns
- High-level WC cisterns.

Most WC pans are manufactured using vitreous china, but WCs used in public places may also be manufactured in stainless steel. These are more resistant to vandalism than the vitreous china version. In cases where a pan forms a part of a WC suite, generally both are made of the same material, but in some situations, for example, low-level and high-level installations, the cistern may be constructed from plastic, and the pan from vitreous china.

Flush pipe

Figure 7.20 Wash down pan

WC pan construction

There are a number of designs for the pan, but there are only two main types:

- Wash down WC pan (see Figure 7.20)

  The wash down pan uses the force of the water from the cistern to clear the bowl.
- Siphonic pan (see Figure 7.21)

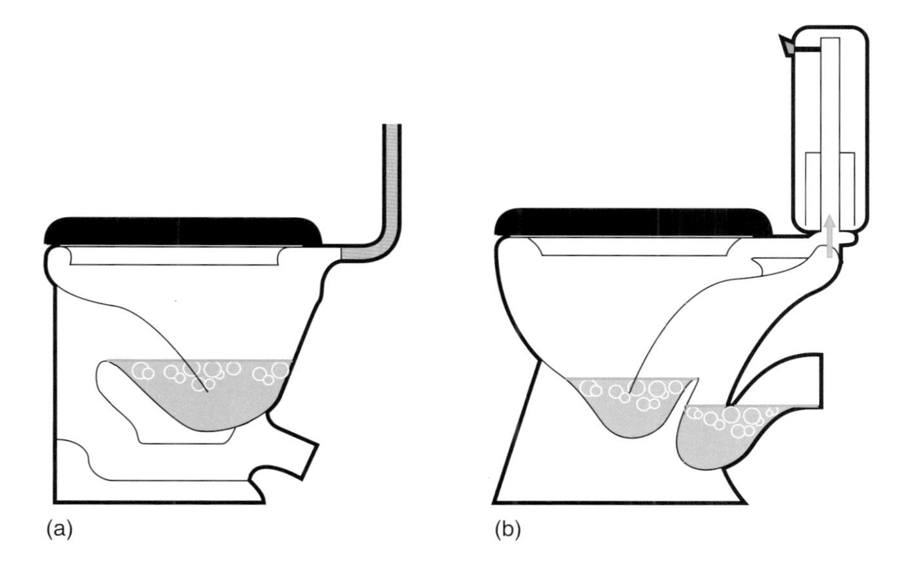

Figure 7.21 Siphonic pan (a) Single trap, (b) Double trap closed coupled

The principle of the siphonic pan is to create a negative pressure below the trap seal. With the single trap pan, this is done by restricting the flow from the cistern and is achieved by the design of the pan.

The double trap close coupled pan uses a pressure reducing device between the cistern and the pan. As the water is released into the second trap, it has the effect of drawing air from the void between the two traps, and siphons the contents from the bowl.

WC cisterns

Water Regulations are covered in the Level 3 Technical Certificate, but you need to be aware of the relatively recent changes to the regulations and how it affects WC cisterns.

Prior to 1993, the capacity of a WC flushing cistern was 9 l. The Water Regulations brought this down to 7.5 l, and from January 2001 reduced it further to 6 l.

In addition, Water Regulations also permit the use of dual flush cisterns; these deliver 6 l for a full flush, and 4 l for a lesser flush. The 6/4 dual flush cistern is now specified more often than the single flush, and on average, will use about 4½ l of water, half that of cisterns before 1993! The Water Regulations allow existing 9 l cisterns to be replaced like for like where the existing WC is not being replaced.

Siphons and flushing valves in cisterns

With the siphon type, when the lever is pressed, the water in the bell of the siphon is lifted by a disc (or diaphragm) up and over into the leg of the siphon towards the WC pan (see Figure 7.22(a)). This creates the siphonic effect, which continues until the water level in the cistern has dropped to a level which allows air to enter the bell.

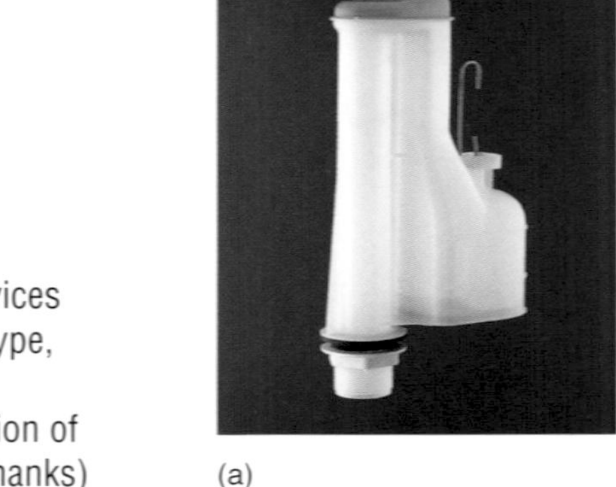

(a)

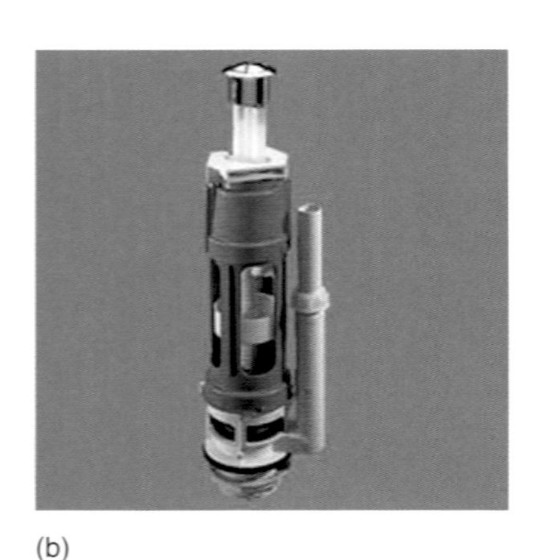

(b)

Figure 7.22 Flushing devices (a) Single flush – Siphon-type, (b) Dual flush – valve-type (Reproduced with permission of Ideal Standard Armitage Shanks)

The dual flush valve is operated by pressing either the full flush or lesser flush buttons, which have to be clearly marked on the chrome button (see Figure 7.22(b)). This operates a valve which releases the water into the WC pan.

In addition to the siphon you will also see these fittings inside the WC cistern (see Figures 7.23(a) and 7.23(b)):

(a)

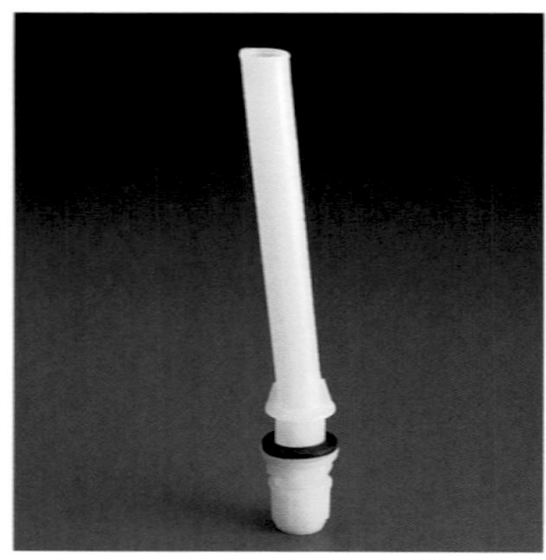

(b)

Figure 7.23 Cistern fittings (a) WC float valve side inlet, (b) WC overflow pipe-bottom outlet (Reproduced with permission of Ideal Standard Armitage Shanks)

Maintenance and replacements

- Cisterns of 9 l and 7.5 l are still available. Water Regulations permit these to be installed as replacements for existing cisterns. This is because the WC pan was designed to work on those capacities.

What is on the market?

On larger housing contracts, or in public buildings, the choice of WC suite may have already been made by the client's architect as part of the contract specification document. However, on one-off jobs the customer or client could ask your advice.

The choice of WC suite depends on a number of factors:

- Location: if it is a public toilet, factors such as durability and ease of cleaning would be a consideration
- Aesthetics (how good it looks) will also play a part, e.g. Victorian versions of the high-level cistern are now very popular
- Cost: how much the customer can afford to pay.

Close coupled WC suite

Hot tips

Here is a list of other types of WC pans and cisterns; use manufacturer's information to find out more about them.

- Wall-mounted WC pans
- Back to the wall WC pans
- Low-level WC cisterns
- High-level WC cisterns.

This arrangement, shown in Figure 7.24, could feature either a washdown or siphonic pan. If a toilet is described as close coupled, then the water cistern (sometimes called the tank or closet) sits directly on top of the toilet pan and is probably the most commonly found type of installation in the UK. The pan is fixed to the floor and the cistern is fixed to the wall.

Some cisterns are reversible, i.e. handles, overflow, outlet and water inlets can be either left- or right-handed. In some cases the cistern is plastic rather than vitreous china. Kits are usually supplied for WC pan soil outlets, so it can be either a 'P' trap or an 'S' trap.

Wash basins

These are categorised into two main types:

- Fixed to the wall
- Counter top

Figure 7.24 Close coupled WC suite (Reproduced with permission of Ideal Standard Armitage Shanks)

Fixed to the wall (refer to Figures 7.25(a)–(c))

These are supported by a pedestal or secured by using purpose-made brackets. Larger type washbasins, fixed using brackets, are mostly found on non-domestic WC installations. The pedestal type is probably the most popular choice for use in homes.

Appearance is the main factor as all the pipework and fittings can be seen with the bracket-mounted basin. The height of the basin is set by the pedestal, but for the bracket-mounted basin they are usually set at 800 mm. The corner basin is also available as wall mounted.

Counter top basins

Are available as:

- Counter top
- Under the counter top
- Semi-counter top.

(a)

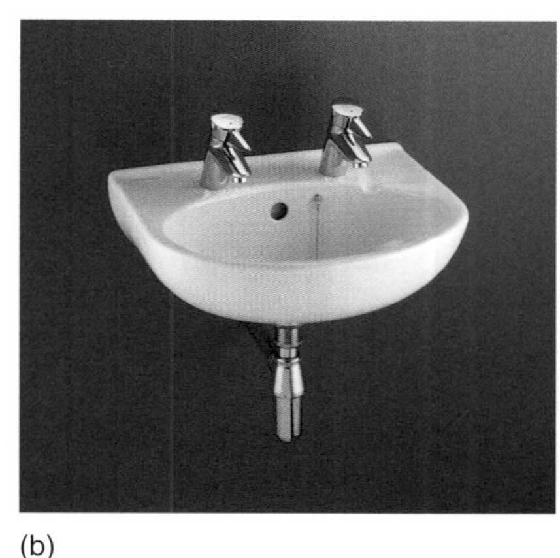
(b)

(c)

Figure 7.25 Wash basins (a) Pedestal, (b) Wall-mounted, (c) Corner basin (Reproduced with permission of Ideal Standard Armitage Shanks)

Hot tips

As with the WCs use manufacturers' information to find out more about the counter top basins.

Waste fittings

Waste fittings for wash basins come in two main types, as shown in Figures 7.26(a) and 7.26(b).

Note how the flange of the waste fitting is tapered so that it fits 'snugly' into the waste hole of the basin.

(a)

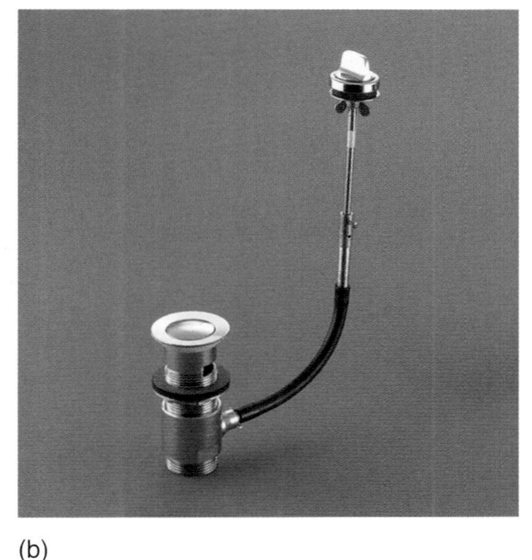
(b)

Figure 7.26 Waste fittings (a) Standard slotted-type, (b) Pop-up waste (Reproduced with permission of Ideal Standard Armitage Shanks)

Baths

Baths can be supplied in cast or formed acrylic sheet, heavy gauged enamelled steel, and vitreous enamelled cast iron. Cast iron baths for domestic use are not as popular these days, but are still available; the Victorian free-standing designs are sometimes specified to give the traditional 'period look'. You will also come across them on maintenance jobs. If you have to do any work on them, do take great care, as cast iron is very brittle. The vitreous enamel provides a smooth hard wearing surface that is corrosion-resistant and easy to clean.

Let us take a look at a couple of examples as shown in Figures 7.27(a) and 7.27(b)

The size of baths varies, depending on the design. The standard one shown measures 695 mm wide by 1695 mm long. Baths should be set at between 480 mm and 540 mm high.

Hot tips

Whirlpool bath and spa bath Again use manufacturer's catalogues to find out more about these appliances. Here are a few examples of bath waste arrangements (see Figures 7.28(a)–(c)).

Shower trays and enclosures

We looked at instantaneous shower and mixer shower appliances in the cold and hot water units so here we will concentrate on the shower trays. Again, there are a number of designs. Shower trays are available in reinforced cast acrylic sheet or fireclay for

(a)

(b)

Figure 7.27 Baths (a) Standard pattern bath, (b) Corner bath (Reproduced with permission of Ideal Standard Armitage Shanks)

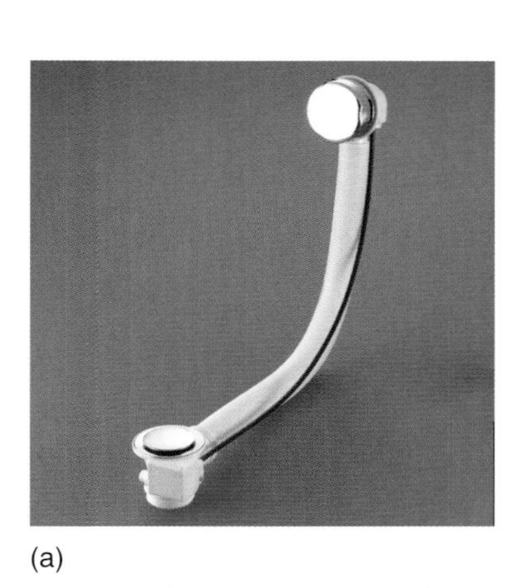
(a)

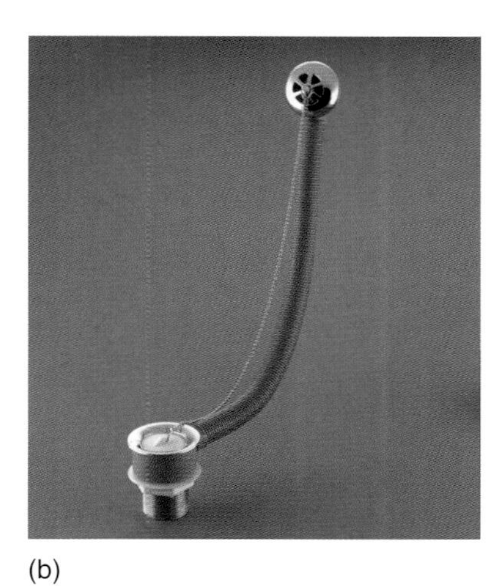
(b)

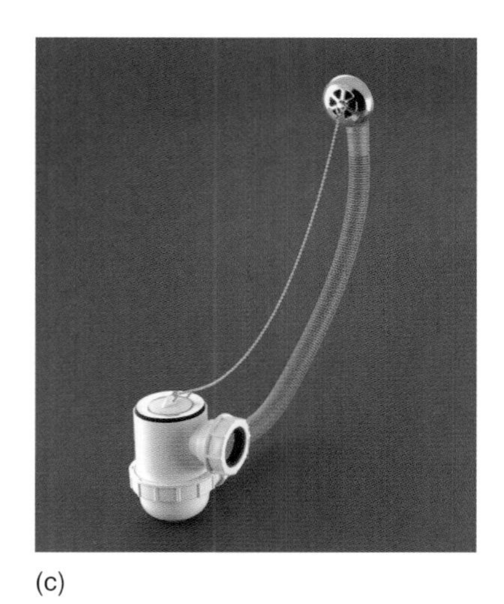
(c)

Figure 7.28 Bath waste fittings (a) Pop-up waste, (b) Combined overflow and (c) Bath waste combined with waste trap (Reproduced with permission of Ideal Standard Armitage Shanks)

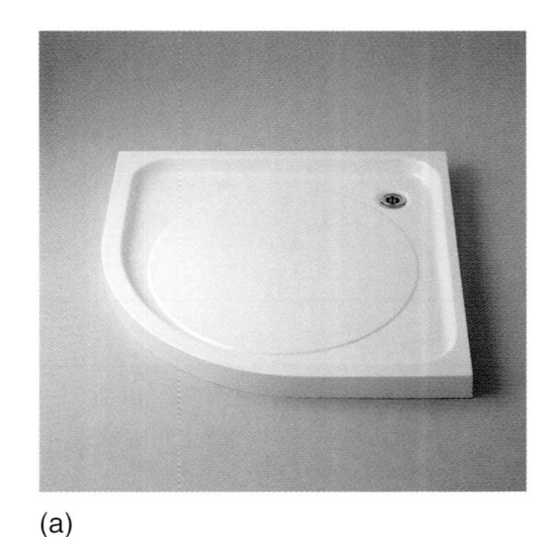
(a)

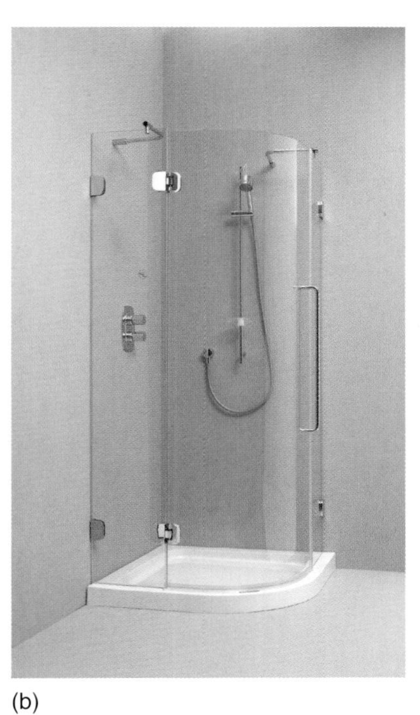
(b)

Figure 7.29 Shower fitments (a) Quadrant pattern in fireclay, (b) Shower enclosure for the quadrant tray (Reproduced with permission of Ideal Standard Armitage Shanks)

heavy duty applications. The outlet is designed to take a 1½″ threaded waste.

Here are a couple of examples, as shown in Figures 7.29(a) and 7.29(b).

Figure 7.30 Bidet (Reproduced with permission of Ideal Standard Armitage Shanks)

Try this

What type of bidet would you say is shown in the illustration in Figure 7.30?

Bidets

There are two types of bidet, one referred to as over the rim and the other termed ascending spray. They are manufactured from vitreous china, and can be floor standing (see Figure 7.30), back to the wall or wall mounted. The tap and waste arrangements are similar to that of wash basins.

The type of bidet shown in the illustration is floor standing.

The ascending spray bidet has to be piped up correctly to avoid risk of contamination to the supply. This is to meet the requirements of the Water Regulations and to prevent the possibility of contamination of the supply. The bidet can be supplied for use with pillar taps, *monoblock* fittings and pop-up waste.

Urinals

It is unlikely that you will encounter a urinal in domestic premises. But you may work in public buildings such as a pub or restaurant where urinals are installed. They are mostly available in vitreous china for urinal or stainless steel (see Figures 7.31(a) and (b)) in restaurant for bowls fixed to the wall, and stainless steel, plastic or fireclay for slab urinals.

Flushing pipes	Waste fitting
This is the pipe-work that feeds the urinal, and in the plumbing industry it is referred to as 'sparge pipes', as shown in Figure 7.32(a)	This is a typical waste fitting for a urinal, as people throw all sorts of things into a urinal, the grill should help to prevent blockages in the trap or waste pipe system, as shown in Figure 7.32(b)

The water supply to a urinal is controlled by an automatic water flow regulator, which can be set to deliver the recommended

(a)

(b)

Figure 7.31 Urinals (a) Urinal bowl installation in vitreous china, (b) Slab urinal in stainless steel (Reproduced with permission of Ideal Standard Armitage Shanks)

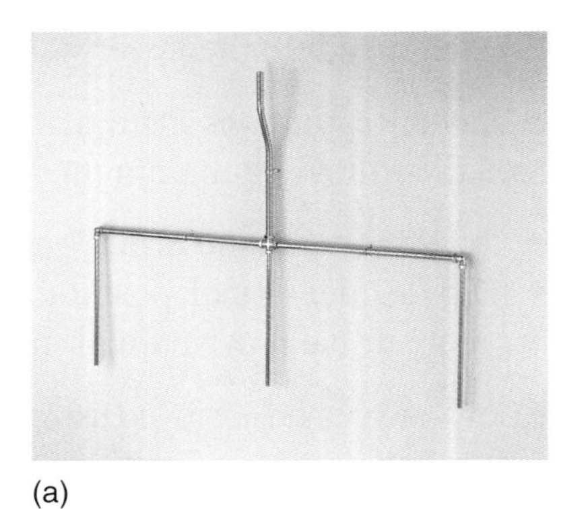

(a)

(b)

Figure 7.32 Pipes and fittings (Reproduced with permission of Ideal Standard Armitage Shanks)

flushing volume within a set time limit. The water supply from the urinal is controlled by an automatic flushing cistern. Both these devices are designed to prevent the undue waste of water.

Figure 7.33(a) is a photograph of a vitreous china cistern, but cisterns are also available in plastic. Figure 7.33(b) in the centre shows a typical auto flushing siphon, and the Figure 7.33(c) on the right is a hydraulic flushing valve.

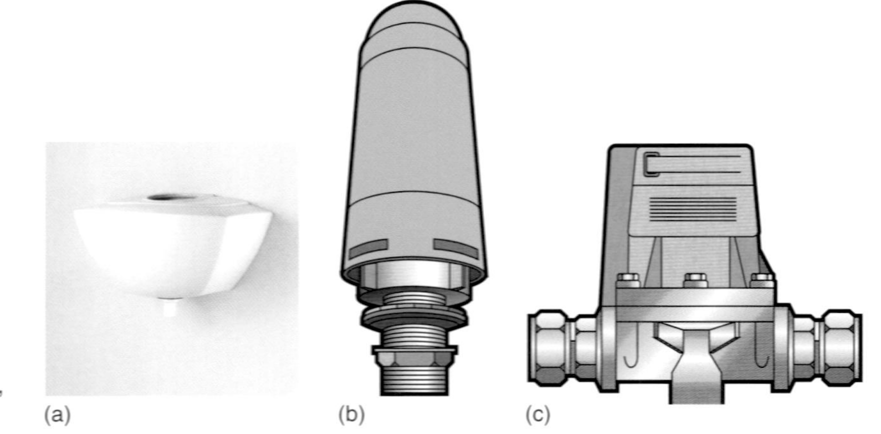

Figure 7.33 Automatic cistern and fittings (a) Cistern (Reproduced with permission of Ideal Standard Armitage Shanks), (b) Siphon, (c) Flushing valve

Design note

The auto siphon's operation is relatively straightforward:

- As the cistern fills and the water rises, the air inside the dome of the auto siphon is compressed
- The increased pressure forces water out of a 'U' tube which reduces the pressure in the dome
- The reduction in pressure causes siphonic action to take place, flushing the cistern
- When the cistern has emptied, the water in the upper well is siphoned into the lower well.

The Water Regulations state that auto flushing cisterns must not exceed the following maximum volumes:

- 10 l/h for a single urinal bowl, or stall
- 7.5 l/h, per urinal position, for a cistern servicing two or more urinal bowls, urinal stalls or per 700 mm slab positions.

The flow rate can be achieved using urinal flush control valves which allow a small amount of water to pass into the system. Timed flow control valves can also be used. These have the additional advantage of being switched off when the building is not in use, i.e. evenings, weekends, factory shutdowns, school holidays, etc. They can be set for 'one-off' hygiene flushes may be once every 24 hours.

Activity 7.7

What do you think the maximum flow rate will be for the three urinal bowl system?

Waterless urinals

Currently, there are two main technologies available, one using fluid treatment and the other using disposable cartridges, as shown in Figures 7.34(a) and 7.34(b).

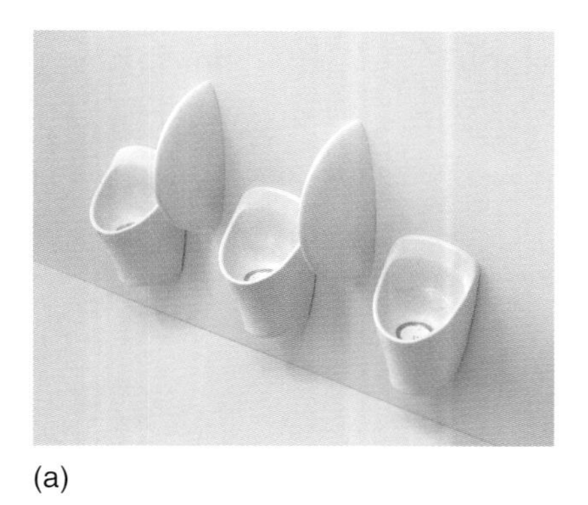
(a)

(b)

Figure 7.34 Waterless urinals (a) Typical urinal bowl installation, (b) Replaceable cartridges (Reproduced with permission of Ideal Standard Armitage Shanks)

The waterless urinal with a siphonic trap which contains a barrier fluid that has to be topped up every 7 to 14 days depending on the intensity of use. Using a wide outlet in the urinal bowl, this contains a pad impregnated with a deodorising agent which needs changing regularly depending on use. In some circumstances, this system can also be retrofitted into an existing urinal bowl.

The system in the photographs is based on the principle of using a deodorising agent in the cartridge. Cartridges last for approximately 7,000 uses. Installers must ensure that urinals are connected to waste pipe systems that are adequately vented. It is estimated that this system would save 236,000 l/year per urinal bowl based on conventional urinal flushing of 4.5 l, 6 times per hour, 24 hours per day, 365 days per year. Installers must ensure that urinals are connected to waste pipe systems that are adequately vented.

Try this

We said on Page 283 that there is a vast range of sanitary appliances, which we can't reproduce here due to space. We strongly recommend that you obtain catalogues or access information from a number of manufacturers, or visit showrooms. Look for new and designer products that are on the market noting their installation details. You might find the Geberit site interesting at www.geberit.co.uk.

Test yourself 7.3

1. State three of the main areas covered by BS 8000 Part 13.
2. List three types of material used in the manufacture of domestic sanitary appliances.
3. State two types of WC/cistern combinations.
4. What is the Water Regulation requirement for the capacity of a WC cistern?
5. What is the Water Regulation requirement for flushing rate for a single urinal bowl or stall?
6. Briefly explain the principle of the operation of a single trap siphonic WC pan.

Installing sanitary appliances

Preparation before you start

We have mentioned it before, but BS 8000 Part 13 gives some excellent guidance on workmanship in relation to above ground drainage and sanitary appliances covering the complete process.

Preparing properly before you start the work is important for any plumbing installation. Preparing to install above ground drainage pipework and sanitary appliances also requires careful preparation and planning before you start in order to ensure there are no hold ups later because:

- Fittings are missing
- Pipe chases/openings in walls have not been prepared (if that is someone else's job)
- Drain connections are not in place
- Sanitary ware is damaged or has not arrived on time.

Storage of materials on site

- If you order materials, get confirmation of delivery date and take delivery in advance of starting the job
- On delivery have your original order with you and cross-check against the delivery
- Handle and transport materials in a safe manner
- When storing the material make sure the store is secure and materials are positioned safely and are not at risk from damage.

Activity 7.8

How would you store vulnerable items such as baths, wash basins and WCs? Jot down your thoughts for your portfolio and check your answer against the one at the end of this book.

Carrying out checks before installing the pipework and appliances

Check list

- The work area must be safe and free from debris
- Drawings or installation information are available, and where appropriate make sure you leave any customer instructions
- If you encounter any problems, for example, the customer wants to change the original specification for the job, notify your boss for advice
- Where appropriate, make sure any preparatory work required by other trades, e.g. cutting holes in brickwork, has been done
- Make provision to protect appliances whilst work is in progress.

Installing all the appliances

So far we have considered sanitary appliances in isolation. In practice, you are often likely to install a complete bathroom suite. Assuming a customer is going for all the appliances it could include:

- Bath
- Pedestal wash basin
- Close coupled WC – wash down pattern
- Shower tray and enclosure
- Bidet.

Hot tips

It is helpful if you have your own manufacturers' catalogues because there will be occasions where a customer asks for your advice so you will be able to show them some actual examples.

'Dressing the suite'

This is a term often used by plumbers, which describes installing the taps, wastes, and in the case of the bath, the cradle frame or feet. It will also include installing float valves, overflows, siphons or flushing valves and the handle assembly to the WC cisterns.

Tap selection

The type of taps may have already been specified or purchased by a customer.

Fitting the taps

This is relatively straightforward and should be done in accordance with manufacturers' instructions. Make sure you do not over-tighten plastic back nuts in case they split, and do not place the jaws of grips directly on the body of the tap.

Waste fittings

These could be either 'slotted' or 'pop-up' and again should be fitted in accordance with manufacturer's instructions, and using washers or fixing kits supplied with the waste fitting.

Baths

Baths are usually supplied with a fixing frame or cradle which has to be assembled to the bath on site. The frame is supplied with adjustable feet to enable the bath to be adjusted to the fixing height and to level. Clips are supplied for fixing the panel to the frame, and grab handles should be fitted to the bath prior to installation.

WC suite

It will be necessary to fit the float valve, siphon, overflow pipe and flushing handle gear to the cistern prior to fixing. In the case of a close coupled suite a bracket is supplied which enables the cistern to be bolted to the pan.

Key point

There are no set rules for the sequence used to install the appliances. Plumbers will work out the best sequence based on the position of the appliances, and the size and layout of the bathroom. They will base their decision on what they think will be the easiest and quickest way for a given situation.

The installation process

On jobs where you are replacing a suite, and the dwelling only has one WC, the first appliance you would install would be the WC suite, so any inconvenience to the customer would be kept to a minimum.

It is not very often you get a bath and shower tray in the same room, unless it is a really large property. Most have shower units installed over the bath with either a shower curtain, screen or part enclosure. A shower tray may be used in an en-suite room, particularly where space is tight, or the occupier has difficulty in using a bath due to age or illness.

Most private customers will have a good idea of how they want their bathroom to look. The layout of the bathroom can also be planned using a scale drawing of the bathroom and manufacturer's dimensions of the appliances to determine the various positions.

There are no hard and fast rules relating to which appliances should be installed first. BS 8000 offers some general guidance in terms of fixing the appliances advising that fixing devices such as brackets and screws that are exposed to wet conditions should be of copper alloy or stainless steel. There is a range of screws and bolts that are manufactured in alloy that are ideal for these fixings.

It also states that appliances are fixed at the required position and height and are fixed plumb and level so that surfaces designed with falls will drain as intended. No appliance should be fixed back to the wall relying on the fixings provided for the installation pipework.

On new work, particularly in small bathrooms, the bath is usually installed first as it is easier to manoeuvre into position without the other appliances in the way. But on a bathroom suite replacement, particularly where there is only one WC suite in the dwelling, a plumber might install the WC suite first so that the customer has access to the toilet if needed.

Hot tips

The Bathroom Manufacturers' Association and its membership via its web site offers guidance, information and practical advice to aid in the process of that 'Special Care Bathroom'. Bathroom manufacturers have a plethora of products for the 'Special Needs' market. It also produces a fact sheet on the topic. The Bathroom Manufacturers' Association offers excellent advice on all aspects of 'bathroom industry', including a number of excellent downloadable fact sheets. They can be contacted on:
www.bathroom-association.org
or
Federation House
Station Road
Stoke on Trent
ST4 2RT
Tel: 01782 747123

Specialist sanitary appliances

Bathing is an integral part of life in ensuring cleanliness and health are maintained. Equally important is retaining a person's self-respect and independence.

Ten percent of today's population suffer from varying degrees of disabilities. In May 2004, the government, via Building Regulations, introduced a revision to Part M which saw a raft of changes not only for disabled people but also including:

- wheelchair users
- ambulant disabled

- people with babies and small children
- people with learning difficulties, visual or hearing impairments
- people who lack tactile sensitivity.

Although these changes to the Building Regulations mostly relate to commercial premises, major refurbishment and new house-build, many of the objectives can be related for inclusive design in the domestic market.

Part M of the Building Regulations takes into consideration the latest requirements of the Disability Discrimination Act 1995 and addresses the main considerations that should be included in the provision of sanitary facilities. Attention to hot taps, WC cubicle doors, opening mechanisms, door widths and position of light switches should all be considered to ensure that bathing facilities meet the needs of all people.

WC Macerator unit

Figure 7.35 shows a WC macerator. It is the box behind the WC pan connected to the trap outlet.

It collects solid waste from the WC and reduces to a 'liquid' state, so that it can be pumped via a 19 mm internal bore pipe into

Figure 7.35 WC Macerator unit

the discharge stack. There are a few things to remember about macerators.

- The units generally accept waste water by gravity. They do not suck in water. All pipework into the units must have a positive gravity fall (1:40).
- To avoid dipping and build up of residual water and subsequent blockage, all discharge pipework should be supported in accordance with manufacturer's specifications and installed with a minimum 1:200 fall to the soilstack
- A drain-off cock should be fitted at the base of a vertical rise
- The macerator can only be fitted where there is another conventional WC in the dwelling
- The macerator has to be vented externally to allow gravity filling and emptying
- The electrical supply to the unit must be via a fused spur connection
- The unit is capable of discharging its contents 4 m vertically and 50 m horizontally
- The unit should be easily accessible and removable in the event of maintenance being required
- All vertical lifts should rise as directly above the unit as possible. Any initial horizontal run from the unit prior to a vertical lift should not exceed 30 cm.

  Only one lift is permitted, at the start of the run, with a 1:200 fall from the high point.
- Where the point of discharge into the soil stack is significantly lower than the base of the unit, an anti-siphon valve will need to be fitted at the highest point in the pipe-run in order to avoid siphonage of the water seal in the unit

  Alternatively, larger bore pipework should be used for the vertical drop to 'break' the siphon.
- Bends should be 'smooth' and not 'elbows'. When utilising uPVC plastic 22 mm pipework it is recommended that two bends are used together to form a right angle. All pipework susceptible to freezing should be adequately lagged.
- Before attempting any maintenance or servicing, the units must be disconnected from the electricity supply. The electrical connection is via an unswitched fixed wiring connector with a 5 amp fuse.

Hot tips

The main advantage of the unit is that it can be sited almost anywhere in a building without the need to connect the WC pan directly to the soil stack.

If you want to know more about WC macerators more information can be obtained from Saniflo at:
Saniflo Ltd, Howard House
The Runway
South Ruislip
HA4 6SE
Tel: 020 8842 0033
Website: www.saniflo.co.uk

- All pipework should be either copper or uPVC conforming to BS 7291
- The use of flexible or push-fit pipework is not permitted
- For long horizontal runs where 22 mm pipework is used this should be increased to 32 mm after approximately 12 m to eliminate the risk of trap siphonage.

Food waste disposal unit

These are usually installed under kitchen sinks to dispose of refuse, such as vegetable peelings, tea bags, leftover food, etc. They require a larger waste outlet than normal, and manufacturers produce sink tops to suit this (see Figure 7.36). The unit should:

- Be connected to the electrical supply via a fused spur outlet
- Be connected to a waste trap
- Be supplied with a special tool to free the grinding blades should they become blocked (do not lose your fingers!)
- Have some form of cut out device that will turn the unit off should it jam.

Figure 7.36 Example of food waste disposal unit

Test yourself 7.4

1. When storing WC pans, what is the recommended maximum number pans you should stack?
   a. 4 pans ☐
   b. 2 pans ☐
   c. 6 pans ☐
   d. not stacked at all ☐
2. What should you do before accepting a delivery of sanitary appliances?
3. When a plumber is 'dressing the suite', what does that mean?
4. When installing a macerator WC pan, what is the recommended fall required for the horizontal pipework?

Case study 7.1

You have been asked by a customer to replace a cracked wash basin and pedestal in a bathroom. You have managed to find a replacement that matches. Describe the procedure from start to finish, including:

- Remove the old wash basin
- Dress the new one
- Mark out and fit the basin and pedestal
- Connect to the hot, cold and waste pipework
- Test the appliance for leaks
- How are you to deal with the customer and how did you make sure the new wash basin was protected during the installation?

Do your report in a concise 'bullet point' fashion. Use the information provided so far in this unit, together with manufacturer's information and any personal experience to cross-check what you have written, we have also covered a few pointers at the end of this unit. Include this piece of work in your portfolio.

Soundness testing

The final part of an above ground discharge system (AGDS) installation is to test it. This includes testing for leaks and making sure that the system discharges correctly without trap seal loss.

Next time you take a bath, think about the amount of water it holds and you will appreciate that sanitary appliances like the bath have to discharge large volumes of waste water. Now think about the mess that is made if you spill a drink and you will realise that it is important that AGDSs are water tight, otherwise we could have serious water damage problems. The test of an AGDS should be carried out in accordance with BS EN 12056. You also need to know what to do to make sure people are aware that testing is going on, and when it is OK for them to use the appliances.

Soundness testing process

Look at Figure 7.37 and the pictures of the test equipment and then read the testing checklist.

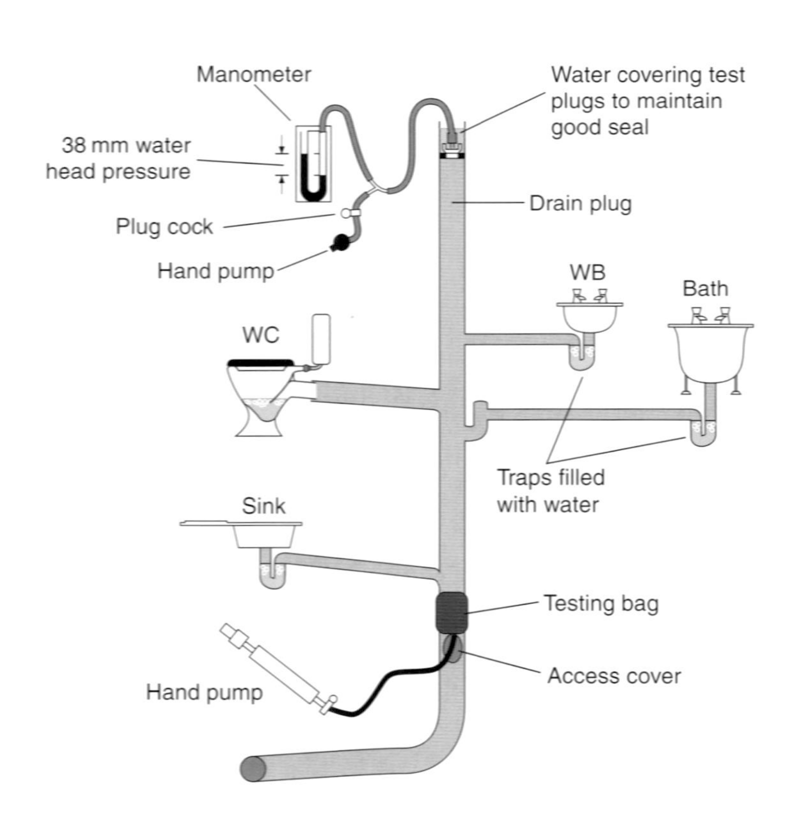

Figure 7.37 Stack system testing procedure

Testing check list

- First, seal the system using hollow drain plugs. Fix a test nipple in the one used for the test. The bottom test plug can be inserted through the access cover at the base of the stack.

  If this proves awkward, a testing bag can be inserted and inflated in position, as shown in Figure 7.37
- Fill the traps with water, and cover the test plugs to make sure they are fully airtight
- The rubber hose, bellows and hand pump are connected to one end. Air is pumped into the system to give a water head of 38 mm. Once this is reached, the plug cock is turned off and the test continued for three minutes
- Where a pressure drop is found all the joints should be tested using leak detection fluid
- Smoke tests are also permissible, but are not advisable, because of the possibility that it may affect the plastic pipework and rubber seals.

Charging procedures

- Once the testing equipment has been removed, all the appliances should be filled to their overflow levels and water released, the WC should be flushed at the same time
- Once you have done all that, check that the trap seals are not less than 25 mm.

  Remember, the depth of the seal is measured, as shown in Figure 7.38.

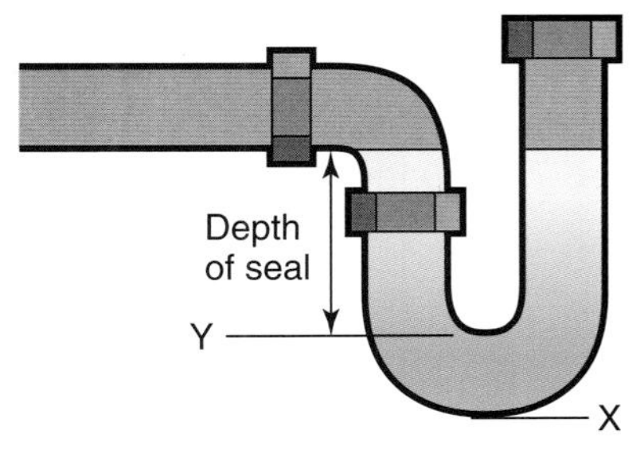

Figure 7.38 Measuring depth of seal

The depth of trap seal can be measured by inserting a thin wooden stick through the waste fitting and down into the bottom of the trap. When you withdraw the stick, the water should leave a mark on the stick. You then deduct the distance from the bottom of the trap X to the bottom line Y to give you the measurement. If you paint the stick matt black, it would give you an even better water mark on the stick.

That completes the soundness testing section, now try the following exercises.

Test yourself 7.5

1. What British Standard is relevant to soundness testing?
2. An AGDS soundness test should be carried out using:
   a. Water ☐
   b. Air ☐
   c. Smoke ☐
   d. Compressed nitrogen ☐
3. The water head required for the soundness test should be:
   a. 18 mm ☐
   b. 22 mm ☐
   c. 32 mm ☐
   d. 38 mm ☐
4. Upon completion of system charging, the trap seals should be not less than:
   a. 25 mm ☐
   b. 38 mm ☐
   c. 50 mm ☐
   d. 75 mm ☐
5. In an occupied dwelling, what should be a plumber's first action before carrying out soundness testing?

Maintenance and de-commissioning

Introduction

Maintenance and de-commissioning of systems is a common feature throughout the Level 2 Technical Certificate, and AGDS is no different, these systems do not have complex controls as, say, water heating systems, but if they are not looked after, eventually they will cause problems.

Older AGDS systems also have to be de-commissioned, and given some of the types of materials used, cast iron and asbestos, brings with it health and safety considerations.

Basic maintenance and cleaning

Traps can be a potential source of problems as they often accumulate hair, soap residue, tooth paste and other objects that are small enough to fall through the grid of the waste hole (as small children will often exploit) and create blockages. Integral overflows, such as the ones on Belfast sinks, are also prone to blockage. The overflow can be rodded with wire, and flushed.

Traps should be cleaned through the access points if fitted, if not the trap should be broken at its joints or removed completely.

Chemical cleaning agents are also available that can be used for both traps and overflows. If you use cleaning agents always follow the manufacturer's instructions, especially with regard to mixing more than one product, as corrosive or explosive mixtures can result.

Make sure that access covers to soil and vent pipes are checked to confirm that they work OK and a visual inspection should be made of waste traps and fittings for signs of leakage.

Blocked pipes and drains

Plumbers who carry out maintenance work will still receive calls to clear blocked sinks. Blockages to discharge pipework on sinks, washbasins and baths can often be cleared using a 'force cup', or 'sink plunger' as most plumbers call it.

Force cup or 'sink plunger'

The blockage is cleared by filling the appliance (say a kitchen sink) with water. Next, press down repeatedly on the handle of the force cup. This creates a positive pressure on the blockage, and a partial vacuum when it is withdrawn, which is usually enough to remove the blockage. Always check the trap for signs of leakage when you have finished. Figure 7.39 shows a force cup.

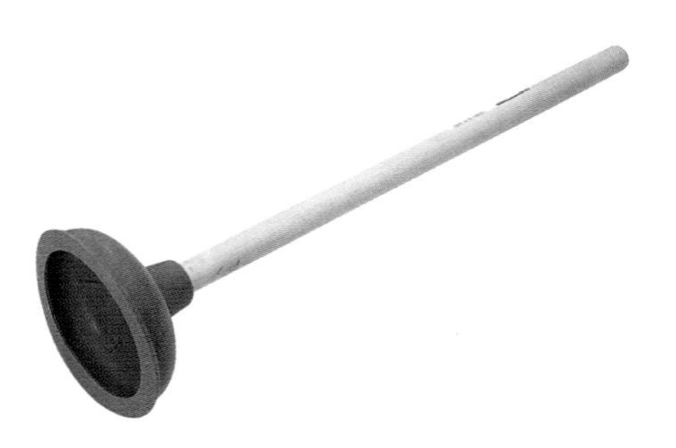

Figure 7.39 Force cup

Figure 7.40 WC plunger

A blocked WC or external gulley can be cleaned using the tool shown in Figure 7.40.

Mechanical equipment

Mechanical equipment is also available for cleaning blockages in pipework systems; a few examples are shown in Figures 7.41(a)–(d), ranging from a simple closet auger to a mechanical version shown bottom right.

Figure 7.41 Range of mechanical equipment for clearing blockages (a) Closet auger, (b) Hand spinner, (c) Hand spinner in action clearing a wash basin waste pipe, (d) Power spinner with autofeed (Reproduced with permission for (b)–(d) from Ridgid Tool)

All these cleaning tools work on the principle of pushing or pulling the blockage using the auger at the end of the pipe.

Hot tips

There is also a wide range of equipment for clearing blockages in below ground drainage, but it is unlikely that you will get involved in this type of work. If you want to find out more about this equipment you could contact Monument tools at: Monument Tools Ltd
Restmor Way
Hackbridge
Wallington
Surrey SM6 7AH
Tel: 020 8288 1100
Fax: 020 8288 1108
Website: www.monument-tools.com.

Other maintenance considerations

Leaking joints on cast iron stacks can be cleaned out, and remade using non-setting pipe jointing compounds.

You will be covering gutters in the next unit, but we will discuss briefly about their maintenance here. Leaking cast iron gutter joints can be repaired by removing the old joint, thoroughly cleaning the jointing surfaces of the gutter and fitting, applying jointing compound, and filling a new gutter bolt.

Gutters should be checked regularly and cleaned as required. They do collect silt residue off roof tile coverings, leaves and bird droppings. The silt can build up and becomes a garden for seeds carried by the wind. Plastic gutters can be stripped in lengths between joints, emptied into a bucket, and wiped clean. Cast iron gutters should be painted to avoid rusting. Word of caution: if you carry out maintenance work, be aware that in the past asbestos-based materials were used for gutters, fall pipes and soil pipe systems. You must take advice if you suspect that the material you could be working with is asbestos.

De-commissioning systems

The same comments about asbestos apply equally here.

De-commissioning an AGDS will normally mean stripping out old appliances and pipework to replace with new ones. When removing appliances, care should be taken not to damage them. Broken vitreous china has edges as sharp as broken glass, so it should be handled wearing strong gloves.

Removing a cast iron bath, particularly if it has to be carried downstairs, requires careful handling. Remember to carry out a risk assessment when handling the bath on things like: will it be a two-handed job? Will you need any mechanical lifting or carrying equipment? Is the access pathway clear and free from any potential risks?

Some plumbers break the bath into four pieces for easier removal. This should be done wearing full face protection, ear defenders and gloves. A club hammer is the best tool for doing this. On the other hand, a cast iron bath in a *reasonable condition* can fetch a high price at an architectural salvage yard, and they can be reconditioned quite readily, so it may be worth getting it out carefully, as it could be worth a good deal more than its scrap value.

Once the appliance has been removed, it should be stripped of any scrap metal which is taken for recycling. In the case of a damaged cast iron bath, this too has scrap value.

Pipework systems could be in lead, so you should take the usual precautions when handling lead. If you are not sure about handling lead you should refer back to the Health and Safety unit.

Where pipework penetrates a wall from the inside of a bathroom to outside, you will need to chop out the mortar between the pipe and the masonry. Take care in doing this, as it will mean less making good if you are installing new pipework.

Removing cast iron soil and vent pipework can be dangerous due to its weight, so it needs careful handling, and it is definitely a two-handed job. It is best to try and take it down in short sections by partially cutting it with an angle grinder and then tapping the pipework with a hammer which will cause it to shear, a rope should be tied to the sections and then the section lowered to the floor.

Make sure no one is in the area where you are working. Fixing lugs can be broken from the joint, and the nails prized out using a wrecking bar. Again, remember your PPE, including a hard hat!

Once the stack is removed, make sure the joint to the drain is covered to avoid anyone falling over it, or debris entering the drain.

Test yourself 7.6

1. Briefly explain the potential source of maintenance problems associated with traps.
2. What is the name of the tool below (Figure 7.42) and what is it used for?
3. What is this tool called (Figure 7.43)?
4. Briefly explain how you would remove a section of cast iron soil stack.

Figure 7.42

Figure 7.43

Now you have completed the maintenance and de-commissioning session, let us see how you are progressing.

Gutters and rain water systems

Introduction

Gutters are used to collect rain water that falls on any type of roof. It then flows down the rainwater pipe and into the drainage systems or soakaways (remember the three types of drainage systems mentioned in Unit 2!) Gutters and rain water pipes are used to prevent damage to gardens caused by water running off the roof and wearing away its surface, it would also wet and stain the external brick wall and in winter this could result in frost damage.

Gutter and rainwater systems

Gutters are classified by their section

There are three main types as shown in Figures 7.44(a)–7.44(c).

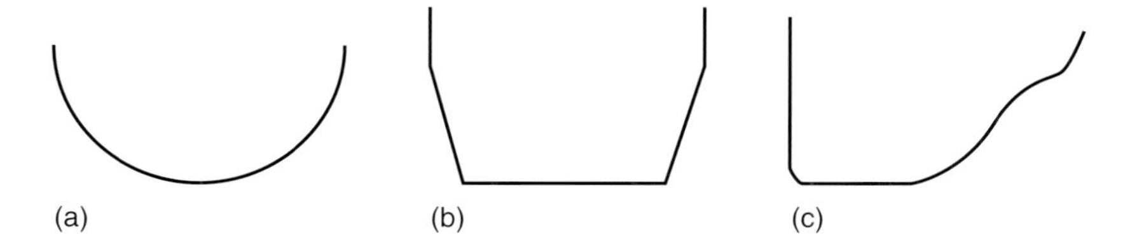

Figure 7.44 Gutter classification (a) Standard Half Round, (b) Square, (c) Ogee

The fall pipes are either square or round. Guttering is supplied in 2 and 4 m lengths, and can be obtained in a range of sizes from 75 to 150 mm; 112 mm is fairly common for most domestic dwellings and fall pipes in lengths of 2.5, 4 and 5.5 m. For most domestic jobs 65 mm square and 68 mm external diameters are used.

Range of fittings – Gutters

Try this

Use manufacturers' catalogues or websites, or look at the fittings in your college or centre to familiarise yourself with the following:

- Angle
- Stop end external
- Running outlet
- Stop end internal
- Union bracket.

Securing gutters and fall pipes

On new work the gutter will be fixed to fascia boards using a fascia bracket. The fascia is the piece of wood (or plastic) like the one shown in Figure 7.45.

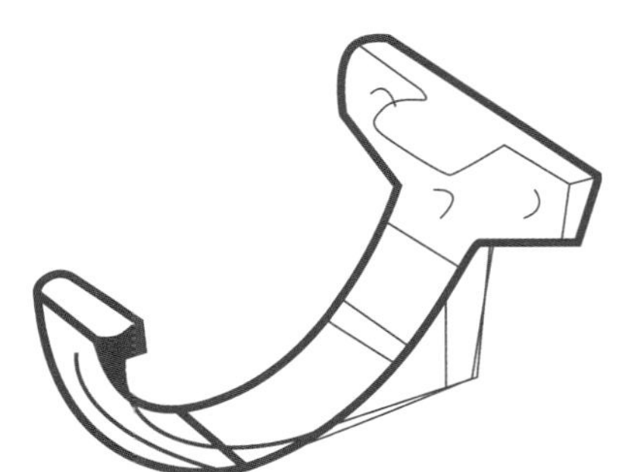

Figure 7.45 Fascia bracket

On some older properties, fascia board may not be fitted. On these jobs metal rafter brackets are used for metal and plastic gutters, these are readily available from gutter manufacturers and are ideal when replacing gutters if you do not want to go to the expense of fitting fascias.

Fall pipe fittings

Try this

Again, use manufacturers' catalogues or websites, or look at the fittings in your college or centre to familiarise yourself with the following:

- Offset
- Tee
- Connector
- Shoe outlet
- Saddle clip
- Barrel clip.

Installing the gutter and fall pipe

Plan your installation using the manufacturer's catalogues to assist you in selecting the correct type and quantity of products required:

  - Fascia Brackets should be spaced at a maximum of one metre apart on straight gutter runs
  - Angles and stopends should have a fascia bracket within 150 mm of the fitting
  - A supporting pipe clip should be used on shoes, branches and bends where necessary
  - Downpipes should be supported below an offset and at maximum intervals of 1.8 m
- On a gutter replacement job, if necessary, remove any old defective gutter and replace old fascia board
- Establish the position of the running outlet(s), usually over an existing drain, and fix securely to fascia board
- Fix a Fascia Bracket 100 mm short of furthest point from the outlet, allowing for a fall to the outlet (1:600 is recommended) using a string line
- Fixings:
  - Fix fascia and union brackets at required intervals
  - Fascia brackets should be positioned so as to avoid the fixing screws splitting the top edge of any timber fascia board. All brackets should be secured to the fascia board, with, for example, two 25 mm × 5 mm (1″ × 10) screws or one 32 mm × 6.5 mm (1¼″ × 14) screw.

- Unions should be fixed using a 25 mm × 5 mm (1″ × 10) screw
- Outlets and angles should be fixed using two 25 mm × 5 mm (1″ × 10) screws
- In areas of heavy snowfall it is recommended that each fascia bracket is secured using two 25 mm × 5 mm (1″ × 10 screws)
- Rainwater downpipe clips should be fixed using two 32 mm × 6.5 mm (1¼″ × 14) screws
- Use round-headed screws which should be brass or zinc-plated.

- Lubricate all gutter seals with silicone spray lubricant to ensure an easy fit and to allow for movement caused by expansion and contraction
- Working from the running outlet insert the back edge of the gutter under the retaining lip of the wrap around clip. Using slight downward pressure on the gutter, snap the front edge of the retaining clip over the front of the gutter. (Ensure that the marked expansion allowance is kept.)
- Use a Union Bracket or Angle to join to next gutter length in order to build up a gutter run. Use a stopend to complete the run.
- Downpipe installation starts at the outlet. If an offset is required use two offset bends with or without a short piece of pipe or soil connector or alternatively use an adjustable offset bend. Ensure a 6 mm gap is left at the top of the downpipe for expansion.
- Pipe Connectors if required should be secured to the wall with a pipe clip
- At the base of the pipe, fit a shoe secured with a pipe clip or connect the downpipe to the underground drainage system using a 110 mm × 68 mm reducer (SP96).

Test yourself 7.7

1. Which of the following materials are you most likely to select for a new domestic gutter and rain water pipe installation?
   a. Copper ☐
   b. Plastic ☐
   c. Cast iron ☐
   d. Asbestos ☐
2. What are the three main types of gutter shown below?

   (a) (b) (c)

3. What is the maximum spacing for fascia brackets on straight gutter runs?
   a. 0.5 m ☐
   b. 0.75 m ☐
   c. 1.00 m ☐
   d. 1.25 m ☐
4. What is the maximum interval for downpipe supports?
   a. 0.80 m ☐
   b. 1.40 m ☐
   c. 1.60 m ☐
   d. 1.80 m ☐

Check your learning Unit 7

Time available to complete answering all questions: 30 minutes

Please tick the answer that you think is correct.

1. Which one of the following covers the design of sanitary pipework systems?
   a. Water Regulations ☐
   b. British Standard 6700 ☐
   c. Construction Regulations ☐
   d. Building Regulations ☐
2. Where would a plumber most likely find information on the type of plastic soil and vent pipe fittings for use on site?
   a. Building drawings ☐
   b. Material specification ☐
   c. Installation schedule ☐
   d. Site layout plan ☐
3. The maximum recommended length of a 32 mm waste discharge pipe from a washbasin when connected to a single stack system is:
   a. 1 m ☐
   b. 1.7 m ☐
   c. 2 m ☐
   d. 3 m ☐
4. What is the recommended gradient on a single stack installation for a 40 mm waste pipe from the basin to the stack?
   a. 8–18 mm/m ☐
   b. 18–28 mm/m ☐
   c. 18–80 mm/m ☐
   d. 18–90 mm/m ☐

Check your learning Unit 7 (Continued)

5. The reason for using a large radius bend at the foot of a single stack system is to prevent:
   a. Wavering out ☐
   b. Back pressure ☐
   c. Self-siphonage ☐
   d. Induced siphonage ☐

6. Which of the following traps would be used in a situation where the Building Regulation design requirements for primary ventilated stacks could not be met?
   a. Bottle ☐
   b. Anti-vac ☐
   c. 'S' trap ☐
   d. Shower ☐

7. In what specific circumstances would the trap above be required?
   a. On installations where their neat appearance is important ☐
   b. Underneath the shower unit where space is extremely limited ☐
   c. Where there is a possibility of seal loss due to evaporation ☐
   d. Where there is a possibility of seal loss due to siphonage ☐

8. Bottle traps tend to be specified because of their compact dimensions, on which of the following appliances would one be used?
   a. Washing machine ☐
   b. Pedestal Wash basin ☐
   c. Plastic Shower tray ☐
   d. Waste disposal unit ☐

9. What is the maximum recommended distance for a horizontal waste discharge pipe from a macerator type WC?
   a. 15 m ☐
   b. 25 m ☐
   c. 40 m ☐
   d. 50 m ☐

10. The normal fixing height for a wall hung washbasin is:
   a. 750 mm ☐
   b. 800 mm ☐
   c. 900 mm ☐
   d. 1000 mm ☐

11. A fitting used to secure a bib tap to a wall surface over a cleaner's sink is called:
   a. Back plate elbow ☐
   b. Straight tap connector ☐
   c. Bent tap connector ☐
   d. Flexible hose union ☐

12. The normal fixing height of an individual wall-mounted urinal is:
   a. 580 mm ☐
   b. 610 mm ☐
   c. 620 mm ☐
   d. 640 mm ☐

13. What is the capacity of a WC cistern when installed together with a new WC pan?
   a. 6.0 litre ☐
   b. 6.5 litre ☐
   c. 7.5 litre ☐
   d. 9.0 litre ☐

(Continued)

Check your learning Unit 7 (Continued)

14. A ventilation stack terminates 300 mm above an opening into a building, how much would it need to be extended to satisfy the regulations?
   a. 600 mm ☐
   b. 700 mm ☐
   c. 900 mm ☐
   d. 1200 mm ☐

15. Water from the wash basin trap sometimes appears in the bowl in the basement flat of a high rise building. What is the most likely cause of the problem?
   a. Induced siphonage ☐
   b. Wavering out ☐
   c. Compression ☐
   d. Self-siphonage ☐

16. An externally mounted horizontal waste pipe is to be extended by 2.4 m from its last clip. How many new clips will be required to complete the installation?
   a. 2 ☐
   b. 3 ☐
   c. 4 ☐
   d. 5 ☐

17. On a WC macerator installation, the electrical supply to the unit should be wired from:
   a. The mains board consumer unit ☐
   b. A switched fused spur outlet ☐
   c. An unswitched fused spur outlet ☐
   d. A pull switch in the WC compartment ☐

18. What type of test is recommended for uPVC sanitation pipework and fittings:
   a. Water ☐
   b. Hydraulic ☐
   c. Smoke ☐
   d. Air ☐

19. What is the permissible pressure drop over the test period in millimetres when air testing sanitary pipework?
   a. Nil ☐
   b. Two ☐
   c. Three ☐
   d. Ten ☐

20. What type of gutter section is shown below?

   a. Half round ☐
   b. Victorian ☐
   c. Ogee ☐
   d. Square ☐

Sources of Information

We have included references to information sources at the relevant points in the text; here are some additional contacts that may be helpful. This list is not exhaustive, you might wish to find additional ones to these.

- Hepworth Plumbing Products
  Head Office
  Edlington Lane
  Edlington
  Doncaster
  DN12 1BY
  Tel: 01709 856 400
  Fax: 01709 856 401
  Website: www.hepworthplumbing.co.uk
- FloPlast Limited
  Sheppy Way
  Howt Green
  Sittingbourne
  Kent
  ME9 8QT
  Tel: 01795 431731
- Bathroom design guide
  The Bathroom Manufacturers Association
  Federation House
  Station Road
  Stoke on Trent
  ST4 2RT
  Tel: 01782 747123
  E-mail: into@bathroom-association.org.uk
  Website: www.bathroom-association.org
- Geberit Ltd
  New Hythe Business Park
  New Hythe Lane
  Aylesford
  Kent
  ME20 7PJ
  Tel: 01622 717811
  Website: www.geberit.co.uk

Try this

If you have access to the internet and a search engine, try typing in words such as:

- Above ground discharge systems
- Sanitary appliances
- Bathroom suite
- Waste traps.

UNIT 8

CENTRAL HEATING SYSTEMS

Summary

This unit covers the requirements of a range of domestic central heating systems, components and controls. **It also covers gas boilers over and above the requirements of Level 2, in order to give you a better insight into typical domestic heating arrangements. Likewise, whilst commissioning central heating systems is not covered until Level 3, it is included here, because even if you do not progress to Level 3, it will give you a very useful insight into commissioning central heating systems for future reference.**

Building Regulation L1A and L1B 2006 which deals with the conservation of fuel and power in buildings, has made a major impact on the plumbing industry, so this is covered in some detail.

Here is what you will cover in this unit:

- Types of central heating systems
- Central heating system components
- Central heating system controls
- Tightness testing and flushing of central heating pipework systems
- Maintenance of central heating systems
- Commissioning and de-commissioning central heating systems.

Central heating systems explained

Central heating systems are closely linked to the hot water system, as in most domestic situations the boiler or heating appliance serves both systems. The central heating system will also be connected to the cold water supply either indirectly or directly depending on the type of central heating selected.

Sealed wet systems, or dry systems such as warm air are covered in Level 3; however, combination boilers, which are classified as sealed systems, are covered separately.

In this unit, you will concentrate on wet-vented systems only – systems that contain water and are vented to the atmosphere. Central heating systems are summarised in Figure 8.1.

Key point

The term heat emitters is usually associated with radiators, but it also describes heat emitting components other than radiators and includes cast iron column radiators, skirting convector heaters, fan-assisted wall convectors, towel rails and low surface temperature (LST) radiators.

Regulations and domestic heating and hot water

Water Regulations do not refer specifically to 'central heating systems', but they are relevant in the sense that the heating system components tend to be connected to the hot and cold water supply. The regulations are particularly relevant when it comes to unvented heating systems which is covered at Level 3, as these are often connected directly to incoming mains. Open-vented systems are also fed by feed and expansion cisterns which are connected to the cold water supply.

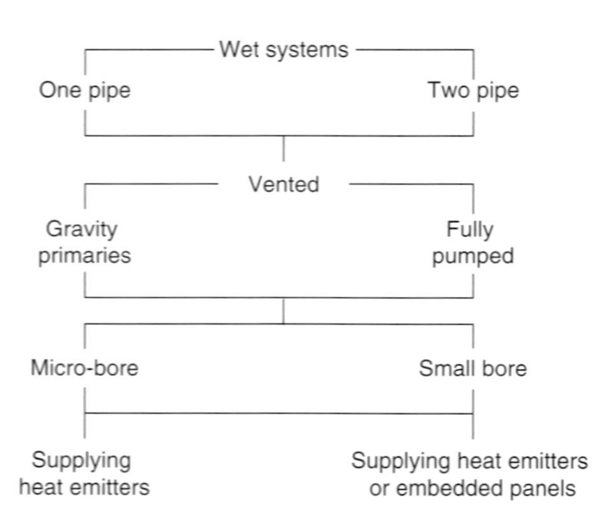

Figure 8.1 Central heating system flow chart

Building Regulation L1A and L1B (2006) and plumbing

Home energy use is responsible for 27% of UK carbon dioxide (CO_2) emissions which contribute to climate change. The Government has reviewed the Building Regulations, introducing requirements intended to make both new and existing buildings more energy efficient and therefore reduce CO_2 emissions.

The regulations aim to reduce CO_2 emissions in dwellings by around 2.8 million tonnes by 2010. Most of this reduction will have to come from improvements to existing homes. The current requirements are set out in a revision to the Building Regulations, which came into force in England and Wales in April 2006. The revision includes modifications to Part L (Conservation of fuel and power) and the introduction of new energy efficiency regulations, for both new and existing dwellings.

Part L is concerned with the conservation of fuel and power in buildings, and for dwellings. Part L is supported by two approved documents:

- L1A
- L1B

L1A gives guidance on how to satisfy the energy performance required by the Building Regulations for new dwellings, and L1B gives guidance for existing dwellings.

We would strongly recommend that you try to access copies of 'The Building Regulations Approved Document Parts L1A and L1B'. These can be obtained from the following website: www.planningportal.gov.uk

Once in the site, click on the Building Regulations link.

Another useful contact is:

The Communities and Local Government
Eland House
Bressenden Place
London
SW1E 5DU
Tel: 020 7844 440
Website on www.communities.gov.uk

Alternatively, you should be able to view a copy in your local library.

In addition to this, I would recommend you source a copy of the 'Domestic Heating Compliance Guide' which is used in conjunction with Parts L1A and L1B of the Building Regulations.

Figure 8.2 Domestic Heating Compliance Guide booklet

It can be downloaded by going to www.communities.gov.uk. Type in 'Domestic Heating Compliance Guide' in their search engine.

This publication is also available to buy from:
RIBA Bookshops (Mail Order Office),
15 Bonhill Street,
London,
EC2P 2EA, UK.
Tel: +44 (0) 20 7256 7222
www.ribabookshops.com

Key point

You'll still hear the term 'Condensing Boiler' used in the industry, but this is correctly termed 'High Efficiency Boilers'

Other useful publications related to regulations

The Energy Efficiency Best Practice in Housing which is managed by the Energy Savings Trust produce the following booklets designed to provide guidance on the various aspects of the regulations together with other related information. The full publication list includes:

- The effect of Building Regulations (Part L1 A & B 2006) on Existing Dwellings
- Domestic Heating and Hot Water – Choice of fuel and system type
- Whole House Boiler Sizing Method for Houses and Flats
- Domestic Condensing Boilers – the Benefits and the Myths
- Controls for Domestic Central Heating and Hot Water – Guidance for Specifiers and Installers
- Domestic Central Heating and Hot water: Systems with Gas- and Oil-fired Boilers – Guidance for Installers and Specifiers
- Best Practice in New Housing – a Practical Guide

They can be contacted on:
Energy Efficiency Best Practice in Housing:
Helpline: 0845 120 7799
E-mail: bestpractice@est.co.uk
Website: www.est.org.uk/bestpractice.

Key point

The Building Regulation L1 covers central heating, and the 2002 version for England and Wales has been amended; the new regulations came into force on 1st April 2006.

Another useful site is:
Heating and Hot Water Information Council
Website: www.centralheating.co.uk
Tel: 0845 600 200
E-mail: info@centralheating.co.uk

Background

All the information related to the Regulations has been applied to the relevant aspects of the central heating systems and controls throughout this unit.

The regulations were brought in because big improvements were required to improve the energy efficiency in existing buildings. Most people will be aware of the greenhouse effect and **global warming** due to carbon emissions. These new measures aim to:

- Reduce carbon dioxide emissions from buildings by 25%
- Save 2.8 million tonnes of carbon emissions per year by 2010.

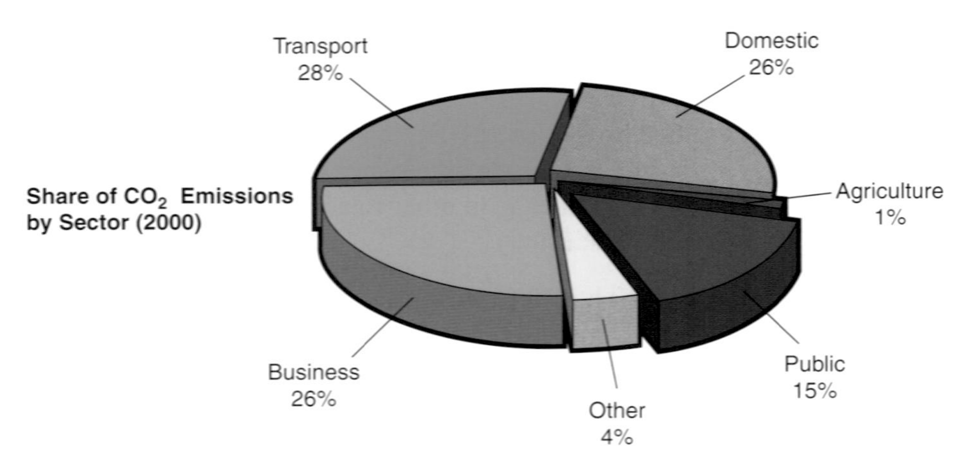

Figure 8.3 CO_2 emissions

What is Global Warming

In simple terms, Global Warming means that the Earth is getting warmer. Global Warming is caused by human activities which are increasing the greenhouse gases in the atmosphere. These gases trap the Sun's heat on the Earth's surface instead of allowing some of the heat to escape back into space.

The Government of the United Kingdom has set a goal for reducing carbon dioxide (CO_2) emissions by 20% below 1990 levels, and hope to achieve this goal by the year 2010. Carbon dioxide is considered to be the leading greenhouse gas and is produced when fuel is burnt. Refer Figure 8.3 for details.

Energy consumed in domestic dwellings accounts for approximately 26% of all the carbon dioxide produced in the United Kingdom. Furthermore, over 70% of this is produced when fuel is burnt to provide heat and hot water.

What the regulations will cover

The Building Regulation Approved Document Part L1A and L1B, 2006, deals with the Conservation of Fuel and Power in buildings, and significant changes have been introduced by the 2006 revisions which apply to new properties and existing homes.

In the Domestic Heating Compliance Guide, the various situations that are applicable to plumbers are:

- Central heating systems in new dwellings, referred to as **'a new system'**.

- Central heating systems in existing dwellings where previously space heating was not provided by central heating, referred to as **'a new system'**.
- Replacement of a central heating system and/or component in existing dwellings where central heating is already installed, referred to as **'a replacement system'**.

This terminology is important as it has a bearing on what is permissible in terms of heating installations in the various categories.

When works are carried out to an existing dwelling, opportunities often arise for efficiency improvements at little extra cost. The regulations require that whenever works are carried out, reasonable provision should be made to improve; the efficiency of the thermal envelope (walls, floors, roof, windows and doors), heating and other fixed services. Improvements are required when undertaking the following types of work:

- Extensions to existing dwellings.
- Change of use (e.g. a barn conversion) or change in energy status.
- Alterations to dwellings (e.g. replacement of an external wall).
- The provision, replacement or renovation of a thermal element. A thermal element is a wall (excluding windows and doors), floor or roof that separates the heated space from the outside or from an unheated space, such as an integral garage. It is defined in regulation 2A.
- The provision or replacement of controlled fittings (windows, external doors and the like).
- The provision, replacement or extension of a controlled service (or part). Controlled services include heating systems, hot water systems, lighting, mechanical ventilation and cooling.

Heating and hot water systems

The Building Regulations apply to new, replaced, or extended heating and hot water systems in existing dwellings. Detailed guidance is provided in the Domestic Heating Compliance Guide (DHCG).

In many cases works will be done under an appropriate competent person's scheme, otherwise a separate application to building control will be required.

For an emergency repair there is no need to delay, but building control is to be notified at the earliest opportunity. The work must comply and commissioning must be carried out accordingly.

For some minor heating works – such as replacement of a pump, a single radiator or controllers using existing wiring – it is not necessary to inform building control, although the works still need to comply and be commissioned, and the user given full instructions. Schedule 2B of the regulations, shown on page 14 in L1B, contains a full list of works not requiring notification.

Heating appliance efficiency

Heating appliances (boilers, etc.) must have an efficiency not less than that shown in the DHCG; the minimum efficiencies depend on the fuel to be used, as shown in the examples below:

Gas: 86% except in exceptional circumstances.

Oil after 1 April 2007: min 86% and to be condensing type. Before this date, efficiencies of 85% or 82% for combination boilers are required.

Solid fuel: min efficiencies are as published by HETAS. For example, independent anthracite boilers: hand-fired (batch fed) 65%, gravity fed 70% to 75% depending on rating.

However, higher efficiencies can be achieved at very little extra cost and guidance on this is available from the Energy Saving Trust.

Boilers have a SEDBUK rating ranging from A to G based on their energy efficiency; more about this later.

System controls

With the exception of solid fuel, new systems must be fully pumped. Where the boiler is being replaced, existing semi-gravity circulation should be converted to fully pumped. Any bypasses installed should be automatic. Most solid fuel systems require a heat leak radiator (e.g. bathroom towel rail) to dissipate heat during slumber periods. In such cases a fully pumped system may not be appropriate.

The DHCG provides minimum provisions for system controls. The following general outline applies to most wet heating systems; however it is important to check the detailed guidance applicable to the fuel used. Existing controls that meet the DHCG provisions would not require upgrading.

Time control of space and water heating should be provided separately to each space and water heating circuit. Where instantaneous hot water is produced, time control is only required for the space heating zone.

All new and replacement systems must include boiler interlock that switches off the boiler and pump when there is no demand for heating or hot water. In the case of solid fuel systems, full interlock may not be required, depending on the type of system installed.

Dwellings should have at least two control zones (one of which is assigned to the living area), each with independent temperature controls. Thermostatic radiator valves in the second zone (e.g. bedrooms) may be used, provided that the main zone is fitted with a room thermostat.

For dwellings over 150 m^2, all zones require separate time and temperature controls. Single-storey open plan dwellings with a living area greater than 70% of the floor area may have a single control zone.

Hot water storage vessels should be insulated as in the DHCG to meet the requirements of BS 1566 if they are vented, or BS 7206 (or acceptably certified) if unvented, and labelled with: type of vessel, capacity, standing heat loss and heat exchanger performance.

Pipework is to be insulated as in the DHCG: Primary circulation pipes for heating and hot water, whenever not in a heated space. Primary hot water circulation pipes, throughout their length.

All pipes (including overflow) connected to hot water storage vessel to 1 m from connection.

Key point

From April 2006, the Building Regulations require that heating appliances need to be 86% efficient so this means a SEDBUK rating of A or B for gas boilers.

The regulation refers to the use of SEDBUK B boilers as 'Basic Practice' and SEDBUK A as 'Best Practice'.

SEDBUK

Reference was made previously to the SEDBUK rating of boilers. SEDBUK stands for Seasonal Efficiency Domestic Boilers in the UK and from 1 April 2006, domestic gas boilers fitted into new buildings were required to have a minimum SEDBUK A or B rating. SEDBUK measures the energy efficiency of the boiler on a scale of A to G with A being the highest.

SEDBUK was developed under the Government's Energy Efficiency Best Practice Programme with the co-operation of boiler manufacturers, and provides a basis for fair comparison of the energy performance of different boilers.

SEDBUK is the average annual efficiency achieved in typical domestic conditions, making reasonable assumptions about pattern of usage, climate, control, and other influences. It is calculated from the results of standard laboratory tests together with other important factors such as boiler type, ignition arrangement, internal store size,

Minimum SEDBUK rating by fuel before April 1st 2007

Central heating system fuel	SEDBUK %
Mains natural gas	86
LPG	86
Oil	85

Band	SEDBUK Range
A	Above 90%
B	86% – 90%
C	82% – 86%
D	78% – 82%
E	74% – 78%
F	70% – 74%
G	Below 70%

Figure 8.4 SEDBUK rating

fuel used, and knowledge of the UK climate and typical domestic usage patterns.

For estimating annual fuel costs SEDBUK is a better guide than laboratory test results alone. It can be applied to most gas and oil domestic boilers for which data is available from tests conducted to the relevant European standards. The SEDBUK method is used in SAP.

Figure 8.4 shows the SEDBUK ranges for boilers; the first table is by fuel type and the second various bands.

Choosing an efficient boiler

A list of Energy Efficiency Recommended 'A' rated boilers is available on www.boilers.org.uk. Alternatively, the Little Blue Book of Boilers can be obtained by calling 0845 727 7200. Both are regularly updated. The Energy Efficiency Best Practice in Housing website also provides tools and further information on financial savings and correct boiler sizing at www.est.org.uk/bestpractice/boiler.

What is the Standard Assessment Procedure (SAP)?

SAP (2005) is the Government's 'Standard Assessment Procedure' for energy rating of dwellings, and is based on the space and water heating running costs. Its value will depend on the type of building, fuel and heating design. SAP ratings are expressed on a scale of 1 to 100 – the higher the number the better the rating.

SAP rating depends on:

- Building insulation
- Building design
- Solar heat gains
- Building ventilation
- Heating and hot water efficiency and controllability (using SEDBUK).

All new buildings have to have a SAP (Standard Assessment Procedure).

What it means to plumbers?

The type of heating system and controls, and the level of insulation and ventilation, influence SAP ratings. Here are a few pointers about SAP:

- Condensing boilers produce higher ratings than traditional boilers and can increase the rating by up to 10 points.

An electric heating system using an economy 7 boiler can reduce the SAP by 11 when compared with a conventional gas-fired boiler.

- The Boiler Efficiency Data File is published as part of the Boiler Efficiency Database scheme. It holds data on domestic boilers, gas- and oil-fired only, current and obsolete, for the purpose of carrying out SAP energy ratings. (www.boilers.org.uk)
- If an existing older boiler is being re-used in conversion, the plumber must ensure that when the pump switches off, the boiler switches off also. The feed to the boiler is not the 'permanent live' but is a switched supply used for the pump. Normally referred to as a 'boiler interlock'.
- Most new and refurbishment designs will require at least a programmer, room stat and thermostatic radiator valves (TRVs) to obtain a good rating, with additional points being added for delayed stats, energy managers, etc
- Cylinder stats are compulsory for a SAP pass (if a cylinder is used)
- Primary pipework insulation is also compulsory and must be insulated within 1.5 m of the cylinder to prevent losses, and to obtain a pass. The depth of the cylinder insulation also has a significant effect on the rating.
- If the pump is fitted in the heated space, it will have a positive effect on the rating, if it is fitted in the unheated space it will have a negative effect.

Activity 8.1

Go to www.bre.co.uk/sap 2005 where you can download a copy of the SAP 2005 specification.

The Benchmark Scheme

The Benchmark Scheme is the Heating & Hot Water Information Council's code of practice for manufacturers of central heating appliances and installers.

Since it was launched in 1999, Benchmark has provided a vital means for improving the quality of installation of heating and hot water systems by encouraging the correct installation and commissioning of appliances and equipment.

Heating and Hot Water Industry Council members can take part in the Benchmark scheme, which is a voluntary code of practice for manufacturers of gas appliances, aimed at raising quality standards in the industry and improving customer satisfaction through a series of codes of practice. It also encourages consumers to have appliances serviced and inspected on a regular basis.

The Government sees Benchmark as a useful vehicle for ensuring that more energy efficient boilers are utilised, cutting CO_2 emissions, in line with Government commitments.

As well as potentially reducing energy consumption and CO_2 emissions, Benchmark also offers a major contribution to the maintaining of safety standards.

The Benchmark Commissioning Checklist

The existing Benchmark log book has been replaced by a new Benchmark commissioning checklist, which is an A4 double sided sheet that will be included in all HHIC gas boiler manufacturer members' installation manuals.

The checklist helps installers record more information about the installation, in order to assist with servicing and repairs, such as details of system cleaners and inhibitors, CO_2 readings, and Gas Work Notification numbers.

The checklist is referred to in the Building Regulations as an example of an effective checklist, demonstrating Benchmark's importance as the standard Code of Practice for the installation, commissioning and servicing of central heating systems.

Further detailed information about the scheme can be obtained from:

Heating and Hot Water Information Council
36 Holly Walk
Leamington Spa
CV32 4LY
Tel: 0845 600 2200
Website: www.centralheating.co.uk

Central Heating System Specification (CHeSS) Year 2005

CHeSS (year 2005) replaces CHeSS (year 2002) to take account of the changes to the Building Regulations in England and Wales.

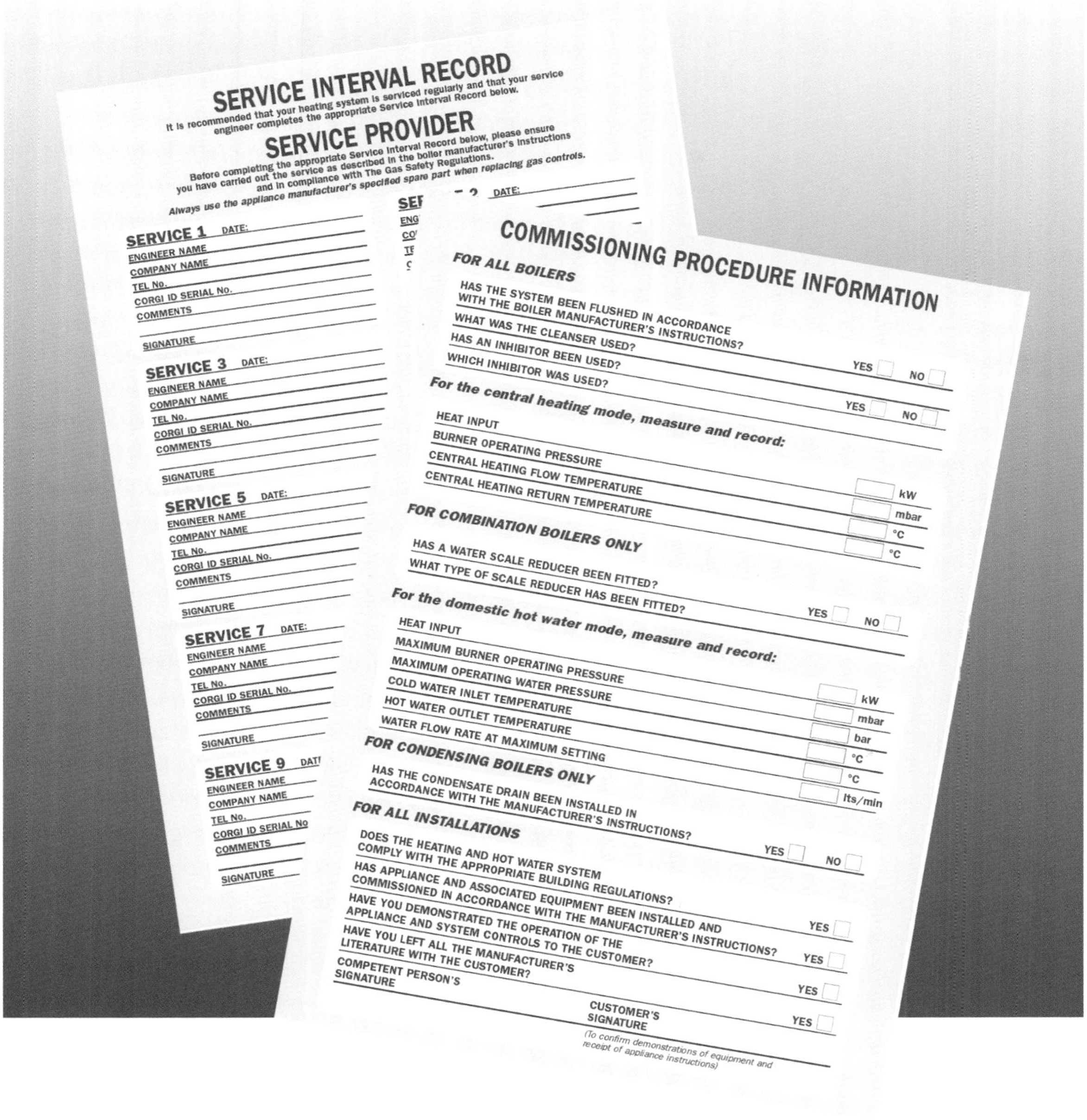

SERVICE INTERVAL RECORD

It is recommended that your heating system is serviced regularly and that your service engineer completes the appropriate Service Interval Record below.

SERVICE PROVIDER

Before completing the appropriate Service Interval Record below, please ensure you have carried out the service as described in the boiler manufacturer's instructions and in compliance with The Gas Safety Regulations.

Always use the appliance manufacturer's specified spare part when replacing gas controls.

SERVICE 1 DATE:
ENGINEER NAME
COMPANY NAME
TEL No.
CORGI ID SERIAL No.
COMMENTS
SIGNATURE

SERVICE 3 DATE:
ENGINEER NAME
COMPANY NAME
TEL No.
CORGI ID SERIAL No.
COMMENTS
SIGNATURE

SERVICE 5 DATE:
ENGINEER NAME
COMPANY NAME
TEL No.
CORGI ID SERIAL No.
COMMENTS
SIGNATURE

SERVICE 7 DATE:
ENGINEER NAME
COMPANY NAME
TEL No.
CORGI ID SERIAL No.
COMMENTS
SIGNATURE

SERVICE 9 DATE:
ENGINEER NAME
COMPANY NAME
TEL No.
CORGI ID SERIAL No.
COMMENTS
SIGNATURE

COMMISSIONING PROCEDURE INFORMATION

FOR ALL BOILERS
HAS THE SYSTEM BEEN FLUSHED IN ACCORDANCE WITH THE BOILER MANUFACTURER'S INSTRUCTIONS?
WHAT WAS THE CLEANSER USED?
HAS AN INHIBITOR BEEN USED?
WHICH INHIBITOR WAS USED?

For the central heating mode, measure and record:
HEAT INPUT
BURNER OPERATING PRESSURE
CENTRAL HEATING FLOW TEMPERATURE
CENTRAL HEATING RETURN TEMPERATURE

FOR COMBINATION BOILERS ONLY
HAS A WATER SCALE REDUCER BEEN FITTED?
WHAT TYPE OF SCALE REDUCER HAS BEEN FITTED?

For the domestic hot water mode, measure and record:
HEAT INPUT
MAXIMUM BURNER OPERATING PRESSURE
MAXIMUM OPERATING WATER PRESSURE
COLD WATER INLET TEMPERATURE
HOT WATER OUTLET TEMPERATURE
WATER FLOW RATE AT MAXIMUM SETTING

FOR CONDENSING BOILERS ONLY
HAS THE CONDENSATE DRAIN BEEN INSTALLED IN ACCORDANCE WITH THE MANUFACTURER'S INSTRUCTIONS?

FOR ALL INSTALLATIONS
DOES THE HEATING AND HOT WATER SYSTEM COMPLY WITH THE APPROPRIATE BUILDING REGULATIONS?
HAS APPLIANCE AND ASSOCIATED EQUIPMENT BEEN INSTALLED AND COMMISSIONED IN ACCORDANCE WITH THE MANUFACTURER'S INSTRUCTIONS?
HAVE YOU DEMONSTRATED THE OPERATION OF THE APPLIANCE AND SYSTEM CONTROLS TO THE CUSTOMER?
HAVE YOU LEFT ALL THE MANUFACTURER'S LITERATURE WITH THE CUSTOMER?
COMPETENT PERSON'S SIGNATURE
CUSTOMER'S SIGNATURE
(To confirm demonstrations of equipment and receipt of appliance instructions)

Figure 8.5 Benchmark service record and commissioning checklist

CHeSS was seen as essential because of the difficulties facing the domestic heating installation industry due to the lack of common standards, and understanding of what should be done to improve energy efficiency. CHeSS gives recommendations for good practice and best practice for the energy efficiency of domestic wet central heating systems, and plumbers should use them as a basis for developing a system specifications and costings.

Further information about CHeSS can be obtained by calling the Energy Efficiency Best Practice in Housing on 0845 120 7799 and asking for:

- Central Heating System Specifications (CHeSS) Year 2005 – Best Practice guide (General Information leaflet 59). This is strongly recommended. It can also be downloaded from: www.est.org.uk/bestpractice

Types of systems

Gravity central heating systems

You definitely will not see new gravity systems installed because they do not comply with current legislation, but you might come across an old system whilst carrying out maintenance work, in buildings such as old Church Halls, where they are designed to heat a large space.

Domestic properties installed with this type of system were referred to as having background heating, and when installed, the installation would consist of a single radiator, normally fitted in the bathroom, close to the hot water cylinder. It would connect into the primary flow and return circulation pipework feeding the cylinder. Generally, these were direct systems, and copper radiators were used to prevent the discolouration of hot water from oxidization. As with any gravity system, it works on the principle of having the pipework installed to the correct gradients to assist circulation. The flow and return circulation pipework to the radiator was normally 22 mm in diameter.

Pumped central heating and gravity hot water systems

Heating installations of this type can either consist of a one-pipe ring circuit or a two-pipe circuit, and are also referred to as Semi-Gravity System.

One-pipe system

Again, you will not install one of these systems a new, but you may come across an old one, either on maintenance work or repair jobs, so you need to know how they look (see Figure 8.6). These

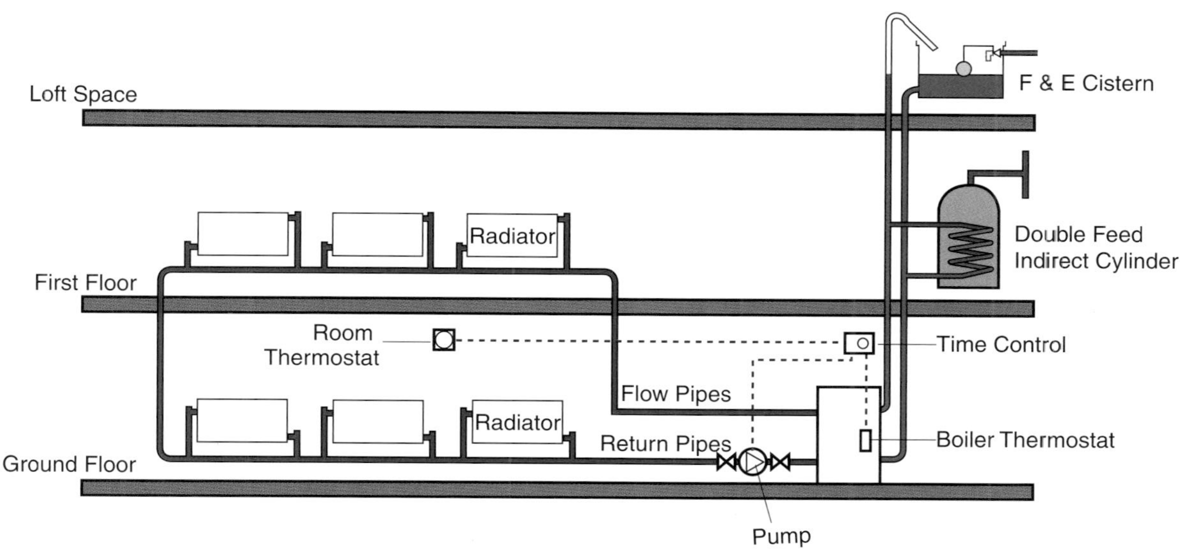

Figure 8.6 One-pipe system

are very early types of gravity or (mainly) pumped heating systems. The radiators are fed with water from a single flow piperun from the boiler, connecting to the radiators by means of short flow and return pipes, returning to the boiler on completion.

This is the basis of a one-pipe system, showing a single flow pipe from the boiler returning back to the boiler. The circulation pump is located on the return pipe. They are sometimes referred to as ring or loop circuits.

Two-pipe system

Two-pipe systems were a common choice, particularly in the late 1970s and 1980s, and you may come across one whilst on maintenance or repair work (Figure 8.7). A typical two-pipe system consists of a main flow pipe from the boiler, connecting into the flow tapping off a heat emitter/radiator, distributing a supply of heated water to the individual radiators with the help of a circulator (pump). A second pipe, connected at the opposite end to the flow connection of the radiator 'returns' the heated supply of water back to the boiler.

The system in Figure 8.8 shows a semi-gravity system design that meets the requirements of Part L.

The system provides independent temperature control of both the pumped heating and gravity hot water circuits. The circulating

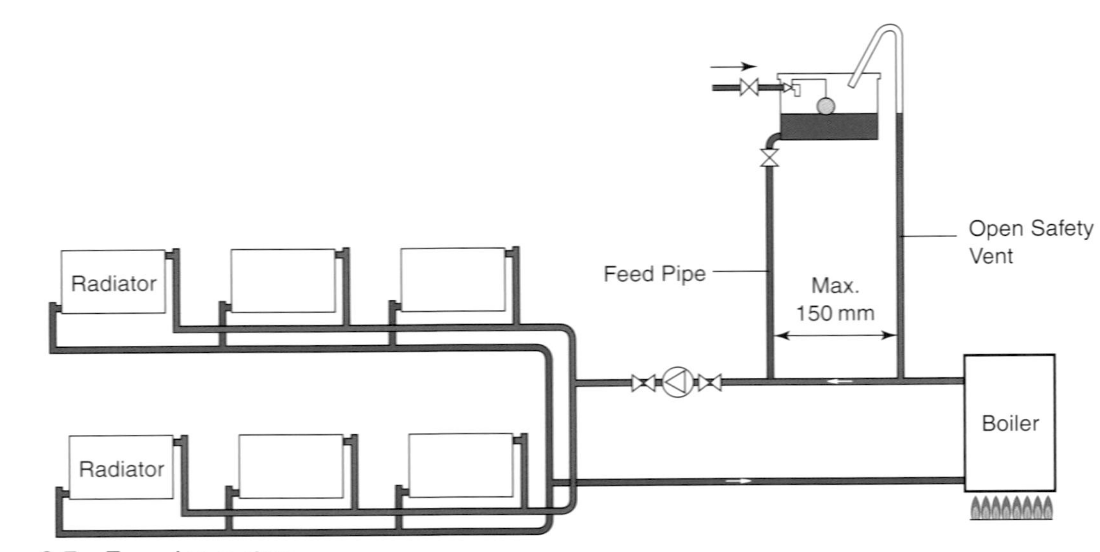

Figure 8.7 Two-pipe system

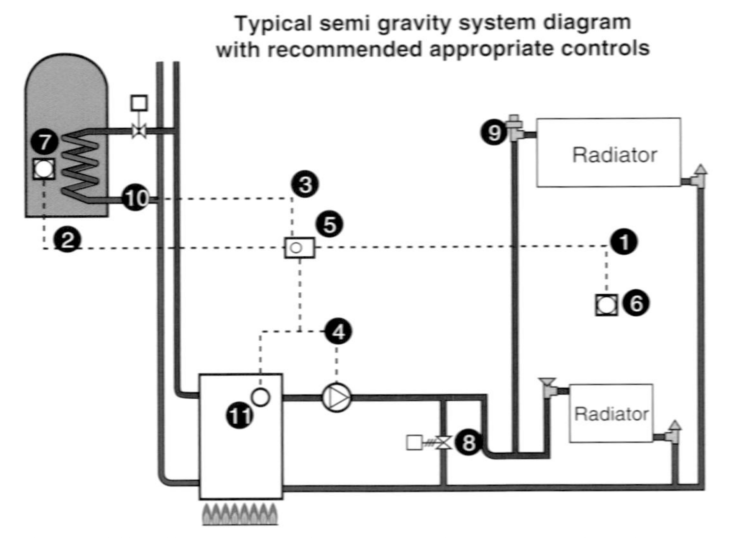

1. Time and temperature control to space heating
2. Time and temperature control to stored hot water
3. Switching of zone valves
4. Boiler and pump interlock
5. Timing control using either a full, standard or mini programmer or separate timers
6. Room (or programmable room) thermostat
7. Cylinder thermostat
8. Automatic by-pass valve to the system
9. TRV's on at least all radiators in the sleeping area
10. Zone valve
11. Boiler thermostat

Figure 8.8 System meeting the requirements of Part L

pump switches 'off' via the room thermostat on achieving the set room temperature, and the two-port zone valve closes, via the cylinder thermostat, on achieving the set water temperature. This means that the heating circulation pump and gravity circulation to the cylinder can be turned off independently by the thermostats. Older systems of this design were commonly referred to as a 'C' plan installation. The boiler interlock ensures that when the demands are met, the boiler does not fire, so as to avoid wasting energy.

The cylinder thermostat should be set at a fixed temperature of not less than 60°C to prevent the possibility of Legionella incubation (a source of water bacteria which can be found in water that has a temperature between 6 and 60°C. Water temperatures between 20 and 45°C seem to favour the growth of the bacteria). Legionnaire's disease is a potentially fatal form of pneumonia that can affect anybody.

Key point

The radiator located in the area of the room thermostat should be fitted with lockshield valves on the flow and return pipework, in lieu of a TRV to ensure that a system bypass of 10% of the minimum heating load is achieved.

Time control can be through either a full-, standard-, mini-programmer or separate timers. TRVs should also be fitted to all sleeping areas but where possible, and acceptable, it is advisable to fit them to all individual rooms to provide additional overriding temperature control.

Ideally, the radiator located in the area of the room thermostat should be fitted with lockshield valves, on the flow and return pipework, in lieu of a TRV.

The system's efficiency could be further improved by the installation of either an additional zone valve or anti-gravity valve, fitted to the heating flow circuit; this will prevent gravity circulation occurring at the upper floor radiators, when only the domestic hot water is in demand.

Fully pumped system

Fully pumped systems shown in Figure 8.9 are probably the most common types used today, on small bore or microbore heating installations. The system works by the flow and return of heating and hot water circuits operating fully under the influence of the circulation pump, and is not reliant on convectional currents to circulate hot water to the hot water storage vessel. This means that as there is no requirement for gravity circulation, the boiler can be sited above the height of the hot water cylinder, giving more design options to the system.

Careful design consideration should be given to a fully pumped vented system, prior to its installation (vented systems are systems that are fed by a feed and expansion cistern, referred to as a F&E cistern) and in particular to the location of the circulation pump. The pump must be positioned to ensure that there is no negative or positive pressure at the open vent pipe, which could result in the system pumping water over into the F&E cistern or air being pulled down the open-ended vent pipe, commonly known as air ingress.

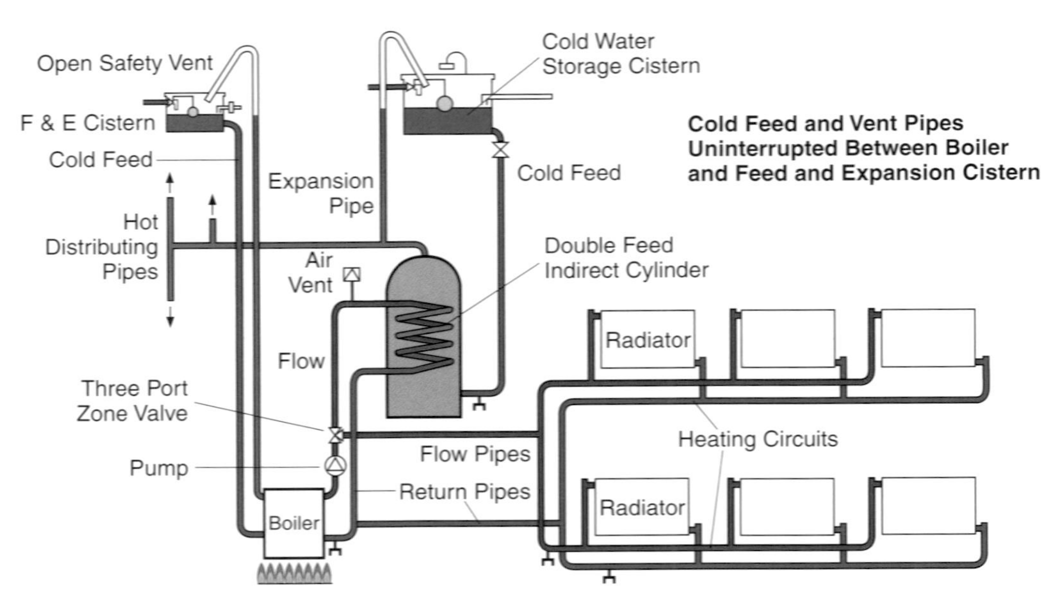

Figure 8.9 Fully pumped system

> **Key point**
>
> Excessive amounts of air or undissolved oxygen in any type of heating system will result in rapid corrosion of the system's ferrous components.

The heating and hot water circuits of newly installed fully pumped systems are controlled by motorised zone valves. A motorised zone valve is an electrically operated valve that can be fitted to the flow or return pipework of a system. The valve opens on heat demand and closes once the demand is satisfied. Activation of the valve can be by either a room thermostat or cylinder thermostat.

There are several types of system designs incorporating either two-port motorised zone valve(s) or a three-port valve (these can be termed as mid-position or diverter) which meet the requirements of the Building Regulations Approved Document L1 A & B 2006.

> **Key point**
>
> Part L1 2006 for domestic dwellings requires that all new heating systems be fully pumped and must provide independent time and temperature control to both the heating and hot water circuits and have a boiler control interlock.

> **Try this**
>
> The following text is an explanation of how the fully pumped system works. We are going to make reference to a number of controls, which we will highlight in italics. As you work through the next few pages please make a note of these highlighted words, because later we will explain how they work and what their role is in heating systems.

Heating temperature control

The heating system must be divided into temperature zones with separate controls. This can be done either by a *room thermostat* in all the zones, a *programmable room thermostat* in all the zones, or by a room thermostat or programmable room thermostat in the main zone with *thermostatic radiator valves* (TRVs) in the remaining zones.

Hot water temperature control

For hot water supplied from a storage vessel, e.g. cylinder, the temperature of the stored water must be controlled by a *cylinder thermostat.*

Time control

Separate independent time control required for the heating and hot water system is achieved using a full *programmer* with individual settings for heating and hot water, or by using two separate *time clocks.*

In instances where the hot water is supplied instantaneously, e.g. by a combi boiler, time control for the hot water is not required.

Where a domestic dwelling has a total usable floor area greater than 150 m^2, it is recommended that the heating circuits be split into a minimum of two zones, having separate timing and temperature controls. This can be done by multiple heating zone programmers or by a single multiple channel programmer.

Why do you think suitability is measured in terms of floor area (150 m^2), rather than by room volume? As it is the volume of air in a room that is being heated, should not a high or low ceiling be taken into account?

Try this

Which part of a dwelling counts as 'usable' for this assessment? Which of the following would you include: kitchen, hall, bathroom, landing, stairs, utility room, porch, conservatory, corridors? Or do you only count the 'rooms' that inhabitants are likely to sit in? Find out from Part L of the Building Regulations.

For practical reasons, area is easier to calculate.

Fully pumped system using a three-port mid-position zone valve

The fully pumped system using a *three-port mid-position valve* provides independent temperature control to both the heating and hot water circuits, via a three-port mid-position zone; older systems of this design were commonly referred to as a 'Y' plan installation.

The design and control layout of this type of system is not suitable for dwellings with a floor area greater than 150 m^2.

Alternatively, the system can be fitted with a three-port diverter zone valve, in place of the mid-position valve. These give priority to the hot water circuit. Older systems of this design were commonly referred to as a 'W' plan (Figures 8.10 and 8.11 installation).

> **Activity 8.1**
>
> With the valve being set for hot water priority, it should not be used in situations where there is likely to be a high hot water demand during the heating season. Why do you think this is so? Write your answer here in your portfolio before checking against the suggested one at the end of this book.

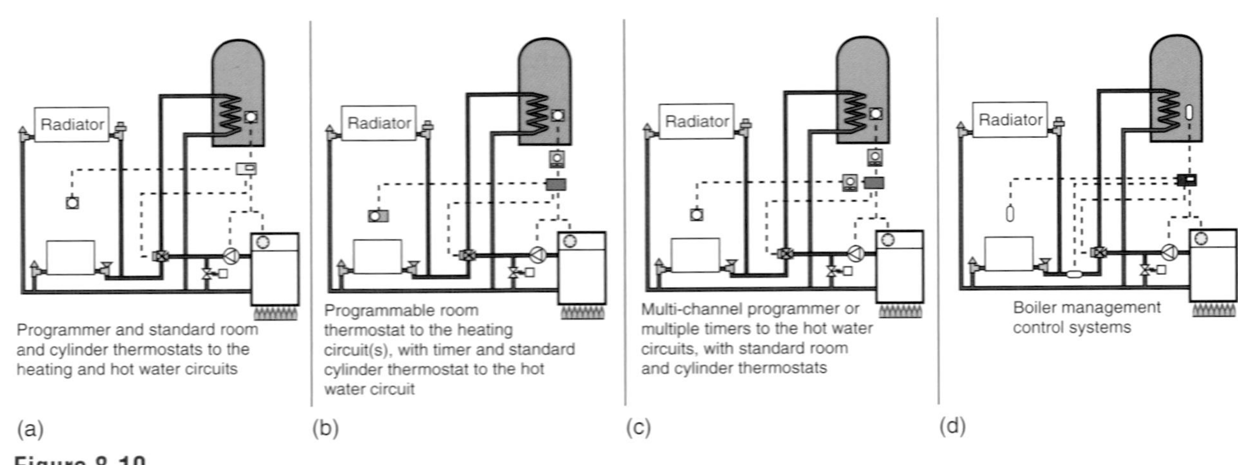

(a) Programmer and standard room and cylinder thermostats to the heating and hot water circuits

(b) Programmable room thermostat to the heating circuit(s), with timer and standard cylinder thermostat to the hot water circuit

(c) Multi-channel programmer or multiple timers to the hot water circuits, with standard room and cylinder thermostats

(d) Boiler management control systems

Figure 8.10

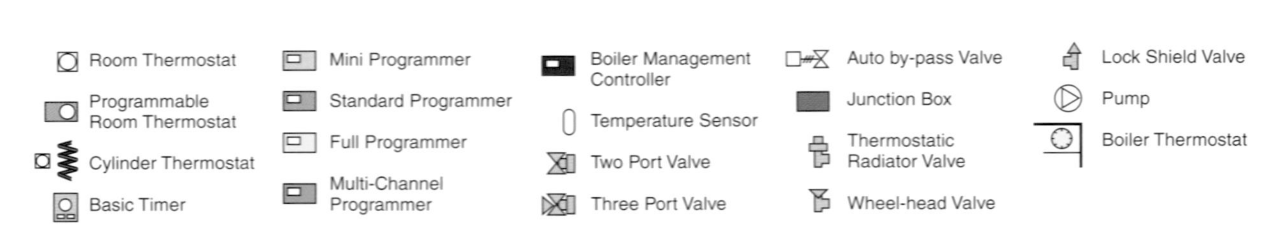

Figure 8.11

A fully pumped system fitted with a three-port mid-position zone valve offers the following features:

- Separate time and temperature control of the domestic hot water
- Separate time and temperature control of the heating circuit
- The circulation pump and boiler switches off when the heating and hot water temperature requirements are achieved. The boiler interlock ensures that when the demands are met, the boiler does not fire, and avoids wasting energy.
- The independent control of the circuits reduces the demand on the boiler at peak periods
- TRVs will provide additional overriding comfort level of temperature control in designated areas.

The boiler manufacturer may specify that the heating installation should incorporate an automatic bypass valve – this valve ensures a minimum flow rate of water through the boiler whilst firing. The valve is particularly useful where the TRVs are installed on a system, as they prevent the system suffering from noise and from a reduced water flow rate as the valves begin to close down.

Fully pumped system using separate two-port zone valves

This system provides separate temperature control for both heating and hot water circuits. Again, the features are similar to the other systems shown earlier. Older systems of this design were commonly referred to as an 'S' Plan installation (Figure 8.12).

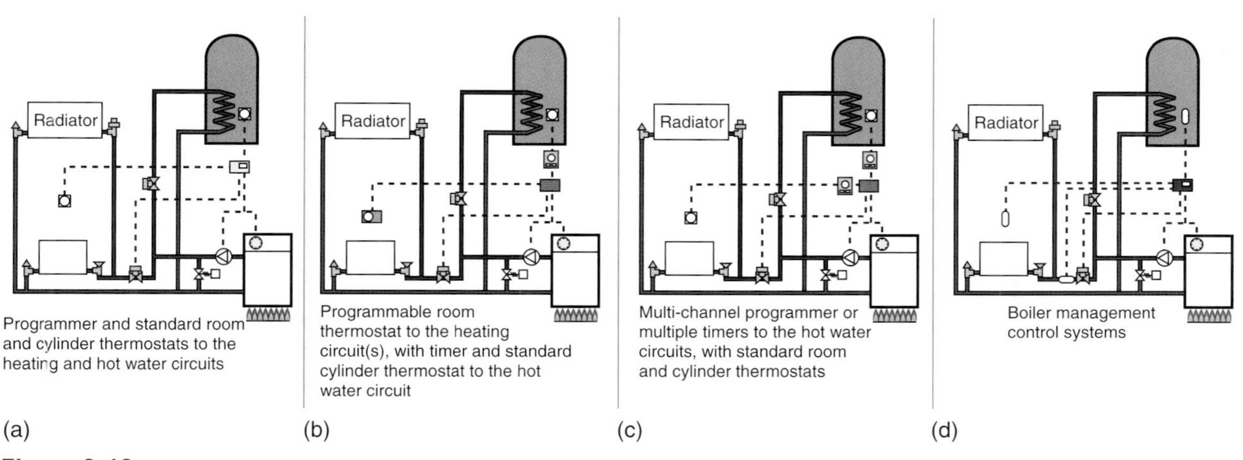

(a) Programmer and standard room and cylinder thermostats to the heating and hot water circuits

(b) Programmable room thermostat to the heating circuit(s), with timer and standard cylinder thermostat to the hot water circuit

(c) Multi-channel programmer or multiple timers to the hot water circuits, with standard room and cylinder thermostats

(d) Boiler management control systems

Figure 8.12

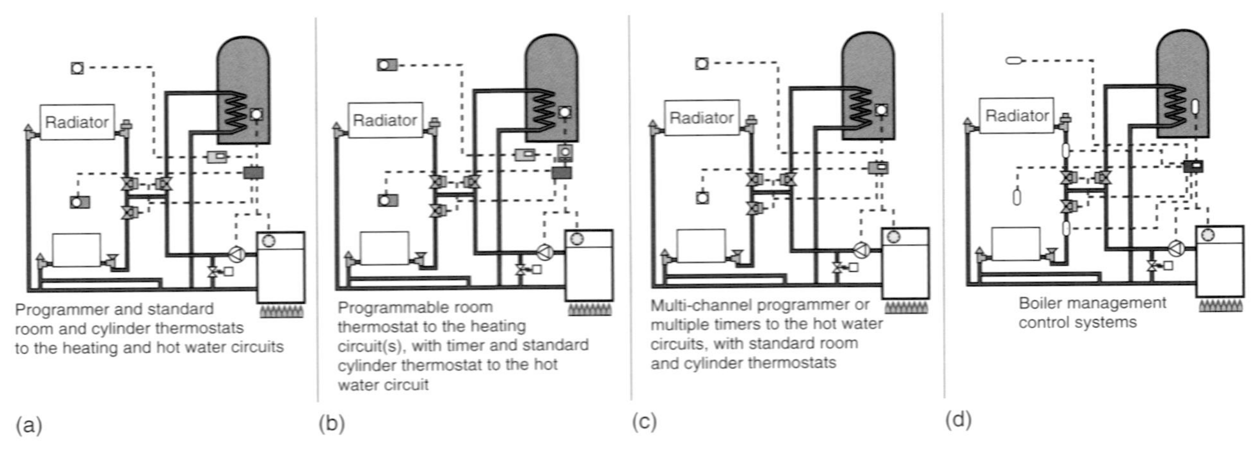

Figure 8.13

Multi-zone installation

This type of installation is recommended for use in dwellings that have a usable floor area greater than 150 m^2. The heating circuits should be divided into a minimum of two zones, with a separate zone for the hot water circuit, having its own independent time and temperature control. The use of the two individual *two-port valves* offers greater flexibility to the design of the heating system, and allows the building to have separate time- and temperature-controlled heating zones (Figure 8.13).

Activity 8.2

Consider the following situation:

A householder has an existing 'Y' Plan system in a house with a floor area less than 150 m^2. They have an extension built, adding a second bathroom and two additional bedrooms occupying a floor area above 150 m^2. The heating system will have to be extended to these areas. What would be your advice? Record your comments in your portfolio and then see what we think at the end of this book.

Key system design components

Whatever system is installed, some of the key components within the pipework system include the:

- Feed and expansion cistern
- Primary open safety vent
- Air separator.

Key point

When filling the F&E cistern the water level should be set low to allow for the expansion of the heated system water, equal to 4% of the total water contained within the system.

Feed and expansion cistern

The feed and expansion cistern (F&E cistern) is used on all open-vented central heating system types and is one of the most important components in a heating system. The cistern serves two purposes:

- It allows the system to be filled with water
- It allows for the expansion of heated water within the system, hence the name feed and expansion cistern.

We remember from the hot water systems unit that when water is heated between 4 and 100°C, it expands approximately 4% (1/24) in volume, and allowances must be made for this expansion. Figure 8.14 shows the space required for the expansion of heated water within the cistern.

It is very important that the cistern is located at the highest point in the system, and it must not be affected by the position and the circulating head of the pump. On fully pumped systems, the cistern should be located at a minimum of 1 m above the highest point of the system.

The F&E cisterns for heating systems (see Figure 8.15), should comply with the requirements of BS 417 and BS 4213 and current Water Regulations.

Try this

The cistern and float valve must also be capable of resisting a water temperature of 100°C. Why would this be necessary? Are there any circumstances when it could come into contact with boiling water? Think about the reason for this, and see if you can find out the answer.

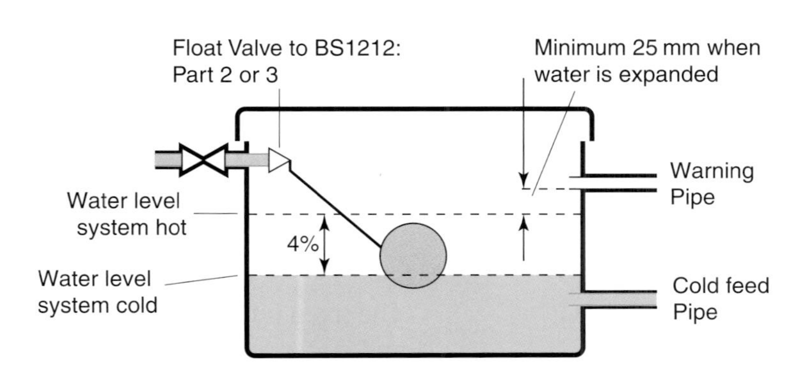

Figure 8.14 F&E details

UNIT 8

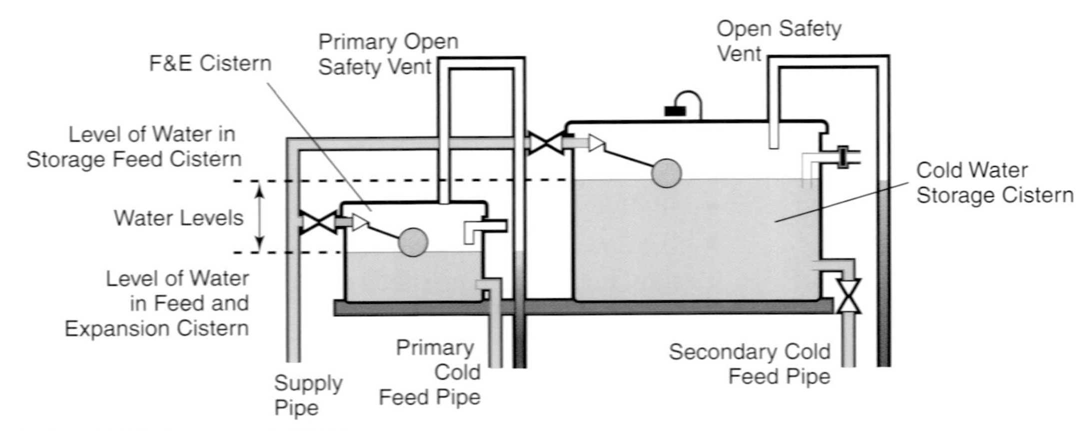

Figure 8.15 F&E cistern and CWSC arrangements

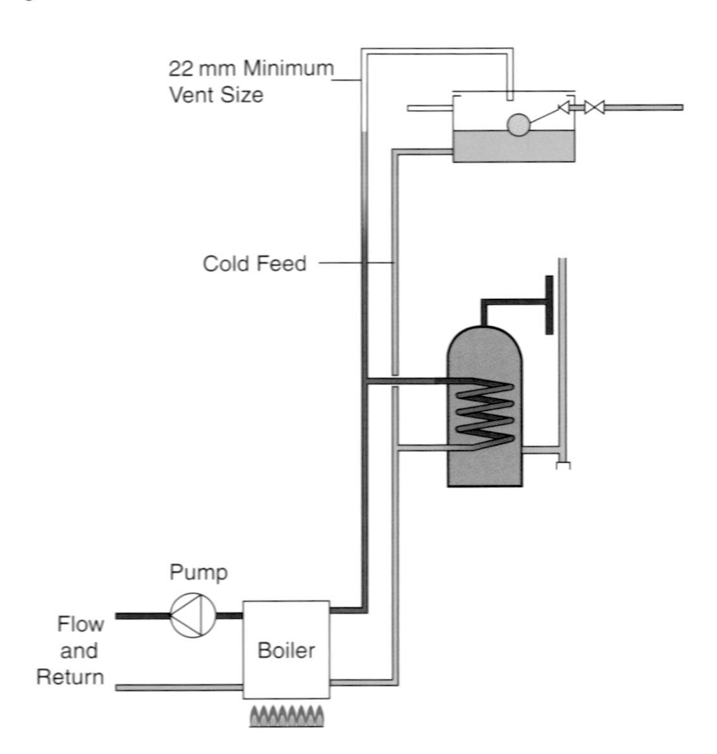

Figure 8.16 Installation of an open safety vent

Primary open safety vent

Try this

Have a look at Figure 8.16. What do you think is the main purpose of the primary open vent pipe?

The main purpose of the primary open vent pipe is to enable steam and water to be *safely vented* away from the system, should the system overheat due to component failure such as a faulty boiler thermostat. This water is discharging at extremely high temperatures, and therefore the cistern and float-operated valve must be capable of withstanding high temperatures.

The open vent will act as an air release in the initial filling and commissioning stages of the system, but it is provided primarily to protect the system for safety purposes, and is referred to as the open safety vent. Very little expansion from heated system water takes place up the open vent pipe and it is incorrect to refer to it as 'the expansion pipe'. The open vent pipe should ideally have a 'dry head' of 450 mm to prevent pump over.

The safety vent pipe must have a minimum diameter of 22 mm, and should never be valved.

Activity 8.3

Why must the safety vent pipe not be valved? Jot down your thoughts for inclusion in your portfolio, and check it out at the end of this book.

Cold feed pipe

The cold feed pipe on heating systems has two functions:

- It feeds water down from the F&E cistern, maintaining the water level in the system
- It allows the system of water, on heating up, to expand into the cistern. The pipe also helps the safety open vent pipe to deal with overheated situations, keeping the system full of water.

The cold feed pipe must have a minimum diameter of 15 mm, and the fitting of isolation valves is not recommended.

Air separators

It is a well-known fact that the presence of air in a system creates a number of problems such as:

- Noisy circulating pump
- Kettling boiler
- Continual venting of radiators.

De-aerators (Figure 8.17) are used to address the above problems, and are designed to remove air from open-vented fully

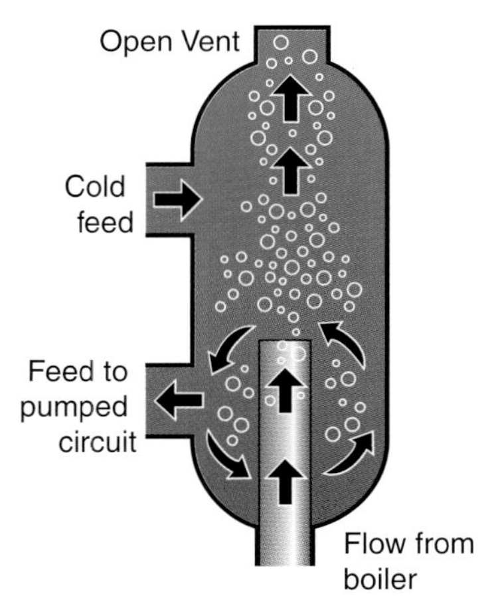

Figure 8.17 De-aerator

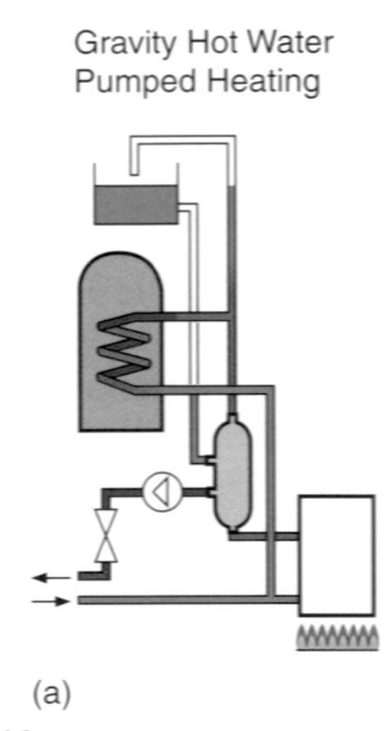

(a)

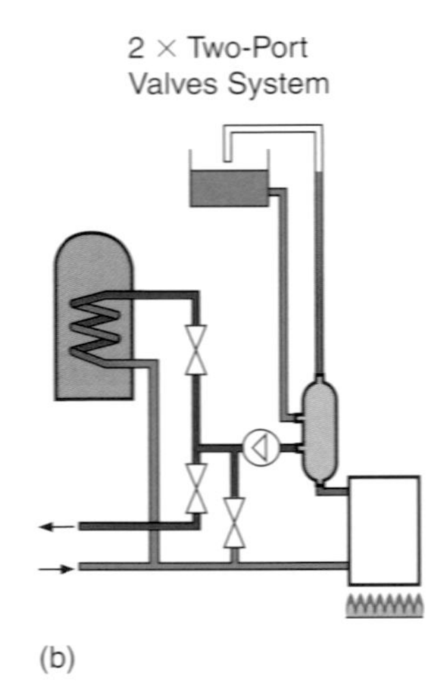

(b)

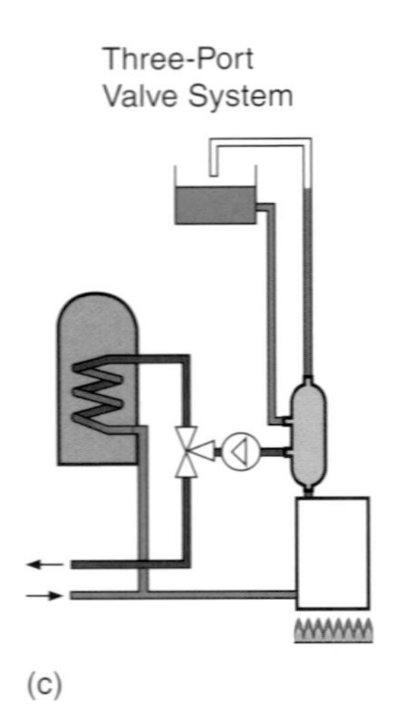

(c)

Figure 8.18

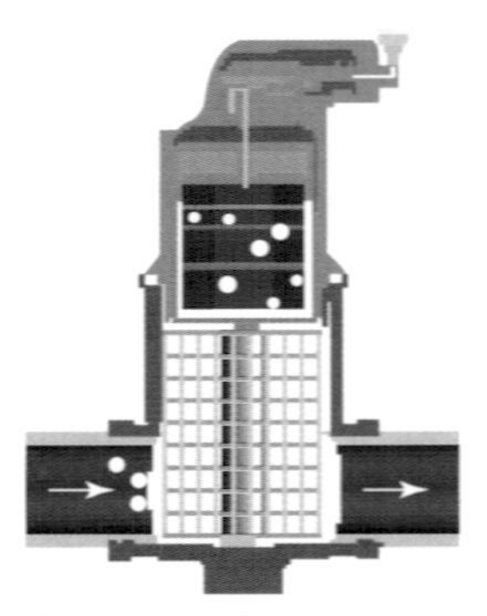

Figure 8.19

pumped or semi-gravity systems. They are ideal for installations where a low head is unavoidable, reducing the possibility of a pump over (this is when the circulator pumps water out of the open safety vent pipe). They also reduce the possibility of air ingress into the system (this is air being sucked down the open safety vent pipe).

The air separator is designed to enable the cold feed and open vent pipe to be joined close to each other. This close connection causes the flow of water to create turbulence allowing air bubbles to formulate and then simply rise out of the system. The separator helps prevent system noise, and reduces the risk of system corrosion.

Figure 8.18 shows examples of how the de-aerators would be installed in different applications.

Other types of de-aerators like the one in Figure 8.19 work by removing the air bubbles that come out of solution when the system water is heated, and before it can be re-absorbed as it cools. De-aerators of this type should be fitted on the flow pipe as close as possible to the boiler (Figure 8.19).

Check your progress by answering questions to see how you are doing. Note this is a self-assessment exercise. Once you have completed answering the questions, check the answer by looking back through the text, then discuss your answers with a tutor.

Test yourself 8.1

1. What Building Regulation document sets out energy conservation requirements for central heating systems?
2. List three specific areas covered under the regulation.
3. An existing pumped system for heating and gravity circulation could be modified to meet the Building Regulation requirements by installing
   a. A two-port valve, cylinder stat, room stat and TRVs ☐
   b. A cylinder stat, room stat and lockshield valves ☐
   c. Gravity check valves, room stat and TRVs ☐
   d. An air separator, two-port valve and cylinder stat ☐
4. Fill in the missing gaps in the following paragraph using the words provided. The paragraph describes a fully pumped system using a three-port mid-position valve.

   This system is designed to provide separate ________ control of the heating and ________ circuits. To fully meet the requirements of the ________, ________ control must be via a ________, and ________ and an ________ valve must be fitted.

   Temperature, gas, heating, Water Regulations, pressure, downstairs, domestic hot water, Building Regulations, time, programmer, TRVs, automatic bypass, automatic air vent, time, upstairs, time clock.
5. Which type of system should be used for dwellings with floor areas above 150 m^2?
   a. Fully pumped system using two × two-port valves ☐
   b. Fully pumped system using three-port mid-position valve ☐
   c. Fully pumped system using two position diverter valve ☐
   d. Pumped heating and gravity hot water using two-port valve ☐
6. What is the main purpose of an F&E cistern?
7. What percentage of the volume of system water should be allowed for expansion in the F&E cistern?
   a. 7% ☐
   b. 6% ☐
   c. 5% ☐
   d. 4% ☐
8. For safety reasons, it is important that a lockshield valve is installed to the primary open safety vent pipe. True or False?

Central heating system components

By now, you should have acquired a good knowledge of the layouts of central heating systems and some of the key system components in terms of the installation design. In this section, you will move on to look at some of the other system components such as:

- Circulating pumps
- Heat emitters
- Radiator valves
- Gas boilers.

Domestic circulators

Circulators are a very important component of heating systems. The purpose of the circulator, also referred to as a pump, is to create a flow of water around the heating system, ensuring that the water is delivered at the desired temperature throughout the system.

Here is what a circulator looks like, and when installed, they are fitted with isolation valves, as shown in Figure 8.20. These allow removal of the circulator for maintenance purposes, without the need to drain the system.

A pump consists of an electric motor which drives a circular veined wheel called an impeller. The rotating impeller pulls the system water through its centre, and then throws it out again by centrifugal force, creating water circulation, as shown in Figure 8.21.

Hot tips

Why not contact a pump manufacturer and find out more about modulating pumps.

Most circulators have a variable three-speed setting, and manufacturers provide performance data for these settings, which show:

- Pressure in Kilo Pascals (kPa) and metres head (m^3/h)
- Flow rate velocity in metres per second (m/s).

Most circulators deliver a 5 or 6 m working head, which is adequate to overcome the resistance of the index circuit (the circuit offering the greatest frictional resistance to flow). The water

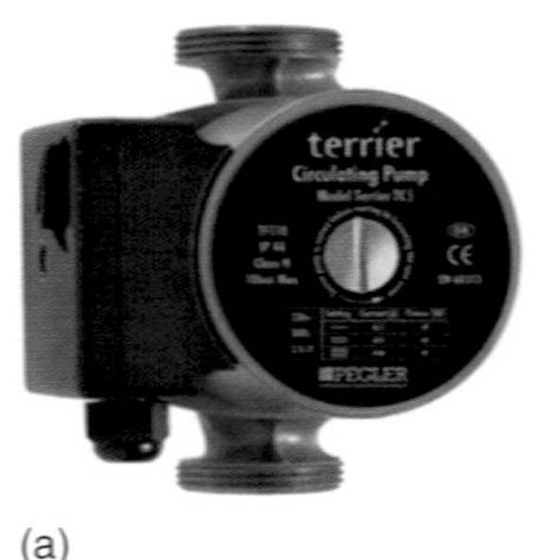

(a)

(b)

Figure 8.20 (a) Pump and (b) Isolating valves (Reproduced with permission of Pegler Limited)

UNIT 8

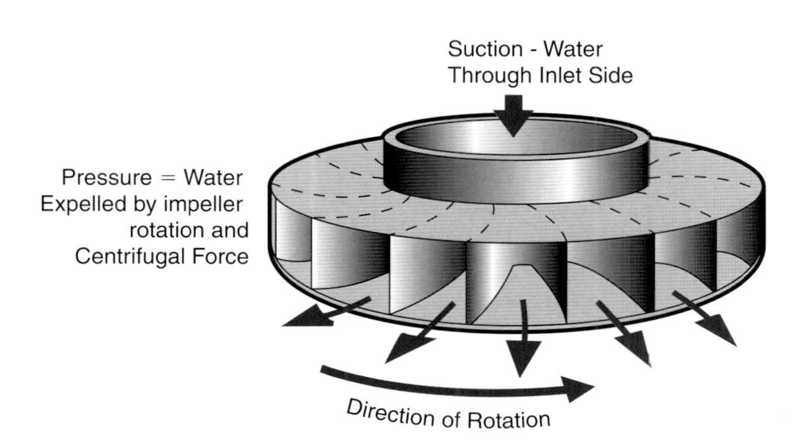

Figure 8.21

Key point

Careful consideration must be given to the location of the circulator on a heating system. It is good practice to install the circulator so that it gives a positive pressure within the pipe circuit. This ensures that air is not drawn into the system through minute leaks from valve gland nuts.

flow rate velocity in heating systems should not exceed 1.0 m/s for small bore systems, and 1.5 m/s for microbore systems, velocities above this rate could result in system noise.

With the ongoing advances in design and technology, energy saving circulators are now available and offer a three-speed operation and modulation mode technology. Circulators of this type (modulating) automatically adjust their operating performance (speed) in terms of pressure in order to match the flow rate demand of the heating system.

The location of the circulator is even more critical on fully pumped systems, in relation to cold feed and open vent pipe, and it must be located in a position to ensure that there is no positive pressure or negative pressure at the open safety vent, which could result in water pumping over, or air being drawn down into the system.

The position at which the cold feed pipe from the F&E cistern connects into the system is regarded as a neutral point, and from this point to the circulator, it would be under a negative influence. Ideally, so as to give a positive pressure, the circulator should be positioned after the cold feed pipe connection, with the open safety vent pipe being a maximum distance of 150 mm away from the cold feed.

Key point

The term heat emitter is used to describe the various designs of heat emitting device, a few examples include radiators, skirting convector heaters, fan-assisted wall convectors, towel rails and low surface temperature radiators (LST).

Figure 8.22 shows the recommended pipework arrangement. There are a few important key principles to note on circulators:

- They must be installed in accordance with the manufacturer's instructions, in the recommended positions
- They must be accessible for inspection, venting and maintenance purposes
- They must be fitted with isolation valves
- Prior to fitting, the pipework system must be flushed, to remove any debris, such as steel wool, solder, dirt, etc

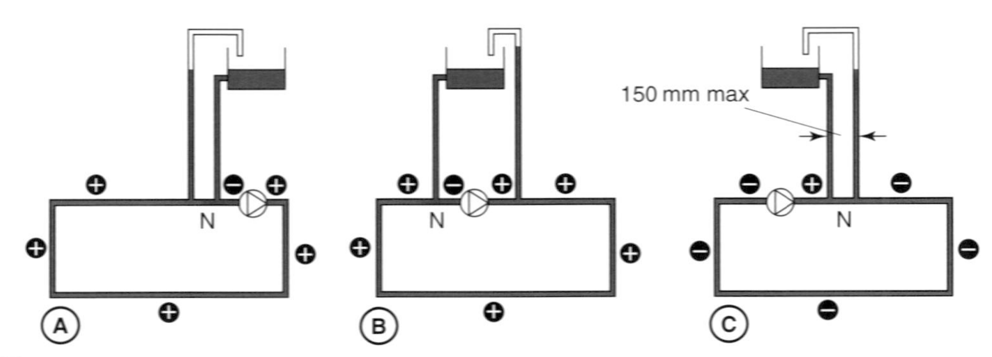

Figure 8.22

- On vertical pipework installations, circulators should be installed to pump upwards, which allows for easy air removal and prevents cavitation (cavitation means pump noise caused by trapped air within the impeller)
- Circulators must be earthed when connected to the electrical supply.

Heat emitters

Radiators

Key point

The fins increase the heat emissions from the hidden surfaces of the radiator, giving improved convection flow, and a higher heat output.

The most common type of heat emitters used on heating systems in the United Kingdom is the steel panel radiator. They are available in several designs and can be single or double panel form, with or without additional convector fins, which are welded directly onto the waterways on the back of the radiator.

Despite their name, a radiator only emits 10% of its heat by radiation, the remaining 90% is emitted by convection.

Radiator types

Try this

Why not take a trip to your local Plumbers' merchants and obtain some manufacturers' catalogues for central heating radiators? These will include the various designs, sizes, and will also give you the heat output for each model. Most DIY outlets will also provide this information. Try to find catalogues that also given tips on installation, and keep copies to include in your portfolio.

Figure 8.23 Cast iron column radiator

Try this

Below is a list of radiator types. Use manufacturers' catalogues to see what they look like, and to find out more about them; keep copies to include in your portfolio.

- Compact radiators
- Combined radiator and towel rail
- Ladder towel rail
- Flat-fronted radiators
- Low surface temperature radiators (LST)
- Skirting convector heating
- Fan-assisted convectors
- Cast iron column radiator (see Figure 8.23)
- See what range of designer radiators are also on the market (www.radiatorfactory.co.uk)

Radiator accessories

Connections

Radiators are available with either top and bottom, or bottom hole ½″ BSP tappings, depending on their design. Domestic installations usually have the radiator valves located at the bottom, and at the opposite ends. Look at the catalogues from the last activity to see examples.

Key point

On some radiators, connections can be placed at the top and bottom of opposite ends, the latter being an ideal location, where the occupant is elderly or disabled making access to the valve easier.

Fixing brackets

These are usually supplied with the radiator, and design may vary depending on the manufacturer. The brackets may have deep hanging slots and corresponding lug positions to provide greater stability to the radiator. Plastic inserts are provided to seat the radiator precisely, and to minimise noise caused by the expansion and contraction of the radiator.

Air vents and plugs

These components form part of the radiator, and are generally supplied with the radiator. Seamless/rolled top radiators generally have the air vent manufactured into the radiator. Radiator packs also contain an air vent key for venting of the radiator.

Radiator valves

With the changes to the Building Regulations Approved Document Part L1, 2002 new heating installations are required to be fitted with TRVs. Older, existing heating installations are often fitted with manual-type radiator valves, so you may be required to install or repair them while carrying out maintenance work.

Here is a selection of valves available from manufacturers.

Manual radiator valves

All the radiator images have been reproduced with the kind permission of Pegler Hattersley, contact details for which are at the end of this unit.

Manual radiator valves are referred to as wheel head valves; they are available in straight or angle pattern, can be of brass or brass with a chrome finish, and pipe sizes range from 8 to 15 mm. The opening and closing of the valve (rotating the head clockwise closes the valve) has to be done manually, by turning the wheel head. Valves of this type do not permit specific temperature control in the same way as a TRV. Figure 8.24 shows the internal parts of a wheel head valve.

Lockshield valve

These are manual valves and are adjusted by the plumber while balancing the central heating system. Balancing means, adjusting the valve radiator by radiator to ensure an even temperature from the first radiator on the system to the last.

Once the valve has been adjusted by the plumber, it cannot be operated by the occupants of the dwelling, thereby preventing tampering. The valve spindle is concealed within a plastic cap; these can be either push fit or fixed with a small screw similar to that shown in Figure 8.25.

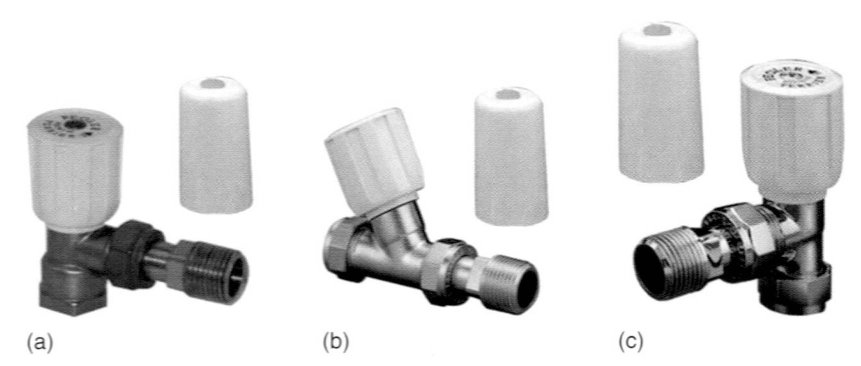

Figure 8.24 (Reproduced with permission of Pegler Limited)

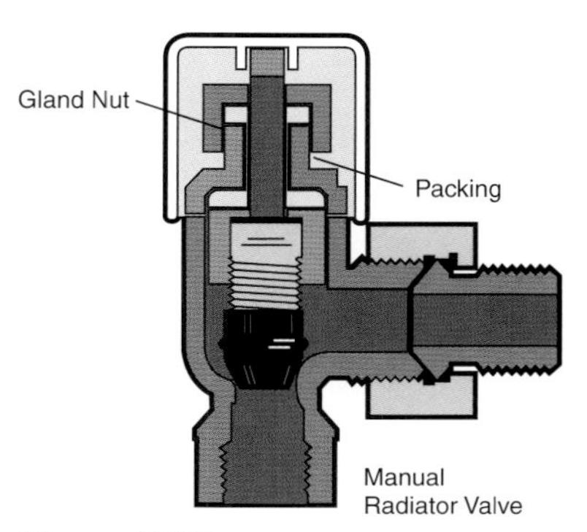

Figure 8.25

Adjustment of the valve is achieved by removing the plastic cover and rotating the valve spindle with a special key, pliers or an adjustable spanner. Figure 8.26 shows they are available in straight or angle pattern, brass or chrome finish, and additionally they can have a built-in drain-off point.

Manual radiator valves conform to the British Standard requirement if they have screwed fixed tops. They are manufactured to either BS 2767-4 or BS 2767-10, the last digit being the pressure that they will withstand, e.g. −10 = 10 bar-rated.

Key point

Adjustment of the valve is only necessary if removing the radiator or balancing the system when commissioning.

Thermostatic radiator valves

Commonly referred to as TRVs, the heads of the valves have a built-in heat sensor. When positioned to the desired setting, the sensor will open and close the valve automatically as governed by the temperature of the room.

The sensors can be either wax- or liquid-filled, dependent upon the type and manufacturer, as shown in Figure 8.27. As the sensor heats up, the wax or the liquid expands and is forced into the bellows chamber within the head. As the bellows expands, it pushes down on the pressure pin within the valve, thus closing the valve.

Thermostatic heads have a number of settings, normally ranging from 0 to 5, this allows variable room temperatures, from 5 to 28°C.

Thermostatic heads are available with remote sensor units, as shown in Figure 8.28, which can be sited up to 2 m away from the radiator. They are particularly useful in situations where curtains may cover the valve, resulting in an inaccurate sensor reading.

In situations where the thermostatic head would be inaccessible for adjustment to the user, or if the user is elderly or has a disability, remote adjustment heads are available (see Figure 8.29).

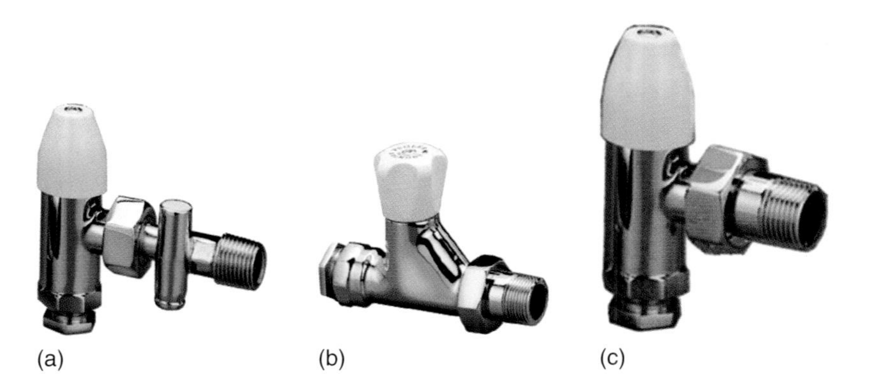

Figure 8.26 (Reproduced with permission of Pegler Limited)

(a)

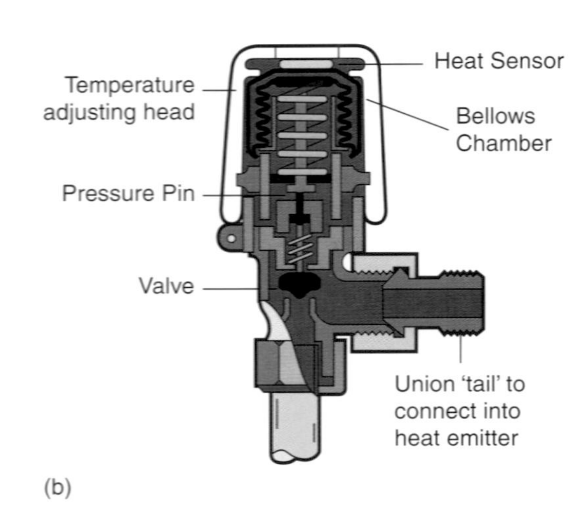

(b)

Figure 8.27 (Reproduced with permission of Pegler Limited)

Figure 8.28 (Reproduced with permission of Pegler Limited)

Figure 8.29 (Reproduced with permission of Pegler Limited)

Here are a few pointers when using TRVs:

- Ideally, when installing TRVs, it is recommended that the system is fitted with a bypass valve to prevent the boiler and pump working against a closed system, should all the TRVs close down once the rooms have reached their required temperature.
- The valves must be installed with the directional flow arrows in-line with the flow of water; failure to do this will result in the noisy valve shut off.
- The valves must be installed in accordance with manufacturers' instructions.
- Remember that the radiator located in the area of the room thermostat should be fitted with lockshield valves on the flow and return pipework, in lieu of a TRV. This will ensure an accurate temperature reading at the room thermostat and also a minimum system bypass load of 10%.
- Certain types of TRVs do not have positive 'shut off ', and if removing the radiator the head must be removed, and replaced with a special manual locking top; this secures the pressure pin down on to its seating, closing the water flow from the valve. Failure to do this on a non-positive shut-off valve will result in water flowing from the valve, should the temperature in the room drop.

Towel rail valves

Decorative period towel rail valves are available in various finishes.

Key point

Because this book covers the requirement of a Level 2 certificate, the coverage of oil and gas boilers is included to provide a basic appreciation.

Central heating boilers

Solid fuel boilers

Although the Level 2 technical certificate in plumbing only covers solid fuel boilers, with oil and gas being dealt with at Level 3, you will also cover gas boilers in this unit to give you a more detailed insight into the industry if you decide not to progress to Level 3.

In terms of a domestic installation you are likely to install either of the two types:

- Independent freestanding boiler, for use internally or externally
- Solid fuel room heater with back boiler.

Independent freestanding boiler – internal

This type of boiler can be obtained in a range of outputs from 13.5 to 29 kW (45,000–100,000 Btu/h). They have a built-in fan to boost heat output on demand. They are usually fed by a hopper which is easy to fill from the top and a typical unit will hold enough fuel to burn continuously for 14 h.

They are designed for installation with a Class 1 flue system.

Try this

Try and find out what is meant by a Class 1 flue system. Here are the typical respective specifications for models with outputs of 13.2 kW, 17.6 kW, 23.5 kW and 29.3 kW

- Rad surface output 21 m^2, 30 m^2, 42 m^2, 53 m^2
- Hopper capacity 30 kg, 37 kg, 41 kg, 48 kg
- Flue Outlet 125 mm, 125 mm, 125 mm, 125 mm
- Weight 125 h kg, 155 h kg, 196 h kg, 230 h kg water flow connections
- 2 flow return 1″BSP 1″BSP 1.5″BSP 1.5″BSP

Independent freestanding boiler – external

This is based on the same principle as the internal boiler. These are installed in purpose-made boiler housings, usually built in the same material to match the exterior of the building. Standards of insulation have to be high.

Standard room heater

This type of appliance is available in two ratings: 33,000 and 45,500 Btu (10 and 13 kW).

The 10 kW version will heat the room where it is located, the domestic hot water, and will provide heat to a total radiator surface of 13.8 m^2 for the 13 kW rising to 20.5 m^2 for the radiators. It fits a standard fireplace opening and is thermostatically controlled.

Flue dimensions are 125 mm (5 in.). It has two flow tappings and two returns for the domestic hot water and heating all 1″ BSP.

Convector room heater

This is designed to provide more direct heat into the room, where the heater is sited. It still provides outputs of either 10 or 12 kW, enough for the central heating requirements of a 2/3 bedroom house. Like the standard room heater it:

- Fits a standard fire place opening
- Has a 125 mm (5″) flue
- Is thermostatically controlled
- Has two flow and return tappings of 1″ BSP.

Try this

This unit only provides basic information about solid fuel boilers; you need to build on this information by finding out more about the various appliances available, and their installation requirements. You can do this by visiting a plumber's merchants and asking for catalogues. Alternatively, you can contact the manufacturers directly. Here is a good contact to get you started:

Trianco Limited
Thorncliffe
Chapeltown
Sheffield
S35 2PZ
Tel: 01142 257 2300
Website: www.trianco.co.uk.

Gas boilers

There are many different makes and models of boilers on the market today like:

- Wall-mounted
- Floor standing
- Gas-fired back boiler (GFBB) units.

Boilers of this type are sometimes referred to as traditional boilers.

All these appliances have various flueing arrangements, and are available with different kilowatt heat outputs. They are designed to be installed on open-vented fully pumped and gravity systems, although some are suitable for sealed pressurised systems.

Wall-mounted

The wall-mounted boiler is popular with installers and customers as it offers flexibility in terms of location. The heat exchanger in wall-mounted boilers can be either lightweight cast iron or prefabricated lightweight metal, which makes them lighter and easier to lift when installing.

Heat exchangers on a cast iron boiler have a close network of waterways, where the alloy tube type passes the tubing through a series of fins. The water contained within the heat exchanger can be as little as 1 l and is referred to as a low water content boiler. The water is heated by hot gases passing through the heat exchanger, when the burner is in operation.

Boilers with low water content heat exchangers generally incorporate a pump overrun device to remove the residual heat from the boiler and must be fitted with some form of bypass. This prevents the boiler from overheating and 'kettling' if the water circulation flow rate through the boiler is slowed down.

Boiler outputs range from 9 to 26 kW, the average SEDBUK rating for these types of boilers is D. The boilers are suitable for most types of systems, but this will be dependent on the model type and manufacturer's recommendations.

Floor standing

Figure 8.30 Floor-standing boiler (Reproduced with permission of Halsted Boilers)

These usually contain a robust cast iron heat exchanger and are designed to sit on a non-combustible floor (e.g. concrete pad), and are usually installed in the kitchen, as shown in Figure 8.30.

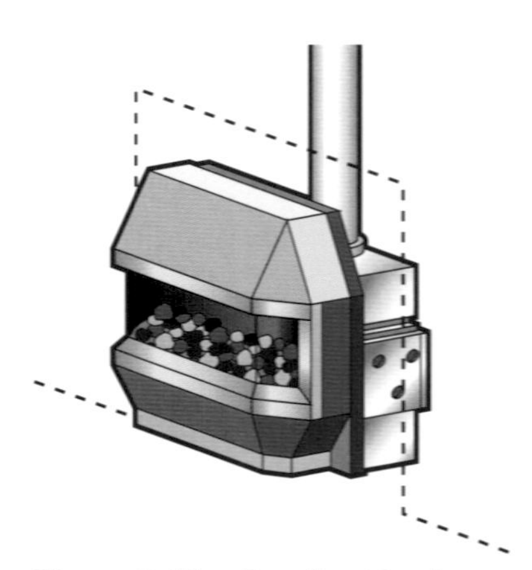

Figure 8.31 Gas-fired back boiler

The boiler has various flue options including natural draught, room-sealed, and even a fan flue version. Heat outputs range from as low as 11.7 kW up to 23 kW. The SEDBUK rating for this appliance is D.

Gas-fired back boiler (GFBB)

These are natural draught appliances that are designed to be installed in builder's openings (fire places) (see Figure 8.31). They were a very popular type of boiler from 1970 up to the late 1990s, slowly being phased out by the replacement of more efficient wall-mounted boilers. Today, they are available with two heat outputs, 13.2 and 16.1 kW, offering a wide choice of fire fronts. The flue and ventilation requirements must be in accordance with BS 5400:2000.

The appliance has the minimum SEDBUK D rating. What is the efficiency of this boiler and would it meet the requirements for an installation into a new building?

SEDBUK D rating is 78–82% and given that the regulations require a SEDBUK rating of A or B as of 1 April 2005, it would not meet the requirements.

Energy efficient boilers

Energy efficient boilers include:

- System boilers
- Combination boilers
- Condensing boilers.

System boilers

Boilers of this type supply domestic hot water indirectly to the storage vessel. The water feed supply to the system is directly from the cold mains, through an approved filling loop, as shown in Figure 8.32.

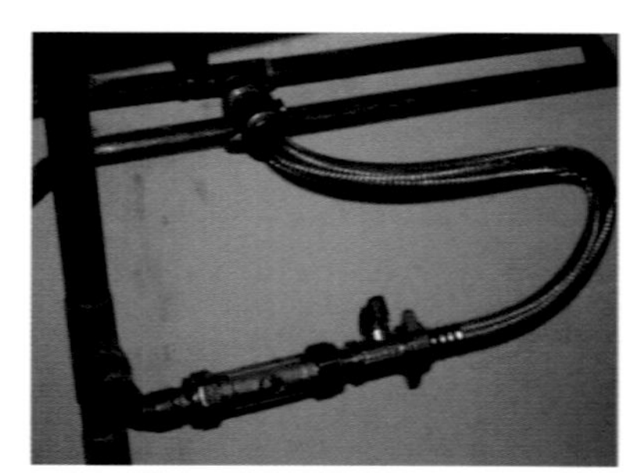

Figure 8.32 Typical approved filling loop

They are referred to as pressurised sealed systems, and do not require a separate F&E cistern. The expansion of the system water is taken up by an expansion vessel, contained within the boiler, usually sited at the rear of the boiler, as shown in Figure 8.33.

Figure 8.34 shows the circulating pump, system bypass, pump overrun, pressure relief valve and other controls are built-in, pre-wired, pre-plumbed and pre-tested. Flue options are generally room-sealed fan flue, heat outputs range from 13 to 24 kW with

Figure 8.33 Example of system boiler (Reproduced with permission of Ravenheat Manufacturing Limited)

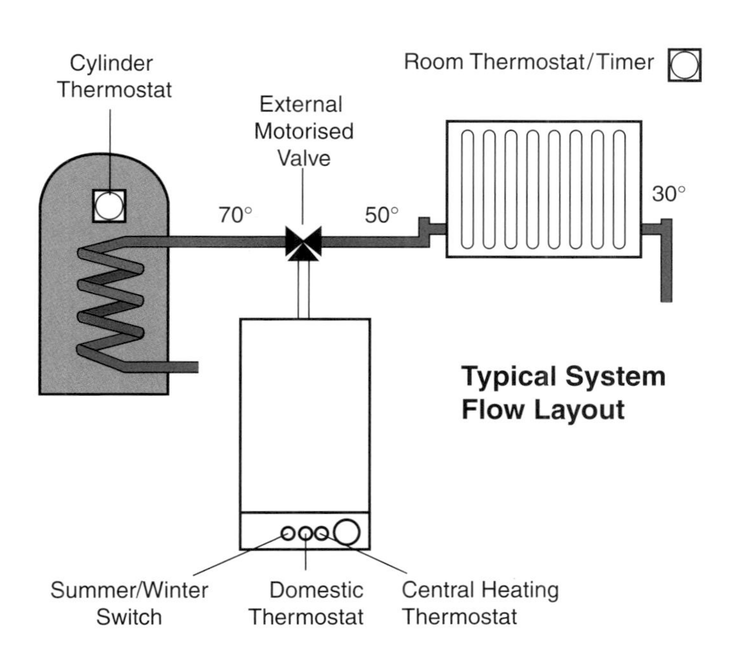

Figure 8.34 Typical system layout

most makes having a modulating gas valve. New, condensing models will achieve a SEDBUK rating A.

Combination boilers

A traditional heating boiler supplies hot water indirectly to the storage vessels, which is usually sited on the first floor level. A combination boiler supplies domestic hot water directly from the boiler via a special heat exchanger (see Figure 8.35).

Combination boilers, referred to as 'combies', are designed to heat up cold water as it passes through the hot water heat exchanger, providing a supply of instant hot water at the tap, on a similar basis to that of an instantaneous water heater.

The boiler supplies heating through an additional heat exchanger within the unit, the systems are generally sealed and require an

Figure 8.35 Combination boiler (Reproduced with permission of Ravenheat Manufacturing Limited)

approved filling loop, although some types will work on an open system. The integral components are very similar to that of a sealed boiler with the addition of a three-way diverter valve and hot water heat exchanger.

Flue options include room-sealed fan flue, and twin flue pipe options. Heating outputs range from 11 to 29 kW with most makes having a modulating gas valve. Condensing combination models will achieve a SEDBUK rating A.

Other types of combination boilers are available that contain an additional small hot water storage cylinder containing approximately 151 litres of stored water; this helps to increase the volume of hot water delivered at the tap.

High efficiency boilers

Higher efficiency (HE) boilers are the most energy efficient boilers currently in the market. The efficiency of a typical traditional boiler is around 78%, whereas current condensing boilers can have an efficiency of over 86% (SEDBUK A&B).

HE boilers have a larger heat exchanger than traditional boilers, and utilise every scrap of energy from the consumed gas, i.e. both latent heat and sensible heat. To achieve this, the boiler incorporates a fan and two separate heat exchangers; as the hot flue gases (having a temperature around 180°C) pass through the heat exchangers, heat is extracted from the hot gases.

This process cools the flue gases to a temperature of around 55°C. The reduction in temperature causes water vapour to form as condensate. This is collected in a receptacle called a condense trap and is taken away from the boiler to a drain or soak away. HE boilers are available as combination boilers, or system boilers.

Flue options include room-sealed fan flues, and twin flue pipe options. Heat outputs range from 13 to 25 kW with most makes having a modulating gas valve. The SEDBUK rating for high efficiency boilers is A to B (Figure 8.35).

Microcombined heat and power units

The boiler provides heat and hot water like a normal boiler. However, it also contains a generator which produces electricity from the same gas to power lights and other electrical equipment, such as

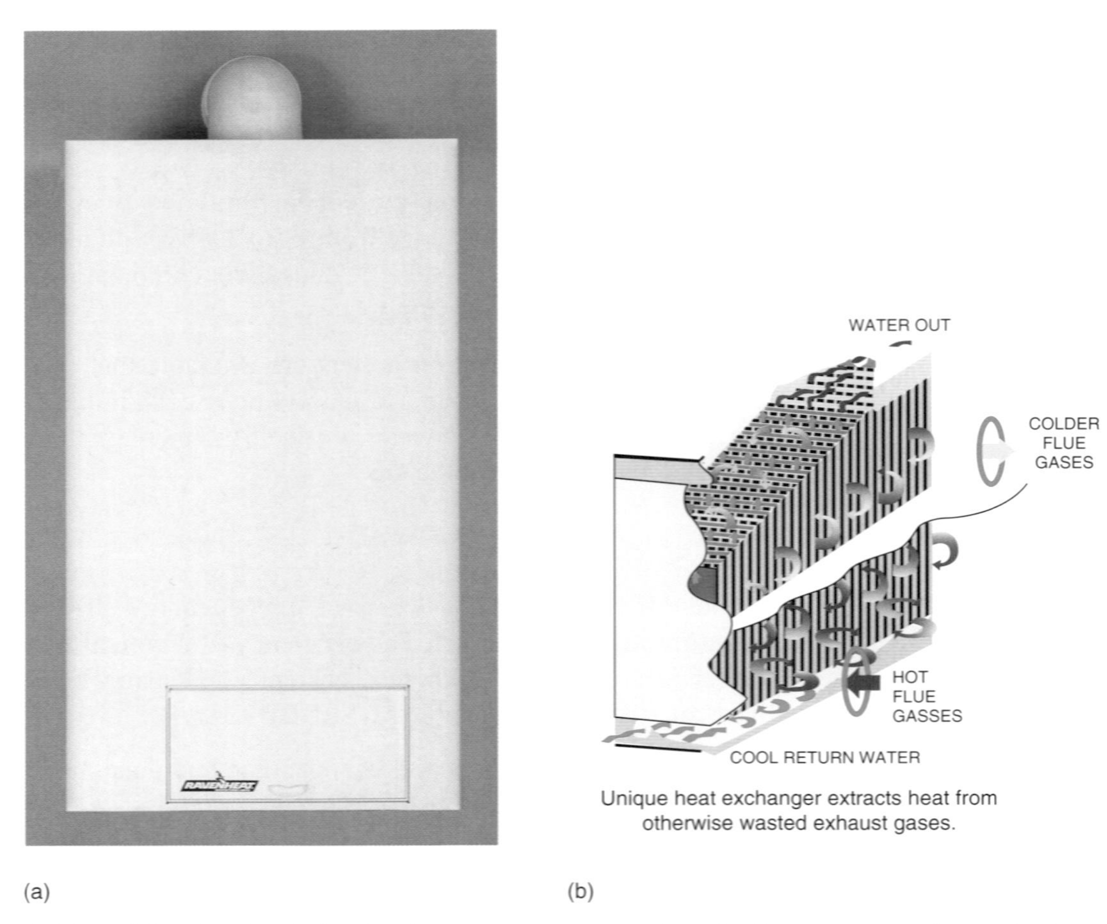

Figure 8.36 (a) Condensing boiler (b) Heat exchanger (Reproduced with permission of Ravenheat Manufacturing Limited)

TVs and fridges, and any surplus electricity is bought back from consumers by the supplier.

This technology is known as 'microcombined heat and power'. Currently these systems are relatively rare in the UK, but they are being piloted and tested in a number of areas. Given the Government's policy to improve energy efficiency, this type of technology could soon become commonplace.

It is always in your best interest to keep abreast of new developments. Make regular visits to your local library to read trade magazines related to plumbing and energy efficiency. One such publication is Energy in Buildings and Industry. Their website is www.eibi.co.uk.

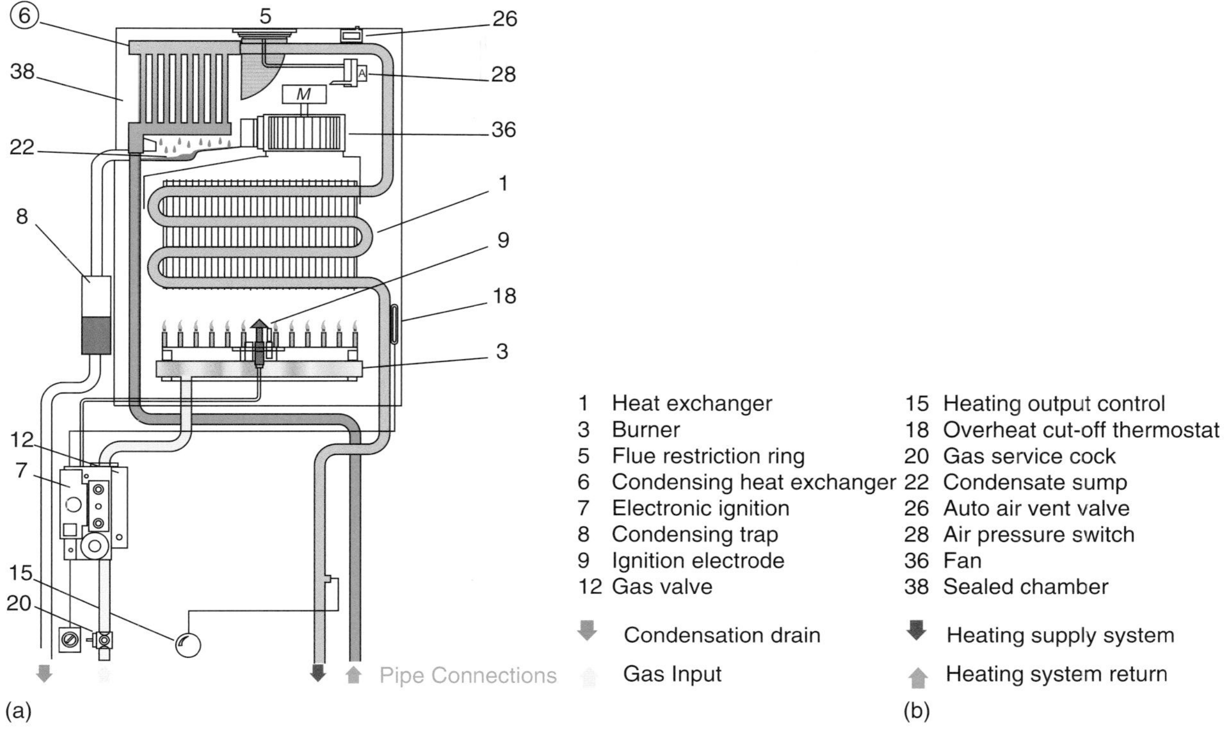

Figure 8.37 (a) and (b) Internal workings of the condensing boiler (Reproduced with permission of Ravenheat Manufacturing Limited)

Boiler efficiency

Now let us look at how boiler efficiency is worked out. Boiler manufacturers give the heat input rates and output rates of the boiler; these rates can be found in the instructions, and on the identification badge plate of the boiler.

We calculate

$$\frac{\text{Boiler output}}{\text{Boiler input}} \times 100 = \text{Efficiency \%}$$

Try one out: A wall-mounted room-sealed boiler has an output of 11.72 and an input of 14.65

$$14.65 = \frac{11.72}{0.08} \times 100 = 80\%$$

The calculations show that the boiler has an overall efficiency of 80%. Therefore the boiler will have a SEDBUK D rating.

Test yourself 8.2

1. In order to avoid noise on a small bore heating circuit, the recommended maximum flow rate in m/s for the pump should be:
   a. 1.0 ☐
   b. 1.4 ☐
   c. 1.8 ☐
   d. 0.4 ☐
2. Radiators give out 90% of their heat by:
   a. Convection ☐
   b. Radiation ☐
   c. Conduction ☐
   d. Evaporation ☐
3. What is the maximum permitted surface temperature in °C for a low surface temperature radiator when the system is operating at design temperature?
   a. 30 ☐
   b. 34 ☐
   c. 43 ☐
   d. 50 ☐
4. What type of boiler is being described below? This boiler can achieve an efficiency of over 86%. It has a larger heat exchanger than other boilers, and utilises every scrap of energy from the consumed gas, i.e. both latent heat and sensible heat. To achieve this, the boiler incorporates a fan and two separate heat exchangers; as the hot flue gases (having a temperature around 180°C) pass through the heat exchangers, heat is extracted from the hot gases.
5. Which type of boiler is most likely to meet the requirements of SEDBUK rating A?

Central heating system controls

Today, everyone has a moral obligation to do what they can to conserve energy. Energy conservation is a high priority of the Government, which has led to the amendments in the Building Regulations. These changes to the Building Regulations Part L1 A and B 2006 have had a significant effect on system design, in particular its control. In this section, we look at mechanical and electrical controls used on domestic heating systems.

Heating controls

Without any form of heating controls, other than the boiler thermostat, a heating installation will provide heating and hot water. But without additional heating and water temperature control, the living environment would possibly become too warm and the hot water too dangerously overheated. Overall, the installation would be uneconomical to run, wasting heat and therefore energy and fuel. Let us take a closer look at the components that make up the control systems.

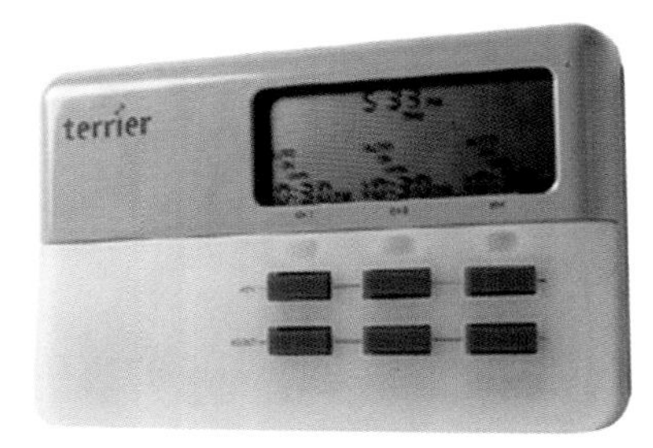

Figure 8.38 Time control programmer (Reproduced with permission of Pegler Limited)

Time switch or time clock

Commonly used on semi-gravity systems, this is a basic switch that is electrically operated, by means of a clock. When set, it offers timed or constant control to both the heating and hot water circuits. If required, it can provide separate hot water, but not independent heating.

Time control programmer

Ideal for fully pumped systems, these put the heating on standby, ready to respond to the requirements of the cylinder and room thermostats. They are available as a 7 day full programmer with independent heating and hot water timings, offering up to 3 on/off periods in a day (Figure 8.38).

The next four figures have been reproduced with the kind permission of Pegler Hattersley.

Figure 8.39 Room thermostat (Reproduced with permission of Pegler Limited)

Room thermostat

The thermostat measures the air temperature within the area where it is located. The occupant sets the desired temperature selected from the dial; normally this is between 18 and 21°C (see Figure 8.39). On achieving this temperature, it turns the heating off, and on temperature drop, turns it on. It can be wired to operate the circulating pump, boiler or zone valve.

Programmable room thermostat

These are room thermostats, as shown in Figure 8.40 that incorporate time control for heating systems; they allow the occupants to set higher or lower room temperatures for different periods of the day. The optional settings can be on a daily or weekly basis, for example a higher temperature may be required from a 3 h period in the evening, or maybe at weekends.

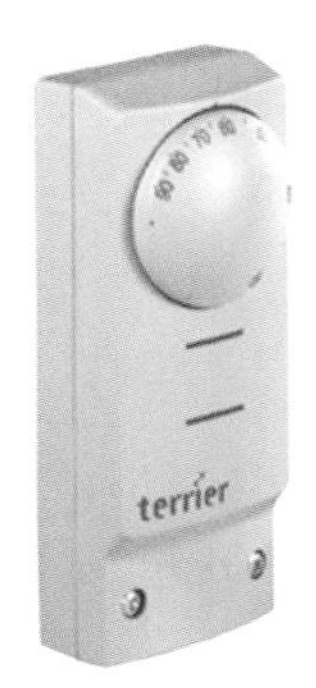

Figure 8.40 Cylinder thermostat (Reproduced with permission of Pegler Limited)

Cylinder thermostat

The cylinder thermostat monitors the temperature of the hot water in the storage vessel, and on achieving its set temperature (typically 60°C) it switches off the flow of hot water to the storage vessel. The thermostat can be wired in conjunction with the boiler, pump or zone valve (Figure 8.40).

Frost thermostat

Frost thermostats give automatic protection to boilers and pipework that are situated in areas at risk from severe cold. A typical example

Figure 8.41 Frost thermostat (Reproduced with permission of Pegler Limited)

Figure 8.42 Bypass valve (Reproduced with permission of Pegler Limited)

is where a boiler and pipework are located in an outbuilding. The thermostat overrides all other auxiliary controls, e.g. time control and temperature control (Figure 8.41).

Circulating pump

This component was covered earlier in the unit; its purpose is to provide a flow of water in the system.

TRVs

These were covered earlier in the unit; you may recall that they offer independent temperature control of the radiators.

Bypass valves (differential pressure)

Figure 8.42 shows a bypass valve. These types of valves are a requirement of Part L1 A and B 2006 and prevent unnecessary increases in circulating pressure if TRVs or motorised valves close, ensuring a flow of water through the boiler. They are ideal for boilers that incorporate a pump overrun.

The bypass valve reduces system noise and valve noise, and helps to increase the life of the circulator, by preventing it from working against a 'dead head'. The valve is marked with a direction flow arrow and is designed to be installed after the circulator.

Key point

This type of valve (Three-port diverter) is not recommended for use in dwellings, where there is likely to be a high hot water demand during the heating season. This could result in a temperature drop to the heating circuits. Installations of this type are sometimes referred to as 'W' plan systems.

Automatic air vents

These devices are designed to remove air automatically from the heating and hot water systems, as shown in Figure 8.43. The vent consists of a vacuum break on the bottom of the valve. This prevents an air lock from forming and encourages air to be released from the water. They should always be installed on the positive side of the system, and positioned where air is likely to get trapped.

Motorised zone valves

The type of motorised zone valve used on an installation will be dependent on the system design. The types of zone valves available are:

- Three-port diverter valve
- Three-port mid-position valve
- Two-port valve.

Figure 8.43 Automatic air vent

Three-port diverter valve

This valve has been designed for fully pumped systems, and provides independent temperature control for the heating and hot water circuits, in conjunction with a cylinder and room thermostat. The valve is a priority valve, and is normally installed to give hot water priority.

Three-port mid-position valve

Figure 8.44 looks identical to that of a three-port diverter valve, designed for fully pumped systems and provides independent temperature control for the heating and hot water circuits, in conjunction with a cylinder and room thermostat. Installations of this type are sometimes referred to as 'Y' plan systems.

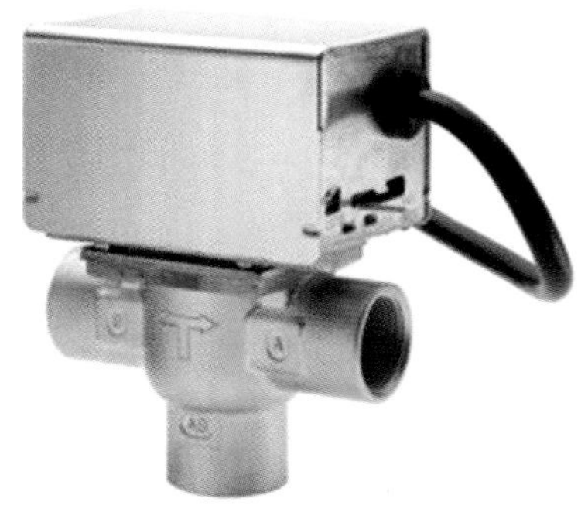

Figure 8.44 Three-port mid-position value (Reproduced with permission of Pegler Limited)

Two-port valve

A two-port valve, as shown in Figure 8.45, is used on gravity domestic and pumped heating systems to enable separate temperature control of both the heating and hot water circuits. Installations of this type are sometimes referred to as 'C' plan systems.

Two-port valves can also be used on fully pumped systems to provide separate control of both heating and hot water circuits. They can also be used to zone different parts of a building, e.g. upstairs and downstairs. Installations of this type are sometimes referred to as 'S' plan systems.

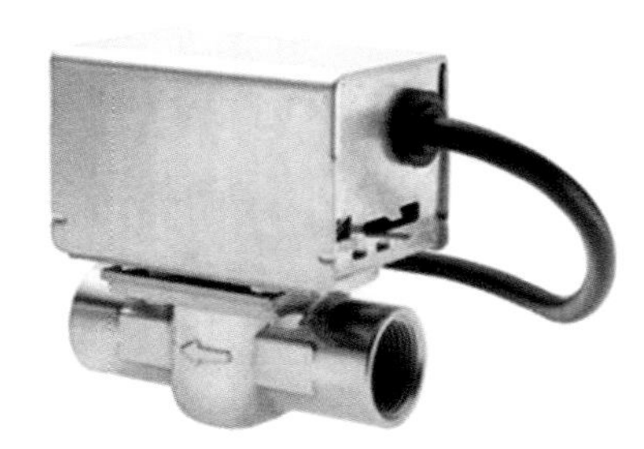

Figure 8.45 Two-port valve (Reproduced with permission of Pegler Limited)

Wiring centre

This provides the connections between the electrical system components and the mains electricity supply. Most manufacturers supply their controls in packs; this usually includes a wiring centre, which is designed to simplify the wiring of the components in the pack. Control packs include zone valves, programmers and thermostats. All the electrical terminal connections are clearly marked, and full instructions are included along with the wiring centre.

Boiler control interlocks

This is a term used in the Building Regulations Approved Document L1. It is not an actual physical control device, but an interconnection of the controls, such as zone valve, room, cylinder thermostats, programmers, etc; connected in an arrangement in order to ensure that the boiler does not fire when there is no demand for heat (refer Figure 8.45).

Key point

It is vitally important that you are aware of Part P of the Building Regulations scheme which covers electrical work in domestic dwellings, and has come into force from April 2005. It is aimed to ensure that any person, including plumbers, who work on electrical systems need to be able to demonstrate they are competent to do so in the area of work that they cover.

Try this

This is a strictly 'look but do not touch' activity. Have a look at the heating systems in your house, that of a friend, or other family members; see how many of these controls are on the systems, and make a note of your findings.

In a system with a traditional boiler this would mean the correct wiring of the room stat, cylinder stat and motorised valve(s). It can also be achieved by more advanced controls such as a boiler energy manager. The installation of TRVs alone is not sufficient for a boiler interlock.

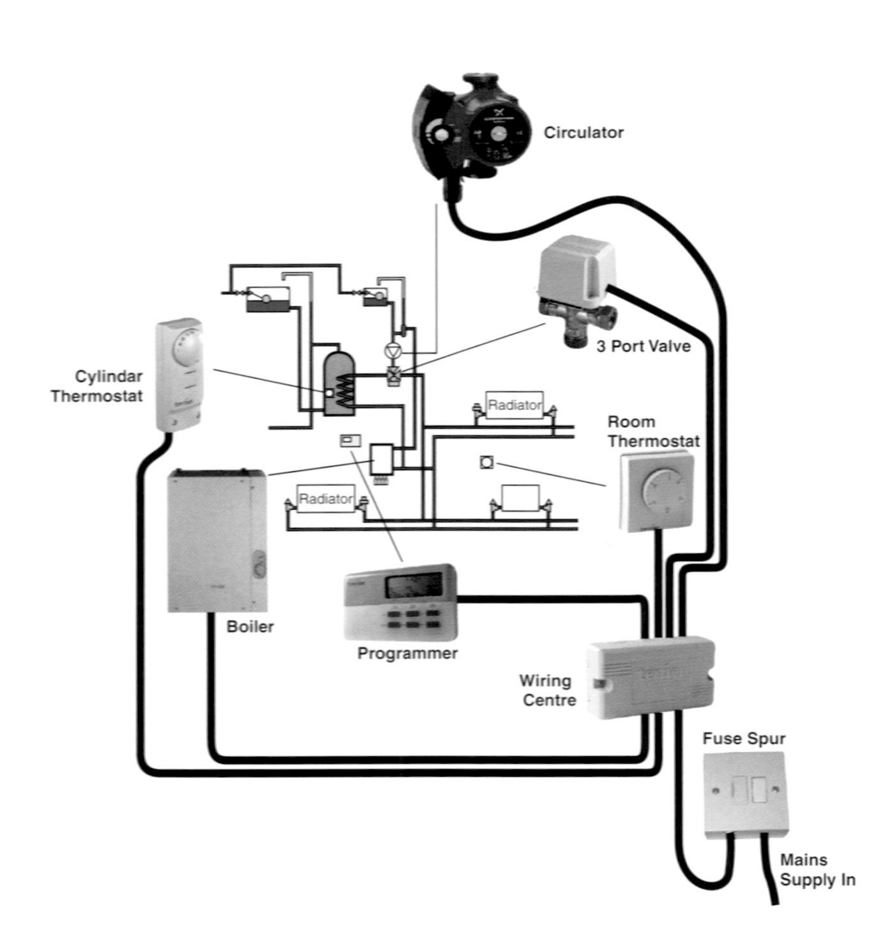

Figure 8.46 Overview of system controls

Basic wiring for controls

Once the heating system has been installed it needs to be wired up. Manufacturers of boilers, pumps and controls provide excellent technical support to guide installers on the requirements of wiring the various components. This will include wiring diagrams and step-by-step notes on what to do.

The wiring system and cables should be run and connected in accordance with the 16th Edition of the Wiring Regulations.

Test yourself 8.3

1. Briefly describe the basic differences between a time clock and a time control programmer.
2. What type of valve is being described below? These types of valves are a requirement of Part L1 and prevent unnecessary increases in circulating pressure if TRVs or motorised valves close, ensuring a flow of water through the boiler. Ideal for boilers that incorporate a pump overrun.
3. What are the three main categories of motorised zone valves?
4. Briefly describe the meaning of the term 'boiler control interlock'.
5. In what position should a cylinder thermostat be located on the external surface of the cylinder?

Soundness testing on central heating pipework systems

On completion of the installation, the heating system should be inspected and tested as appropriate; the tests should be carried out prior to any pipe insulation being applied and the covering of any pipe ducts.

We have already looked at the testing requirements for hot and cold water.

You may recall, this included:

- Visual inspections
- Testing for leaks and flushing the system
- Pressure testing
- Final checks.

We will look at these in the context of central heating as well as flushing the central heating system pipework and using additives and in-line corrosion protectors.

Testing procedures

Visual inspection

All the pipework should be visually inspected to ensure that:

- All connections are tight
- They are free from jointing compounds, flux, etc
- In-line valves and radiator valves are closed to allow stage filling
- The inside of the F&E cistern is clean and free from swarf
- All the air vents are closed
- It is fully supported; checks should include the bases of F&E cisterns, and storage vessels
- Before filling and testing, it is recommended that the circulating pump is removed and replaced with a section of the pipe. This will prevent any system debris entering the pump, damaging the impeller.
- The customer or other site workers are advised that soundness testing is about to commence.

Testing for leaks

- Turn on the water supply to the F&E cistern; allow the system to fill
- Turn on the radiator valves fully, and vent each radiator (starting on low levels first, working up to upper floors where applicable)
- Visually check all the joints for signs of leaks
- Drain down the system flushing out all the debris, wire, wool, flux, etc
- Replace the circulation pump
- Refill the system and check for leaks again
- Make sure the water level in the F&E cistern is set at a level allowing for expansion of the system water when hot.

Pressure soundness testing

Soundness testing of the system is carried out hydraulically using a test pump and the system must be sealed from the atmosphere. On larger contracts, this testing procedure can form a part of the specification contract.

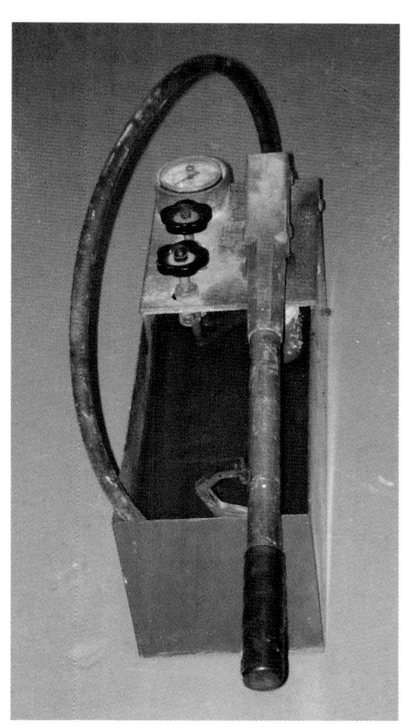

Figure 8.47 Example of a hydraulic test pump

With the system filled with water, having the pump attached to a suitable test point on the system and any open ends of pipework sealed, the installation is tested to 1½ times its working pressure, and allowed to stand for 1 h.

Hydraulic test equipment

Hydraulic test pumps (Figure 8.47) of this type have easy fill water reservoirs (to maintain system make-up water), and are complete with test hose and pressure gauge. The pump is capable of pressurising a system up to 50 bar.

Final system checks

On completion of a successful pressure test (no pressure drop on the gauge) perform a final check to ensure that everything is sound, closing the test point and removing any capped ends. Advise the customer, or other site workers, when system testing is complete.

Corrosion protection

Corrosion will take place to some extent in a domestic heating system. The severity will depend on factors like:

- Types of metals used in the system
- Nature of the water, hard or soft (acidic)
- The degree at which air is drawn into the system; this is generally associated with poor system design, e.g. pumping over or air being pulled down the open safety vent pipe.

It is recommended and is a good practice to add corrosion protector in a liquid form or install an in-line corrosion protector to maximise the life of the system and its components. They should be added or installed in accordance with the manufacturers' instructions. The liquid types can be added as the system is refilled after flushing, or injected through a radiator vent.

In-line corrosion inhibitors can be installed to either the heating flow or the return pipework, as close as reasonably practical to the boiler (refer to Figure 8.48).

Here are some advantages the corrosion protection offers:

- Prevents the build up of 'black sludge' (magnetite (Fe_3O_4) and haematite ($6Fe_2O_3$) – a major cause of heating problems), which can result in the pin holing of radiators
- Reduces fuel costs

Figure 8.48 Range of corrosion inhibitors (Reproduced with permission of Scalemaster)

Key point

It's important not to use inhibitors on systems using Primatic cylinders as loss of seal could result in the inhibitor mixing with the hot water supply.

Try this

Check out the website of manufacturers for the various additives available for heating systems. Write down the purpose of at least three products, noting any advantages or disadvantages of the product.

- Prevents frequent venting of the radiators
- Reduces system noise and reduces scale build up
- Non-acidic neutral formation, hence safe to use
- Harmless to the environment.

Test yourself 8.4

1. A new heating installation requires a pressure test. What is the test pressure and the time duration of the test?
   - **a.** 1½ times the working pressure and 1½ h ☐
   - **b.** 1 time the working pressure and 1 h ☐
   - **c.** 1½ times the working pressure and 1 h ☐
   - **d.** 2 times the working pressure and 1 h ☐
2. State two factors that can cause corrosion in a heating system
3. State the three advantages offered by corrosion protection

Maintenance of central heating systems

Domestic heating systems require periodic maintenance and plumbers need to understand the principles of maintaining the components within the system. The Level 2 Technical Certificate also requires you to be able to change a defective radiator valve or circulation pump, so you will look at these two components.

By now, you should have a good insight into the importance of a system's energy efficiency. Maintenance plays an important role here; it ensures that the system components operate effectively and efficiently. For example, a pump that is not performing to its design specification could make the system work inefficiently, thus wasting energy.

You can also check for system leaks that may have gone undetected where pipework is located under ground floors.

When carrying out maintenance work to a heating system the procedures may consist of:

- Locating the fault or defective component that is to be serviced
- Isolating the water and electrical supply (you must be competent to work on the electrical supply)
- Stripping and repairing a defective component, if possible
- Replacing a defective component
- Reassembling the components (if applicable)
- Turning on the water and electrical supplies, testing the component for correct operation on completion (again, you must be electrically competent).

Replacing a radiator valve

Irrespective of the type of radiator valve, replacement follows the same procedure.

Draining down the system

This method can make what is a relatively simple task into a long job. It is not so much the draining down of the system, it is the refilling and making sure the system is working correctly, and removing any air locks that may form in the system that takes the time.

Here is a checklist:

- Ensure that the electrical supply to the heating system is isolated, turn off the boiler

- As a precaution, remove the fuse from the spur outlet to the wiring centre or junction box for the controls
- Advise the customer what you are doing and not to touch the controls. Only work on the system when the system water has cooled
- Turn off the service valve to the F&E cistern, if one has not been fitted tie the float valve up so you do not have to turn all the cold water supply off
- Locate the nearest drain valve at the lowest point below the radiator valve to be changed
- If working on an upstairs radiator, do not drain all the system down, just drain the first floor level
- Draining off a system is done by connecting a hose pipe to a drain valve located on the system, running the other end of the hose to an outside drain
- Once drained, the installation of the valve can begin, but one of the two things could happen:
  - The valve can be fitted without the need to alter the pipework, this is because the valves are of the same size, or there is sufficient movement in the pipe to enable it to fit
  - The dimensions of the valve are different so the pipework has to be altered.
- If the valve can be changed without altering the pipework, remove the valve leaving the existing nut and compression ring on the pipe (provided the threads are compatible)
- Check whether the new valve body fits the existing valve tail in the radiator. If it does not fit, change the tail. If it is of the type with an external nut, this can be done with an adjustable spanner
- If not, you will need a radiator spanner/Allen key which goes inside the tail to remove or fit. When fitting the new tail, wrap PTFE tape around the threads 5 or 6 times in the direction that it is being tightened (clockwise)
- Remove the nut and compression ring from the new valve and place the new radiator valve body in position. If it is a TRV, remove the thermostatic head first, and as the compression ring has been used before, wrap the PTFE tape around the compression ring and the valve body
- If you have to alter the pipework, either by shortening or extending using a fitting, complete the process, and fit the new nut and compression ring to the pipe, then follow the subsequent steps

Key point

Remember, on a sealed system, it needs to be re-pressurised when filling the central heating system.

- Once the valve has been secured to the pipe, fit the valve body to the radiator tail and tighten, using an adjustable spanner/grip
- Fit the thermostatic head to the TRV when installing this type of valve
- Turn off the drain tap, turn on the water supply to the F&E cistern and begin to vent the system. This will involve venting all the radiators, circulation pump and any other manual air vents on the system. So, do not forget your radiator air key
- On completion of filling, remember to add a liquid corrosion inhibitor if an in-line type is not fitted.

Note: This does not apply on systems with a self-venting cylinder, e.g. a primatic.

- Finally, check your 'handiwork' for water tightness, removing the hosepipe on successful completion.

Replacing a defective circulator pump

Remember, it is important before commencing the pump replacement to advise the customer of what you are doing, both in advance and during the actual job.

Key point

On a maintenance job, the circulator could be on any type of heating system, located almost anywhere in a dwelling. For example:

- On fully pumped systems it will be located close to the domestic hot water storage vessel
- On older systems with gravity primaries it could be located close to the boiler
- On system boilers; combies it will be located inside the boiler casing.

Activity 8.4

By now you should have an understanding of systems and components to enable you to tackle this activity, but there is nothing stopping you from doing some additional research on pumps to help you with this task.

Have a go at writing a short report/checklist on how you would go about replacing a central heating pump. In this case, the report can be done using a series of bullet points rather than long-winded paragraphs. Do it on a separate sheet(s) of paper and then include it in your file. When you have completed your report, check it against our version at the end of this book.

General system maintenance

Prevention is better than cure, and many heating firms offer planned maintenance packages for domestic heating systems. This could include:

- Boiler and other heating appliances servicing at least once a year (a legal requirement for gas appliances in tenanted properties)
- Maintenance of circulators and radiator valves
- Cleansing and flushing of heating systems.

A Domestic Service and Maintenance Contractor says that the majority of maintenance and repair jobs are caused by:

- Failure of boiler parts, heating controls, motorised valves and thermostats
- Leaks caused by occupiers, typical examples are by the removal and replacement of radiators for decorating purposes
- Leaks or burst pipes.

One of the contractor's main concerns was that customers believe there is a fault with the system, but it often turns out that the customer does not know how to operate the system correctly. So, it is vitally important to ensure that the customer fully understands how to operate the system efficiently and effectively.

Motorised valves are generally very reliable devices, but eventually they will reach the end of their working life, and require replacement. At one time, to carry out this task meant draining the heating system. However, some manufacturers have now made it possible to replace the defective motors or the actual power head of the valve.

Try this

Try to obtain manufacturer's literature about the various types of pipe freezing equipment available; use the information to decide which equipment you might select and why.

Using pipe freezing equipment

There are several types of kits on the market, and some kits are available for hire from hire companies.

Freezer kits can be used on other systems including cold and hot water supply. They vary in specification and cost, and if purchasing one, you will have to decide the best kit to suit your needs in terms of the type of work undertaken and the frequency of use. It may be more economical to hire the kit for one-off jobs.

Most kits are suitable for use on copper, plastic, lead and even iron pipework; this will depend on manufacturer's recommendations. On obtaining a kit, whether brought or hired, make sure you can use it competently, reading the manufacturers' instructions prior to use.

System cleansing

When working on older heating installations, it is recommended that the system is flushed out to remove any system sludge; this helps system efficiency and gives longer life to radiators. Flushing can be done on a yearly basis in conjunction with the servicing of the appliance, or when carrying out system repairs.

There are additives available on the market that assist in the cleansing of heating systems, old and new, ranging from pre-commissioning cleaners to scale and sludge removers.

Most boiler manufacturers insist that before installing a new boiler to an existing heating system, the system is thoroughly flushed out, and may even make it part of the terms in the appliance warranty.

Portable high pressure cleansing machines are available that can rapidly cleanse a system with 20 or more radiators connected to it.

Approved cleaning agents should be used with this type of equipment. The unit comes complete with hoses and valved connections. The machine can also be used for descaling combination boilers and water heaters. When used for cleansing a system, ideally a ground floor radiator should be removed, connecting the hose connections to the radiator valves. Forced water containing the cleaning agent is then passed through the entire system. The machine incorporates a flow reverse device that helps dislodge and mobilise deposits, effectively removing them from the system. On completion, remember to include a system additive when re-filling, if an in-line type is not fitted.

Note: This does not apply on systems with a self-venting cylinder e.g. a primatic.

Test yourself 8.5

1. In terms of ongoing maintenance, how often should a central heating system be flushed out?
2. What is an alternative method to draining down a heating system in order to carry out repair job such as replacing a radiator valve?
3. It is permissible to use system additives on primatic cylinders. True or false?

Commissioning and De-commissioning of central heating systems

On completion of the installation, the system and appliance must be commissioned in accordance with Benchmark 30. This will include a certificate of compliance on completion.

Try this

You have encountered Benchmark earlier. Write a short summary explaining what the term means and why it is relevant to commissioning heating systems. Check your summary against the text in the section on Central Heating Systems explained, at the beginning of this unit.

Key point

System flushing should be carried out with the circulating pump removed, bridging the gap between the circulator valves with a make up pipe. Ensure all radiator valves are fully open.

Filling

After a successful water tightness test, the system can be filled and flushed to minimize the presence of solid particles, e.g. wire wool, swarf, solder, and chemical residues, e.g. fluxes, jointing compounds, which may cause corrosion and damage within the system.

Fill the system, venting all the high points and radiators, set the water level in the F&E cistern, allowing space to accommodate expansion equal to 4% of the water in the system. Check the system for leaks, drain down the system, and refit the circulation pump.

Wherever recommended by the boiler manufacturer, add a chemical cleanser into the system, at dilutions recommended by the cleanser manufacturer. Refill the system, venting all the high points, radiators and circulating pump. Set the hot water and room thermostat to the correct temperatures, turn on the boiler thermostat.

The boiler must be commissioned in accordance with the manufacturers' instructions, checking electrical polarity, electrical fuse ratings (remember Part P), gas pressure rates (standing and working pressure), and water flow rates where applicable. When commissioning is complete, its safe operation should be confirmed including safe removal of the products of combustion.

Once in operation, allow the system to heat up to normal operating temperature, check the system again for any leaks, if a chemical cleanser has been added allow sufficient time for it to work (this will be recommended by the cleanser manufacturer's instructions).

On reaching the temperature, switch off the boiler and circulating pump, drain down the complete system whilst still warm. If a chemical cleanser has been used, the system must be re-flushed, to ensure that the cleanser has been removed.

Refill and vent the system as done earlier. If introducing a corrosion inhibitor into the system, this should be added in accordance with the manufacturer's instructions. Labels should be fixed to the boiler or drain tap, stating that the system contains corrosion inhibitor, naming the type and the date of application.

Recheck the system for leaks, refixing any cistern lids and any insulation material. Where a system has been installed and commissioned in unoccupied properties, at a time of the year when freezing conditions can be expected, it is advisable to leave the system drained. Leave warning notices and all manufacturers' instructions in the dwelling for the future occupier.

Balancing

The heating circuit and radiators now require balancing. To complete the task you will require either two clamp-on pipe thermostats or a digital 'touch thermometer'. The system should be balanced by regulating the flow rate of the circulation pump to ensure the design temperature difference across the boiler.

Key point

The design flow temperature of a domestic heating installation must not exceed 82°C; the design return temperature must not be less than 66°C unless the boiler is of condensing design (High efficiency), and then a lower return temperature is acceptable. A temperature difference of 20°C is recommended in order to keep the boiler on 'condensing mode'.

The water flow rate through the individual radiators should be regulated to ensure a mean water temperature at each radiator according to design. In practice, the system flow temperature may be lower than the design flow temperature; this is due to the boiler cycling and intermittent operation. Hence, ideally there should be a 10°C temperature drop between the heating flow and return pipework.

Balancing the radiators is achieved by closing all the lockshield valves on the radiators. With the boiler thermostat set on high, start at the lowest radiator, closest to the boiler. Connect the thermostats and balance the radiator to each room slowly opening the lockshield valve to give a mean water temperature to each radiator, ensuring that the design temperature drop (10°C) is maintained at the boiler.

Handing over

On completion of the commissioning procedure and testing the operation of all the controls, the working of the system, including all the controls, should be demonstrated to the user with the explanation of best methods of economic and efficient use.

The user should be advised about the method of summer and winter operations of the system, be left with all the relevant manufacturer's instructions relating to the components of the system, together with an 'air vent' key for venting the radiators and any lockshield key required for the installation. The installer should also leave a permanent card attached to the boiler detailing:

- The name and address of the installer
- Date of installation.

The user should be offered, or made aware of, a regular service contract to ensure that the equipment is maintained in an efficient and safe operating condition. Finally, the user should sign the benchmark book to confirm the demonstration of the equipment and receipt of appliance instructions.

De-commissioning of a heating system

De-commissioning includes draining down the system and could be required in the following situations:

- Where an old system is going to be completely stripped out of a dwelling and replaced by a new system, e.g. an old gravity system replaced by fully pumped condensing boiler system
- Where a system is going to be stripped out permanently, for example, prior to the demolition of a building
- Where a dwelling is going to be unoccupied for a period of time.

It is mostly common sense by making sure you:

- Keep customers or other site workers informed of what work is being carried out and when
- Make sure all supplies are isolated and capped off (where necessary) and that electrical supplies are safe
- Fix notices on taps, valves, boilers, etc, e.g. Not in Use, 'Do Not Use', etc.

Test yourself 8.6

1. What code of practice is being described in the paragraph below?
   This is a series of codes of practice that will encourage the correct installation, commissioning and servicing of domestic heating boilers and system equipment.

   The code of practice places an obligation on installers and manufacturers.

   Installers are required to fit the equipment in accordance with the manufacturer's instructions and then confirm this by completing a log book. These are now provided with virtually every type of gas boiler and major piece of heating equipment.

2. Referring to the log book in question 1, list at least three things that should be recorded by the installer before handing it over to the customer.

3. When balancing a central heating system, what is the design temperature drop between the flow and return pipework?
   a. 5°C ☐
   b. 10°C ☐
   c. 15°C ☐
   d. 20°C ☐

Check your learning Unit 8

Time available to complete answering all questions: 30 minutes

1. The statutory requirements of space heating systems controls are covered under the:
   a. British Standards ☐
   b. Construction Regulations ☐
   c. Building Regulations ☐
   d. Water Regulations ☐

2. One disadvantage of using a one pipe heating system is that:
   a. It is not possible to use TRVs ☐
   b. The central heating pump must always be on the return ☐
   c. The pipework to the radiator valve must always be 22 mm ☐
   d. Accurate balancing of the heat emitters is a must ☐

3. What is the maximum velocity rate for a circulating pump on a small bore central heating system?
   a. 1 m/s ☐
   b. 1.4 m/s ☐
   c. 1.6 m/s ☐
   d. 1.8 m/s ☐

4. Which one of the following systems would a three port valve, programmer, room thermostat, cylinder thermostat, and TRVs normally be used on?
   a. Pumped heating and gravity hot water for a two-pipe system. ☐
   b. Fully pumped heating system (hot water priority) using a two position zone valve. ☐
   c. Pumped heating and gravity hot water for a one-pipe system. ☐
   d. Fully pumped heating system using a three-port mid-position zone valve. ☐

5. The minimum size of vent pipe from the primary flow on a pumped central heating system while using gravity primaries is:
   a. 15 mm ☐
   b. 25 mm ☐
   c. 28 mm ☐
   d. 22 mm ☐

6. A boiler that works on the principle of extracting heat from otherwise wasted flue gases is known as:
   a. Thermostatic ☐
   b. Combination ☐
   c. Condensing ☐
   d. Independent ☐

7. Which of the following is the correct flue classification for a solid fuel boiler (excluding wood burning) when connected to a brick flue?
   a. 1 ☐
   b. 2 ☐
   c. 3 ☐
   d. 4 ☐

8. On an open-vented fully pumped system, positioning the cold feed and vent to the rear of the pump would:
   a. Avoid the need to install a sealed expansion vessel ☐
   b. Drastically decrease the efficiency of the boiler ☐
   c. Create a negative pressure in the system ☐
   d. Create a positive pressure in the system ☐

Check your learning Unit 8 (Continued)

9. Which of the following gas boilers alleviates the need for a CWSC, F&E cistern and hot water storage vessel?
   a. Combination ☐
   b. Back-boiler ☐
   c. Condensing ☐
   d. Room heater ☐

10. The recommended position of a cylinder thermostat from the base of the cylinder is:
   a. $\frac{1}{4}$ height ☐
   b. $\frac{1}{3}$ height ☐
   c. $\frac{1}{2}$ height ☐
   d. $\frac{3}{4}$ height ☐

11. What is the recommended distance from the nearest heat source when positioning a single room thermostat?
   a. 1.5 m ☐
   b. 2 m ☐
   c. 3 m ☐
   d. 3.5 m ☐

12. The purpose of the primary open vent on a central heating system is to:
   a. Provide a safety outlet should the pressure vessel fail ☐
   b. Increase system pressure and thus reduce pump speed ☐
   c. Provide a safety outlet should the system overheat ☐
   d. Reduce system pressure and thus increase pump speed ☐

13. When setting the water level in an F&E cistern, what allowance, in volume, should be made for the expansion of the water?
   a. 4% ☐
   b. 8% ☐
   c. 10% ☐
   d. 12% ☐

14. The recommended pressure when testing heating systems pipework is:
   a. $1\frac{1}{2}$ times the working pressure ☐
   b. Twice the working pressure ☐
   c. $1\frac{1}{2}$ times the standing pressure ☐
   d. Twice the standing pressure ☐

15. If TVRs or zone valves become closed on a system, the main purpose of an automatic bypass valve is to make sure:
   a. Water can flow through the cylinder ☐
   b. Air cannot enter the heating system ☐
   c. Water can flow through the boiler ☐
   d. Water cannot pump over the vent ☐

16. An existing two-pipe system installation can be confirmed by checking that:
   a. There is no return heating pipework to the boiler ☐
   b. The main circuit pipework is 28 mm diameter minimum ☐
   c. The primary flow and return pipes are fully pumped ☐
   d. Both radiator valves are fed by separate pipes ☐

(Continued)

Check your learning Unit 8 (Continued)

17. When removing a single radiator on the first floor of a dwelling, what is the most efficient way of draining the radiator?

- **a.** Drain down the system as far as the first floor ☐
- **b.** Completely drain down the whole of the system ☐
- **c.** Use freezing equipment on the pipes to the valves ☐
- **d.** Isolate the valves and drain down the radiator ☐

18. If hot water discharges from the vent pipe to a F&E cistern while the heating system is running, the most likely cause is that the:

- **a.** Pump has been installed between the cold feed and vent ☐
- **b.** F&E cistern has been installed higher than the CWSC ☐
- **c.** F&E cistern has been installed lower than the CWSC ☐
- **d.** Pump has been installed horizontally and not vertically ☐

19. One of the requirements of the revised regulations Part L1 when replacing the boiler and/or controls is that the system should be:

- **a.** Gravity hot water and pumped heating ☐
- **b.** Heated by combination boilers only ☐
- **c.** Fully pumped with a boiler interlock ☐
- **d.** Installed by CORGI-registered plumbers ☐

20. The pipework to a bathroom radiator is being altered, what should be done to minimise the safety risk to the plumber?

- **a.** Apply temporary earth continuity bonding ☐
- **b.** Turn off the cold water supply stop tap ☐
- **c.** Isolate the cold water supply at the float valve ☐
- **d.** Turn off the electricity supply at the mains ☐

Sources of Information

We have included references to information sources at the relevant points in the text; here are some additional contacts that may be helpful. This list is not exhaustive; you might wish to find additional ones to these.

- Caradon Stelrad Ltd
  Stelrad House
  Marriott Road
  Mexborough
  Rotherham
  S64 8BN
  South Yorkshire
  General enquiries, Tel: 01709 578 950
  Technical information, Tel: 01480 498 663
  Website: www.stelrad.com
- Pegler Limited
  St.Catherine's Avenue
  Doncaster
  South Yorkshire
  DN4 8DF
  Brochure hotline 0870 1200284
  Technical support 0870 1200285
  Website: www.peglar.co.uk
- Baxi
  Brown Edge Road
  Bamber Bridge
  Preston, PR5 6UP
  Tel: 08706 060780
  Fax: 01772 695420
  Website: www.baxi.co.uk

Technical Services, Training and Marketing

- Halstead Boilers Limited
  5 Titan Business Centre
  Spartan Close
  Tachbrook Park
  Leamington Spa
  Warwickshire
  CV34 6RR
  Tel: 01926 834800
  Website: www.halsteadboilers.co.uk
- Ravenheat Manufacturing Limited
  Chartists Way, Morley
  Leeds, LS27 9ET.
  Tel: 0113 252 7007
  Fax: 0113 238 0229
  Website: www.ravenheat.co.uk
- Honeywell Control Systems Ltd
  Honeywell House
  Arlington Business Park
  Bracknell
  Berkshire
  RG12 1EB
  Tel: 01344 656000
  Website: www.honeywell.com/uk
- Ideal Boilers
  PO Box 103
  National Avenue
  Kingston upon Hull
  East Yorkshire
  HU5 4JN
  Tel: 01482 492251
  Fax: 01482 448858
  E-mail: enquiries@idealboilers.com
  Website: www.idealboilers.com.

UNIT 9

ELECTRICITY

Summary

Recall from Unit 8 Central heating systems, the importance of Part P of the Building Regulations and how it impacts on the plumbing industry. Here is the text again for your reference and information.

It is vitally important that you are aware of Part P of the Building Regulations Scheme which covers electrical work in domestic dwellings, which came into force in April 2005. It aims to ensure that any person, including plumbers, who work on electrical systems needs to be able to demonstrate that they are competent to do so in the area of work that they cover.

There are several accreditation bodies that have been approved by government to run the scheme. Basically, a plumber will need to be qualified in their own trade, and then be able to demonstrate competence in the areas of electrical work they carry out. This would be achieved by carrying out a course of training and assessment approved by the accreditation body.

Part P gives guidance on areas of competence, the category appropriate to plumbers being Level B.

Level B

Electrical work in relation to design, installation, inspection and testing of defined electrical installation work. This includes working on existing circuits and can also include the installation of a new circuit.

More information can be found at: www.partp.co.uk.

More information on the requirements of Part P can be found at: www.planningportal.gov.uk.

Plumbers work on a wide range of appliances and components which form a part of the various plumbing systems. This includes appliances like boilers, pumps, showers and systems controls powered by electricity.

It is therefore essential for your safety and the safety of others, that you are competent to work on the electrical supply to a particular plumbing appliance or component. If you are in a situation where you are not fully competent, you must seek the services of a qualified electrician.

This unit will provide the underpinning knowledge to help you understand basic electricity. It is not in itself a qualification or proof that you are competent to carry out electrical work. For this reason, we have not covered the requirements of testing and de-commissioning, as this should be covered in a college or centre.

This unit covers:

- Principles of Electricity:
  - Electrical flow
  - Resistance
  - Fuses and Fuse Rating
  - Circuits.
- Electricity Supply:
  - Domestic Circuits
  - Earth Continuity and Bonding.
- Domestic electrical installation requirements:
  - Electrical systems
  - Electrical safety and isolation
  - Electrical installation skills
  - Practical requirements for central heating controls.
- Earth continuity:
  - Earth continuity systems for domestic appliances and controls
  - Temporary earth continuity bonding to permit work on piped systems.

Principles of electricity

Perhaps the simplest way of describing how electricity works to a plumbing student is to compare it to water flowing through a pipe. Both water and electricity can do 'work' as they flow between two points. For example, flowing water can turn a water wheel or a turbine; electricity can produce light as it flows through a lamp, or rotation (which can drive something) if it flows through an electric motor. The flow is the result of *pressure* in each case – the higher the pressure, the greater the flow.

But acting against this pressure, there is something called *impedance* or *resistance*. A similar situation applies to electricity in that some materials allow electricity to flow easily, whilst others resist it to a greater or lesser degree and may completely prevent it. Materials which permit a flow of current are called *conductors* and those which prevent it are called *insulators*.

Conductors are like the water pipe; a big one allows current to flow more easily than a small one, but the material of the conductor also affects the freedom with which the current can flow through it – the resistance. Most metals have low resistance and are good conductors, whilst many non-metallic materials such as ceramics, plastics and rubber have extremely high resistance and are used as insulators.

Like water, electricity needs somewhere to flow – in other words, a circuit. A circuit starts at the source of supply, in the case of your house effectively at the meter. It then travels through the conductor (the *phase* wire is often referred to as the live wire, but since both supply connections are effectively '*live*' the term 'phase wire' is more appropriate) to wherever it is needed to do work. Then it flows back through another conductor (the *neutral wire*) to its source.

The flow of an electric current is measured in *amperes* (A or amps for short), and is driven by the pressure difference (called potential difference) between the ends of the circuit; this is measured in *volts* (abbreviated to V), and on domestic installations a potential difference of 230 V is the norm (nominal 230 V to harmonise with other European Community countries). The potential difference drives the flow of electricity against the circuit's impedance (measured in *ohms*, often given the symbol Ω).

As electricity flows through the circuit, the amount of work it does is measured in *watts* (W for short) and in practical terms the watts consumed are given by the product of the supply voltage and the current drawn; watts = volts × amps. So, an electric fire taking a current of 10 amps from a 230 V supply will consume 2300 W or 2.3 kilowatts (2.3 kW). The time factor must also be taken into account when measuring the amount of electricity we use, so our fire taking 2.3 kW and burning for an hour will use up 2.3 kilowatt-hours (kWh) of electricity. This is recorded by your home's electricity meter as 2.3 kWh or 2.3 units (a unit is 1 kWh).

The wiring through which the current flows possesses resistance and the pressure of the flow against the resistance creates heat. The smaller the cable, the higher the resistance, so it is important that any cable is large enough not to cause heat to be generated which could damage the cable or anything in contact with it.

Table 9.1 SI units relevant to electricity

SI Unit	Measure of	Symbol
coulomb	Charge	C
joule	Energy	J
ohm	Resistance	Ω
volt	Potential difference	V
watt	Power	W
ampere	Electric current	A

Basic circuit theory

At its simplest, an electric current is a flow of electrons. For an electric current to flow in a simple circuit, two requirements are necessary:

- a source of chemical energy
- a continuous loop of a conducting material which will allow the transfer of that energy.

Key point

Batteries are relatively safe because they produce small amounts of electricity and cause a 'one-way' electron flow – they are said to produce a 'direct current' as opposed to an 'alternating current'.

The source of chemical energy most commonly used is the electric cell (a group of cells form a battery).

All matter is made up of atoms which arrange themselves in a regular framework within the material. The atom is made up of a central, positively charged nucleus, surrounded by negatively charged electrons. The electrical properties of a material depend largely upon how tightly these electrons are bonded to the central nucleus.

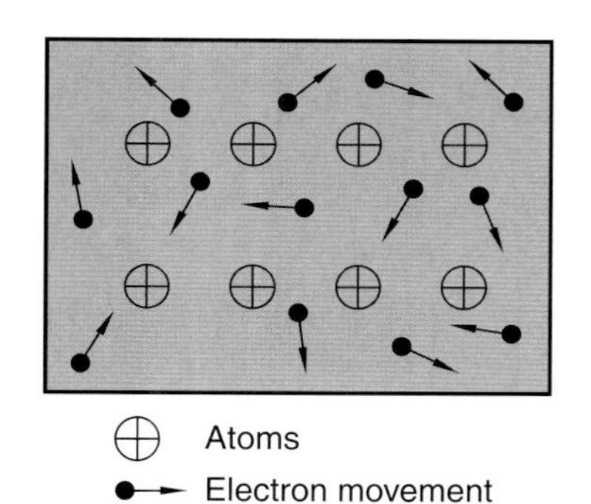

Figure 9.1 Electrons and conductors

A conductor is a material in which the electrons are loosely bound to the central nucleus and are, therefore, free to drift around the material at random from one atom to another, as shown in Figure 9.1. Materials which are good conductors include copper, brass, aluminium and silver.

An insulator is a material in which the outer electrons are tightly bound to the nucleus, and so there are no free electrons to move around in the material.

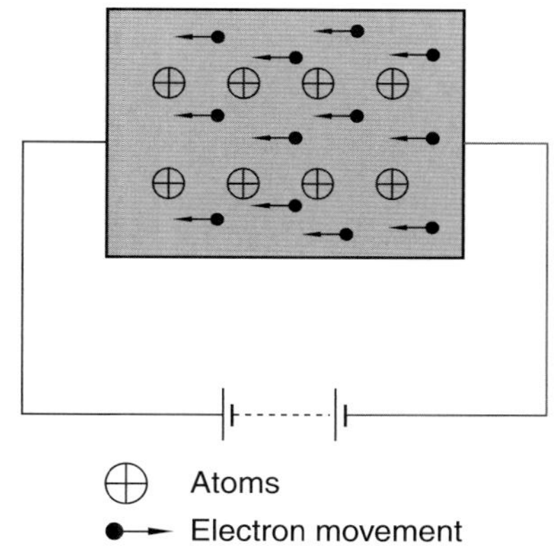

Figure 9.2 Battery attached to conductor

If a battery is attached to a conductor, as shown in Figure 9.2, the free electrons drift purposefully in one direction only. The free electrons close to the positive plate of the battery are attracted to it, since unlike charges attract, and the free electrons near the negative plate will be repelled from it. For each electron entering the positive terminal of the battery, one will be ejected from the negative terminal, so the number of electrons in the conductor remains constant.

This drift of electrons within a conductor is known as an electric current, measured in amperes and represented by the symbol A. For a current to continue to flow, there must be a complete circuit for the electrons to move around. If the circuit is broken by opening a switch, for example, the electron flow and therefore the current will stop immediately.

To cause a current to flow continuously around a circuit, a driving force is required, just as a circulating pump is required to drive water around a central heating system. The driving force is the *electromotive force* (abbreviated to emf). Each time an electron passes through the source of emf, more energy is provided to send it on its way around the circuit.

An emf is always associated with energy conversion, such as chemical to electrical in batteries and mechanical to electrical in generators. The energy introduced into the circuit by the emf is transferred to the load terminals by the circuit conductors. The *potential difference* (abbreviated to p.d.) is the change in energy levels measured across the load terminals. This is also called the volt drop or terminal voltage, since emf and p.d. are both measured in volts. Every circuit offers some opposition to current flow, which we call the circuit resistance, measured in ohms (symbol Ω), in honour of the famous German physicist Georg Simon Ohm, who was responsible for the analysis of electrical circuits.

Ohm's law

In 1826, Ohm published details of an experiment he had done to investigate the relationship between the current passing through and the potential difference between the ends of a wire. As a result of this experiment, he arrived at a law, now known as Ohm's law, which says that:

> *The current passing through a conductor under constant temperature conditions is proportional to the potential difference across the conductor.*

This may be expressed in the formulae:

- $V = A \times \Omega$ or
- $A = V \div \Omega$ or
- $\Omega = V \div A$

Where V = volts, A = amps (current) and Ω = ohms (resistance).

You may see in some textbooks that the symbol used for measuring current (ampere) is I, and that for resistance (ohms) is R. We have avoided using these here in order to avoid confusion.

Example 1

An electric heater, when connected to a 230 V supply, was found to consume a current of 4 A. Calculate the element's resistance.

Using $\Omega = V \div A$

$$\Omega = \frac{230\ \text{V}}{4\ \text{A}} = 57.5\ \Omega$$

Example 2

A 12 V battery is used to serve a circuit in which the total resistance is 3 Ω. Calculate the electrical current.

Using $A = V \div \Omega$

$$A = \frac{12\ \text{V}}{3\ \Omega} = 4\ \text{A}$$

Example 3

When a 4 Ω resistor was connected across the terminals of an unknown d.c. supply, a current of 3 A flowed. Calculate the supply voltage.

Using V = A × Ω

$$V = 3A \times 4\Omega = 12\,V$$

Resistivity

The resistance or opposition to current flow varies for different materials, each having a particular constant value. If we know the resistance of, say, 1 m of a material, the resistance of 5 m will be five times the resistance of 1 m.

The *resistivity* (symbol ρ – the Greek letter 'rho') of a material is defined as the resistance of a sample of unit length and unit cross section. Typical values are given in Table 9.2. Using the constants for a particular material we can calculate the resistance of any length and thickness of that material from the equation.

$$R = \frac{\rho l}{a} = (\Omega)$$

where

ρ = the resistivity constant for the material (Ωm)
l = the length of the material (m)
a = the cross-sectional area of the material (m^2).

Table 9.2 Resistivity values

Material	Resistivity (Ωm)
Silver	16.4×10^{-9}
Copper	17.5×10^{-9}
Aluminium	28.5×10^{-9}
Brass	75.0×10^{-9}
Iron	100.0×10^{-9}

The resistivity values table gives the resistivity of silver as $16.4 \times 10^{-9}\,\Omega$m, which means that a sample of silver 1 m long and 1 m in cross section would have a resistance of $16.4 \times 10^{-9}\,\Omega$.

Fuses

Why fuses are used?

See Keypoint.

Key point

Fusing is a safety measure which aims to prevent high electrical current passing through wires that are not designed to carry such large charges. This is very important because if a current which is too high for the wire is passed through it, the overheating of the wire causes a serious risk of fire.

How do fuses help?

All the various types of fuses that exist contain fuse wire (see Figure 9.3 for examples of cartridge and rewireable fuses). The fuse wire will melt or 'blow' if electric current above the specified amount is passed through the wiring. Fuses come in different sizes to protect against different levels of current.

You will also come across the miniature circuit breaker (mcb), a device which will trip a switch to break the electrical current if excessively high current is detected. These are more expensive than fuses, but are re-settable and are commonly found in consumer units within newer domestic properties.

Fuse Rating

To ensure that the appropriate size fuse is used, the *fuse rating* can be worked out by using the simple formula *watts* ÷ *volts* = *amps*.

Frequently, in domestic environments, fuses are overrated, i.e. a fuse with too high rating is used. For example, if a lamp contains a 100 W light bulb, the fuse rating would be:

100 ÷ 240 (the voltage of domestic mains supply) = 0.416 amps.

Key point

If the wiring in the appliance is faulty, the fuse should 'blow', thus breaking the circuit. But, if a very high fuse rating is used, the current will continue to flow, and the appliance may 'blow' instead.

Therefore a 1 amp fuse would be sufficient for use, though usually a 3 amp fuse would be fitted.

Many people, out of ignorance, would fit a 13 amp fuse to all appliances. Why is this not a good idea?

There is potential for confusion here, because 1 amp of current is more than sufficient to kill at 240 V. Remember that the purpose of the fuse is to protect the wiring; and that of the earth is to protect you – the user.

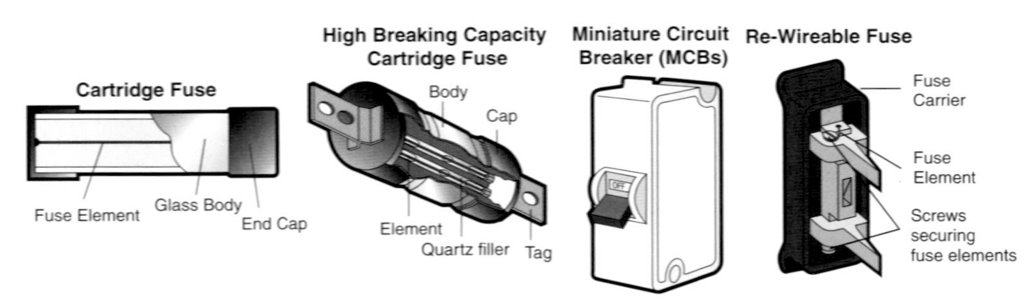

Figure 9.3 Types of fuses

Activity 9.1

Work out the correct fuse to use for a dishwasher which is rated at 1005 W. Check your answer at the end of the book.

Circuits

There are two basic types of electrical circuits. These are:

- series circuits
- parallel circuits.

Series circuit

Figure 9.4 describes a system where the current flow is made to pass through each component (e.g. a bulb) in a circuit. The current should be the same in any part of the circuit, but the voltage will vary depending upon the resistance of each component. The total voltage of all the components must not exceed the total available voltage, otherwise the bulbs will not glow sufficiently.

Parallel circuit

In a simple parallel circuit as shown in Figure 9.5, however, there are alternative routes open to the flow of electrons, and the current will flow along both. This results in a very different effect from a series circuit, when two electric lamps are connected in parallel.

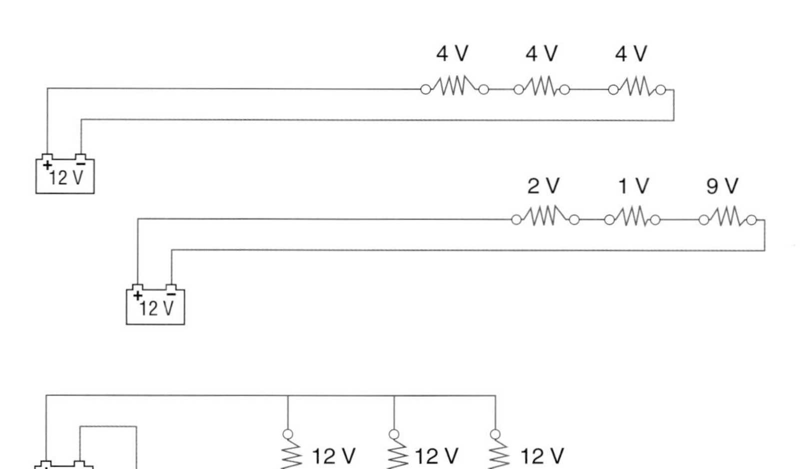

Figure 9.4 In series circuit

Figure 9.5 Parallel circuit

In this system, when one bulb blows, the others stay alight; whereas in a series system, when one bulb blows, it breaks the circuit and all the bulbs go out, for example, the lighting in a house is wired up in parallel and decorative Christmas tree lights are normally wired up in series.

Direct and alternating current

Direct current (d.c.)

In a d.c. electrical circuit, the electron flow is in the same direction all the time. One example would be from the anode to the cathode of a battery around a simple circuit, as shown in Figure 9.6.

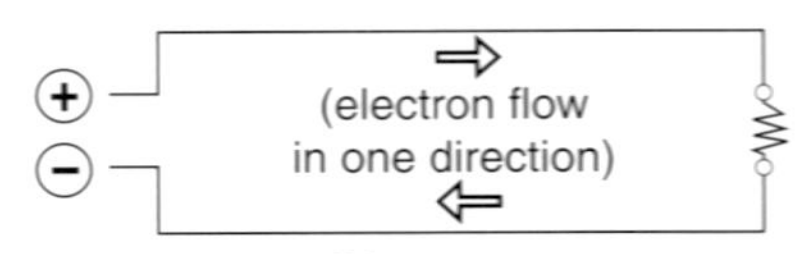

Figure 9.6 Direct current

Alternating current (a.c.)

Alternating current is found in the majority of domestic property, the usual rate 'at socket level' being 230 V a.c. Within the alternating current, electrons travel continuously back and forth, as shown in Figure 9.7. The reason for this is a result of the way the current is produced. Alternating current is produced as a result of *electromagnetism*.

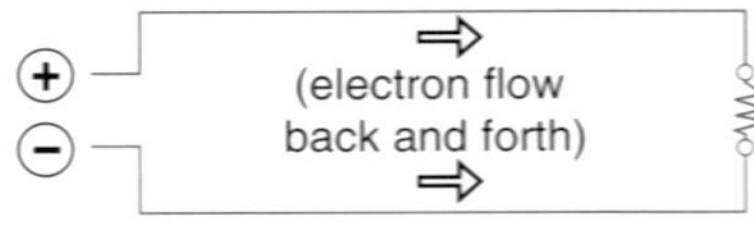

Figure 9.7 Alternating current

- All electrical currents produce a magnetic force; this is the basic fact that underpins the creation of almost all the electricity used in today's world
- The application of this fact was first demonstrated by Michael Faraday in the 1830s who discovered that electricity could be generated by moving a magnet in and out or around a coil of wire, which is wound on to a soft iron core
- Electric generators at power stations still produce a.c. electricity on this principle today.

Key point

When a.c. electricity is used, it is essential that appliances are 'earthed' as this completes the formation of a circuit necessary for current flow. The way in which this works is that the current flows to an appliance from the phase (live) wire and then from the neutral wire (which is, in effect, connected to earth); the current flows continuously back and forth at a rate of 50 times a second (50 hertz).

Basic domestic circuits

The final journey of electricity is from the consumer unit to familiar domestic output devices, such as sockets and electric lighting. As a plumber you will usually come into contact with three types of wiring circuits: Lighting, Ring main and Spur outlets.

Lighting circuit

This is a radial circuit which means it feeds each overhead light or wall light in turn. To stop the light being on continuously, the live or *phase* wire is passed through a wall-mounted switch, used by the property owner to turn lights off and on at will. Two-way switches are also used (usually on stairways) and these require special switch controls.

The lighting circuit is usually fed by a 1.5 mm^2 twin and earth PVC insulated cable and is protected by a 5 amp fuse or mcb at the consumer unit. You will often find that the lighting circuit in domestic houses is split into two – upstairs and downstairs.

Ring main circuit – 13 amp socket outlets

The sockets you will see in domestic properties feeding televisions and stereos will normally be 13 amp sockets fed from a continuous ring circuit. As with the lighting circuit, cables circulate from the consumer unit round each socket and then return to the consumer unit (see Figure 9.8).

The ring main circuit is fed using a 2.5 mm^2 twin and earthed PVC cable and is protected by a 30 amp fuse or mcb.

Spur outlets

Spur outlets are used where it is inconvenient to place a socket from the ring main. The spur is connected to the ring main through a joint box, or is wired directly from the back of an existing socket. Spurs can be either fused (Figure 9.9) or non-fused.

Figure 9.10 shows examples of all the features described. Take a moment to closely examine the diagram and try to recognise as many of the features described as possible.

Figure 9.8 Lighting switch and socket outlets

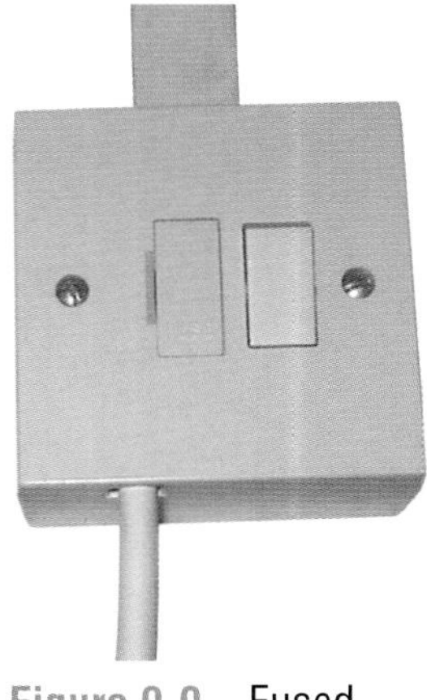

Figure 9.9 Fused spur outlet

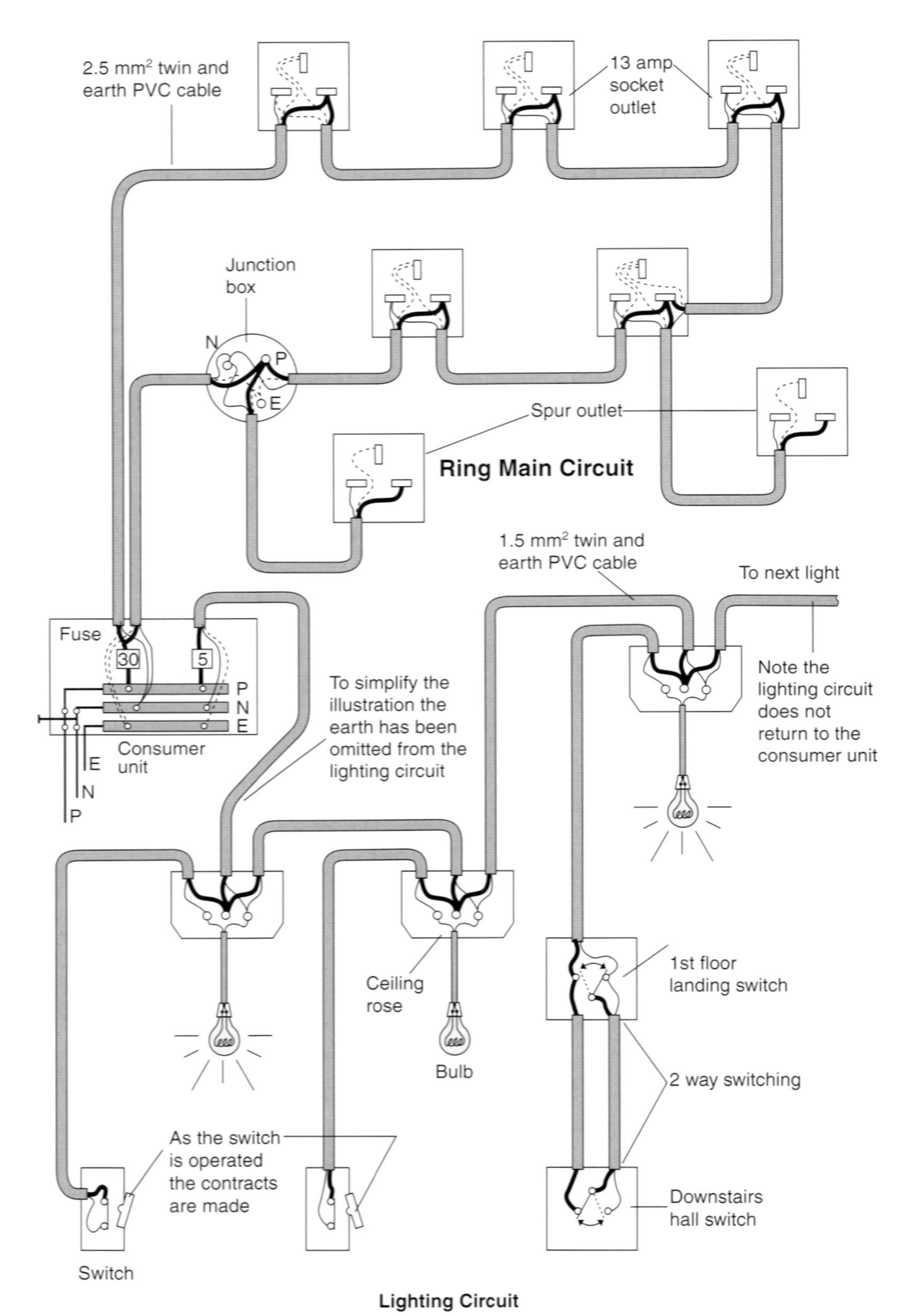

Figure 9.10 Lighting and ring main circuit

Test yourself 9.1

1. Which of the following is the unit of measurement for the resistance in an electrical circuit?
   a. ohm ☐
   b. coulomb ☐
   c. volt ☐
   d. hertz ☐
2. The change in energy levels measured across load terminals is referred to as:
   a. Electromotive force ☐
   b. Impedance resistance ☐
   c. Potential difference ☐
   d. Fuse rating ☐
3. Ohm's law is concerned with which factors?
   a. Power, voltage and resistance ☐
   b. Voltage, current and power ☐
   c. Work, power and resistance ☐
   d. Voltage, current and resistance ☐
4. Which of the following electrical protection devices is most commonly used in newly built domestic consumer units?
   a. Cartridge fuse ☐
   b. Fuse wire ☐
   c. Miniature circuit breaker ☐
   d. High rupturing capacity ☐
5. Alternating current (a.c.) electricity is produced as a result of what?
   a. Atmospheric pressure ☐
   b. Electromagnetism ☐
   c. Potential difference ☐
   d. Chemical energy ☐
6. The electrical voltage in a domestic property is approximately?
   a. 230 V ☐
   b. 110 V ☐
   c. 400 V ☐
   d. 1100 V ☐
7. A consumer unit supplying a ring main circuit feeding socket outlets should be protected with what size of fuse or mcb?
   a. 3 amp ☐
   b. 5 amp ☐
   c. 13 amp ☐
   d. 30 amp ☐

Now check the answers at the end of the book.

Electricity supply systems installation

Scope of electrical work carried out whilst working on domestic plumbing installations

Plumbers carry out work on electrical systems, which in the 'main' is restricted to connecting supplies to appliances and controls. You are unlikely to be required to install electrical circuits in either new or existing dwellings, but it is important that you have a basic understanding of how a circuit works.

When working on domestic plumbing installations, you will be expected to prove competence in the electrical installation requirements and safe systems related to domestic installation work in the following areas:

- Understanding the electrical hazards and risks to which you may be exposed during your work, and the precautions to prevent electrical injury
- The procedures for preventing electrical injury during installation, commissioning, maintenance, diagnostic testing, fault finding, modification, and de-commissioning of domestic plumbing installations. This must include the safe isolation and the safe connection/reconnection of electrical circuits and components in domestic plumbing systems, and the safety requirements relating to working on, or near, live electrical systems where there is a risk of electrical injury.
- The installation requirements for electrical circuits and components in domestic plumbing installations, as specified in BS 7671 'Requirements for Electrical Installations'
- The requirements relating to the earthing and bonding of exposed and extraneous conductive parts in domestic plumbing systems, including the requirements for temporary or permanent earth continuity bonding of pipework where necessary during work on any domestic plumbing installation.

Anyone working on electricity supplies *must be competent*. The aim of this section is not to turn you into an electrician, but to give you sufficient knowledge to do your part of the job competently. In this section, you will look at safety, and then look at practical requirements of system installation.

Safety requirements when working with electricity

The two most important things here are:

- Circuit protection
- Safe isolation.

Circuit protection

The Wiring Regulations require that an electric current is disconnected automatically should a greater current flow through a conductor (cable) than what the circuit was designed for. This should happen within 0.4 s for appliances with a socket outlet, and 0.5 s for fixed appliances, e.g. immersion heater.

Activity 9.2

What do you think are the most common causes of faults in circuits? Jot your thoughts down in your portfolio and then check them out at the end of this book.

The faults identified in *Activity 9.2* are usually down to badly designed circuits, faulty appliances, and/or poor installation. Circuits are protected against overload and short circuits using either fuses or circuit breakers. Circuit protection against earth faults includes fuses, circuit breakers, or a residual current device (RCD).

Think about the electrical components of any water heating appliances you are familiar with. What kind of faults can occur, and what might cause these faults?

Electric shock

Key point

There are two ways a person can receive an electrical shock:

- By direct contact with a live supply
- By indirect contact with a live supply.

Direct contact

There are different methods to prevent against direct contact. Insulating the live parts of the system which carry the current with non-conductive material, e.g. conductors in a cable or flex are protected with a plastic coating, is one such method. Components such as pumps and valves using electric motors are also insulated internally.

Another method is to ensure that, under normal circumstances, live components are placed within an enclosure such as a junction box, which means the exposed cabling cannot be touched.

Think about the effect that an infestation of mice or rats (or an escaped hamster) could have, given a rodent's liking for gnawing on anything resembling a treeroot (which includes plastic-coated cabling). What telltale signs might alert you to a rodent's presence?

Indirect contact

Indirect contact is prevented by a method called Earth Equipotential Bonding. This ensures that all exposed metalwork in a building is bonded together and connected to the earthing block in the consumer unit. Should any exposed metal work become live due to an electrical fault, then the current will be discharged to earth. The fault will also be detected by the RCD, which will automatically cut off supply.

> **Key point**
>
> Before working on any electrical supply, you must make sure that it is completely dead and that it cannot be switched on accidentally without your knowledge. Not only is this a requirement of the Electricity at Work Regulations, but it is essential for your personal safety, and that of your customer or co-workers.

Activity 9.3

Write down some situations where an RCD will protect against electric shock. Some suggested answers are given at the end of this book – but you may come up with your own, which may be just as valid.

Safe isolation

The approved method of testing if a circuit is live is by using a voltage indicating device. This can be done using a pocket-sized voltage detector, as shown in Figure 9.11.

Figure 9.12 shows a tester with multiple functions, such as voltage, continuity and current.

> **Key point**
>
> The use of a home-made test lamp or neon screw-driver is not acceptable, as this will not show a supply that has low voltage.

Inspection and testing

Before starting work on any electrical circuit, you should make sure that it is completely isolated from the supply. Most fatal accidents

Figure 9.11 Pocket-sized voltage detector

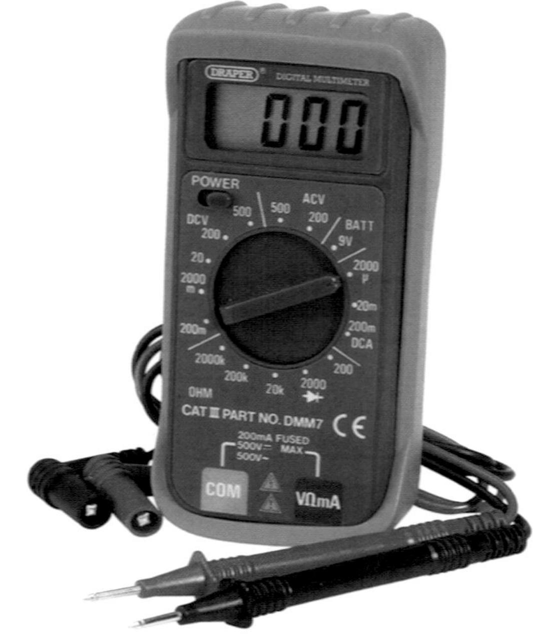

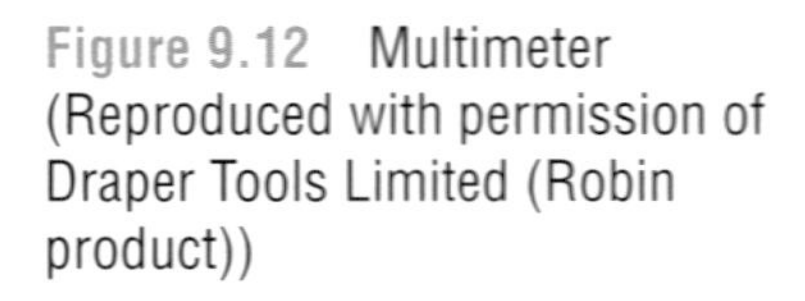

Figure 9.12 Multimeter (Reproduced with permission of Draper Tools Limited (Robin product))

involving electricity occur at the isolation stage. It is at times like this when you must be really careful and know what you are doing, as you may have no idea of the type of supply you may be confronted with. Do not take any risks, if you are in any doubt, seek assistance from a qualified electrician.

Activity 9.4

Put the steps for inspection and testing below in the correct order:

1. Test if the equipment system is dead
2. Secure isolation
3. Isolate
4. Begin work
5. Identify sources of supply.

Check if you were right by checking at the end of this book. Standardised procedures have been drawn up by the Electrical Contracting Industry; these are used to set the standard for safe systems of working in the Plumbing industry.

Identify sources of supply

It is not only important that you identify the *type* of supply, but you also need to know the *source* of supply. In domestic dwellings, this will be a single phase 230 V supply to the circuits.

Do you remember the various types of circuit? The main types of circuit are:

- 32 amp ring main for 13 amp socket outlets
- 6 amp lighting circuit.

In addition to these circuits, you may also encounter:

- 16 amp radial circuits to a 13 amp socket
- Spur outlets.

The 32 amp ring mains were covered earlier in terms of layout and cable sizing. As a plumber you are unlikely to work on lighting circuits, but you need to know the circuit layout and cable sizes. You will look at the wiring of fused spur outlet from the ring main.

The 16 amp radial circuits supply the socket outlets, but unlike the ring main, the circuit terminates at the last socket and is not returned to the consumer unit. This is not as popular as the ring main, because you are limited in the number of sockets you can install.

Another point to remember is that supplies to electric cookers and showers are in 6 mm cable, immersion heaters are still 2.5 mm, and are all wired directly to the consumer unit and use amp ratings higher than 13 amps for the fuse.

Isolate

Now the type and source of supply have been identified, they need to be isolated. Regulations require that a means of isolation must be provided to enable skilled persons to carry out work on, or near parts which would otherwise normally be energised (live).

Isolating devices (fuses, miniature circuit breakers, residual current devices) must comply with British Standards and the isolating distance between the contacts must comply with the requirements of BS EN 60947 – 3 for an isolator. The position of the contacts must either be externally visible or be clearly, positively and reliably indicated.

Secure isolation

To prevent the supply being turned on accidentally by the customer or other co-workers the fuse or circuit breaker should be removed and kept in your pocket, or the isolator locked off. As an extra precaution, a sign saying 'work in progress and system switched off' could be left at the consumer unit or the area in which you are working.

Test equipment and system is dead

Any circuit you work on *must* be tested to ensure that it is dead.

Test equipment

Test equipment must be regularly checked to ensure that if it is in good and safe working order. You must ensure that your test equipment has a current calibration certificate, indicating that the instrument is working properly and providing accurate readings. If you do not do this, test results could be inaccurate.

Equipment check list

- Check the equipment for any damage. Check to see if the case is cracked or broken. This could indicate a recent impact, which could result in false readings.
- Check that the batteries are in good condition and have not leaked, and that they are all of the same type
- Check that the insulation on the leads and the probes is not damaged. Check that the insulation is complete and secure.
- Check the operation of the meter with the leads to both open and short-circuit
- Then zero your instrument on the ohm scale.

If you have any doubt about an instrument or its accuracy, ask for assistance. These instruments are very expensive and any unnecessary damage caused due to ignorance should be avoided.

Test the voltage indicator on a proven supply before you start, to confirm that the kit is working. For single-phase supplies, this is done by testing between phase and neutral and phase and earth.

Remember, all live conductors must be isolated before the work can be carried out. As the neutral conductor is classified as a live conductor this should also be disconnected. This may mean removing the conductor from the neutral block in the consumer unit.

Begin work

Make use of the warning notices, 'Plumber at Work'. It may be helpful to put your name and contact number on the note, so if you have to leave the job whilst the customer is out, they can contact you to find out why power has been turned off.

The flowchart in Figure 9.13 shows the procedure for isolating an individual circuit or item of fixed equipment. Work through each step – does it make sense to you?

Electrical installation

Fused spur outlets

You saw what a fused spur outlet looked like in Figure 9.9. It can be surface-mounted, or mounted flush to the wall finish. Ones that are flush with the wall are fitted to a metal back box.

Do you remember when you looked at the layout of a domestic house wiring in Figure 9.10. It showed the ring main and lighting

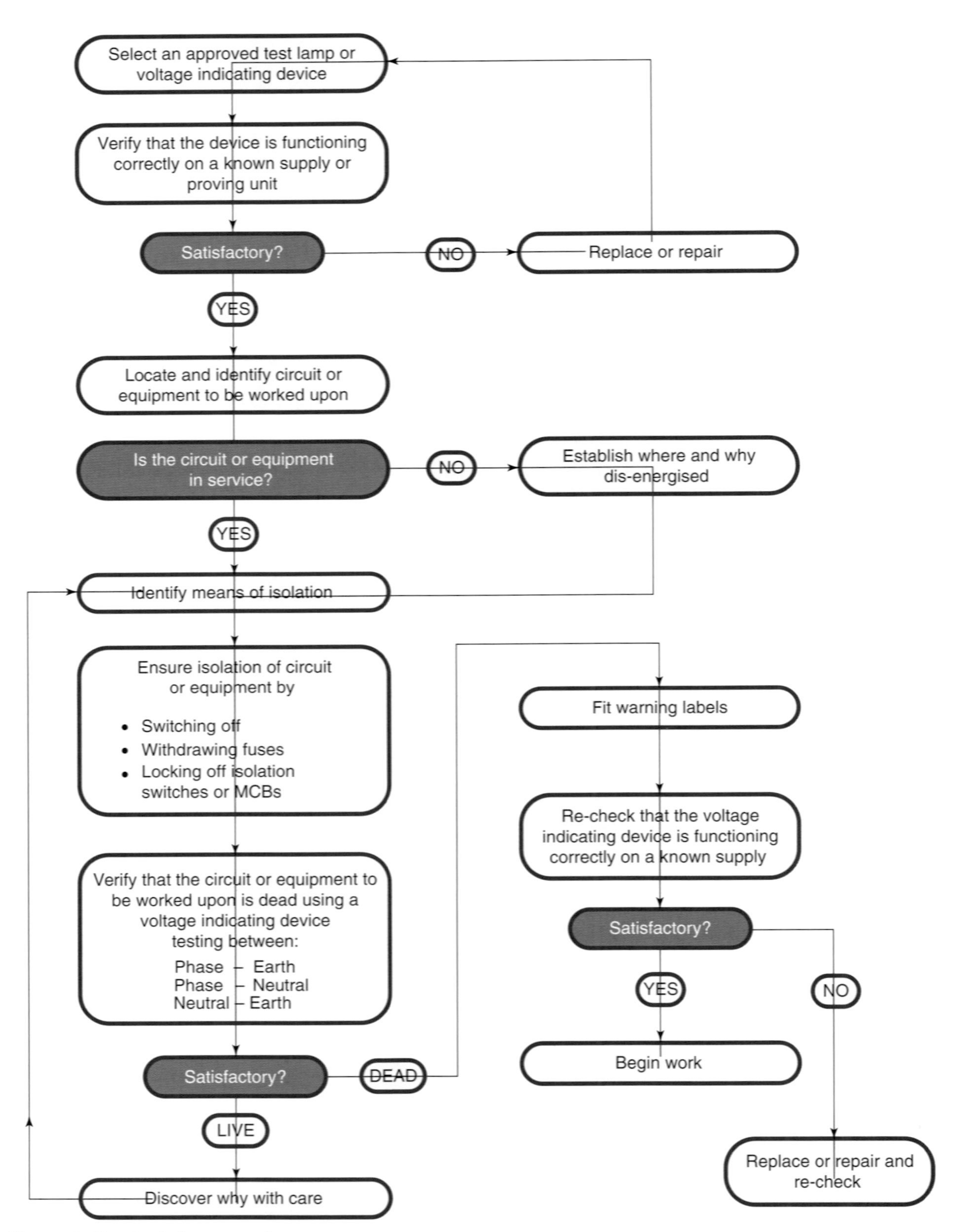

Figure 9.13 Isolation procedure

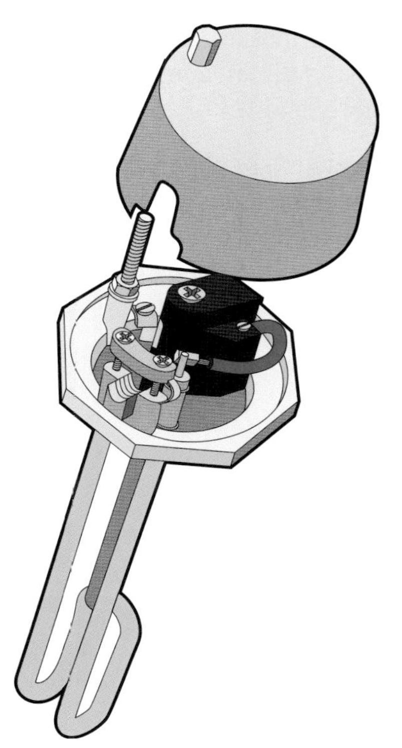

Figure 9.14 Immersion heater connections

circuit from a consumer unit? Get your tutor to show a practical example of how the appliances and components are connected to the circuit using a fused spur outlet.

Immersion heaters

Immersion heaters are not wired directly from the ring main, but from the consumer unit in 2.5 mm^2 twin and earth PVC cable, and have a fuse rating of 15 amps.

It is fed to a double pole switch, which has to be located at a maximum of 1 m from the connection to the immersion heater. The cable between the pole switch and the immersion heater is 1.5 mm^2 heat resistant flex, usually butyl. A double pole switch is safer than a single pole version, in that when it is turned off both the live and neutral are isolated.

It is most likely that you would be required to disconnect and reconnect the supply when replacing a defective immersion heater (see Figure 9.14 for connections).

Test yourself 9.2

This concludes the section on electricity supply systems installation. Try the following questions to see how you are progressing.

1. The main requirement for a plumber working on an electrical system is that they must be:
   a. Competent ☐
   b. Over 21 years old ☐
   c. CORGI-registered ☐
   d. Confident ☐
2. Briefly summarise the differences between circuit protection and safe isolation.
3. State how people are protected against indirect contact from an electrical supply.
4. What piece of equipment would you use to check whether an electrical circuit was dead?
5. What size cable, ring main amp rating and socket amp rating would you expect to see on a ring main?
   - Cable size
   - Ring main amp rating
   - Socket amp rating.
6. How would you make sure a system had been safely isolated?

There are no answers to this progress check, because this section is quite important, we would ask that you check your answers by going back through the session text. Make any corrections or add additional information that you think are necessary.

Earth continuity

We discussed earth continuity bonding in the last section, where we referred to it as protection against indirect contact. In this section, we will look at the topic in more detail.

Earthing and bonding is a very important area of electrical wiring, so you should be able to tell if it has been installed correctly. You may also have a cause to carry out work on these systems, but should do so only if you are competent. Never carry out work of this nature if you are not sure. If that is the case, contact a qualified electrician.

Equipotential bonding

A standard method of protecting against indirect contact using equipotential bonding and automatic disconnection of supply is explained below. This meets the requirements of the Wiring Regulations (see for example Figure 9.15).

Metal pipework can provide a route for stray electric current to earth. This could cause an electric shock for someone touching the live pipework if the metal is not properly earthed. It can also cause the pipework to corrode.

Note the size of the bonding conductors, and the need for the clamp to be labelled. The bonding conductor from the main terminal to earthing clamp is 10 mm^2. The bonding to gas, water, or other services should be as close as possible to the point of entry, and for the gas supply within 300 mm of the meter.

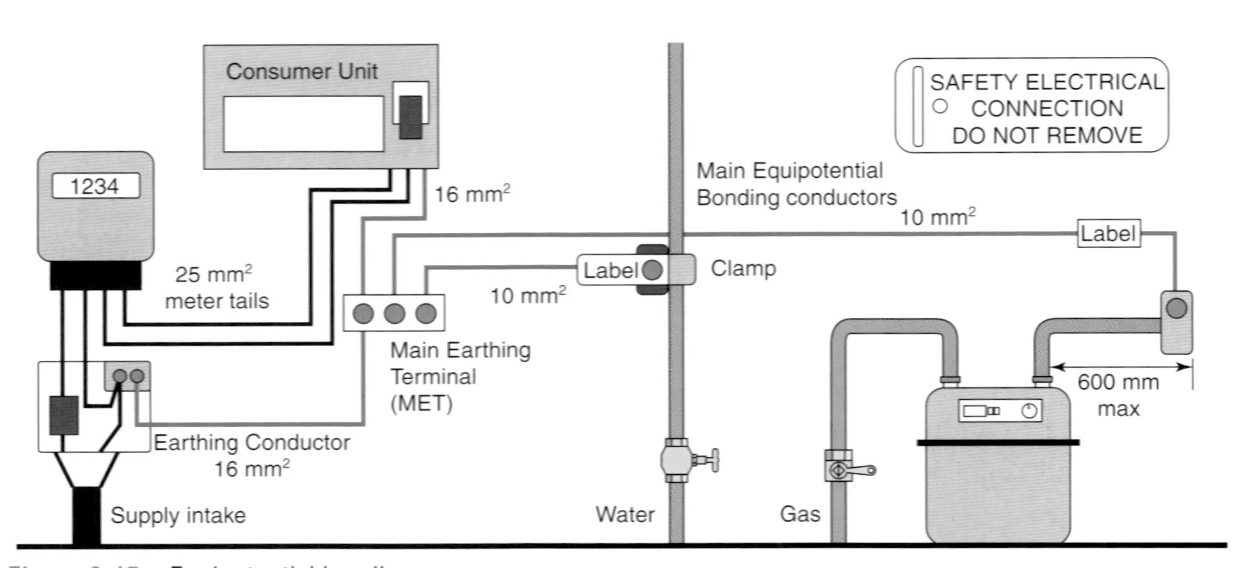

Figure 9.15 Equipotential bonding

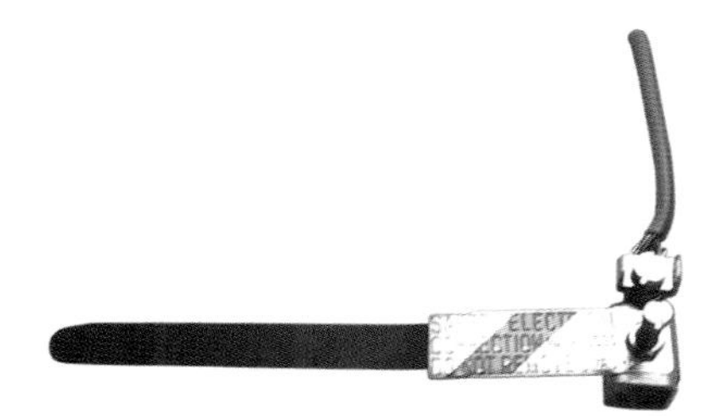

Figure 9.16 Earthing clamp

Bonding the pipework, as shown in Figure 9.16, will provide a safe route to earth.

Supplementary bonding

You should be able to tell if a plumbing installation is correctly bonded. You can see from the previous figure that the equipotential bonding only connects to one point of the pipework.

There are other exposed metal parts within the domestic hot, cold and central heating systems within a dwelling, which may not be protected because they have been isolated from the earth by plastic fittings, cisterns, etc used in the system. This affects the conductivity of the pipework to the earth. To maintain the earth continuity, supplementary cross-bonding is used; a typical layout is shown in Figure 9.17.

The bonding wire is indicated by the striped cable, this would be green and yellow in the workplace.

Key point

The clips should be placed in a position to bridge the gap of the pipe or fitting that is going to be removed. Work can then safely take place. Remove the clamps once the job is complete.

Temporary bonding

Plumbers are in constant contact with metal pipework and metal surfaces when carrying out their job. If you are required to work on a repair or maintenance job, it is important that you check the pipework you are going to work on is bonded correctly. If you are going to remove a section of metal pipework or fittings it is essential that earth continuity is maintained before any cutting or disconnection takes place. Otherwise, if there was a fault and the

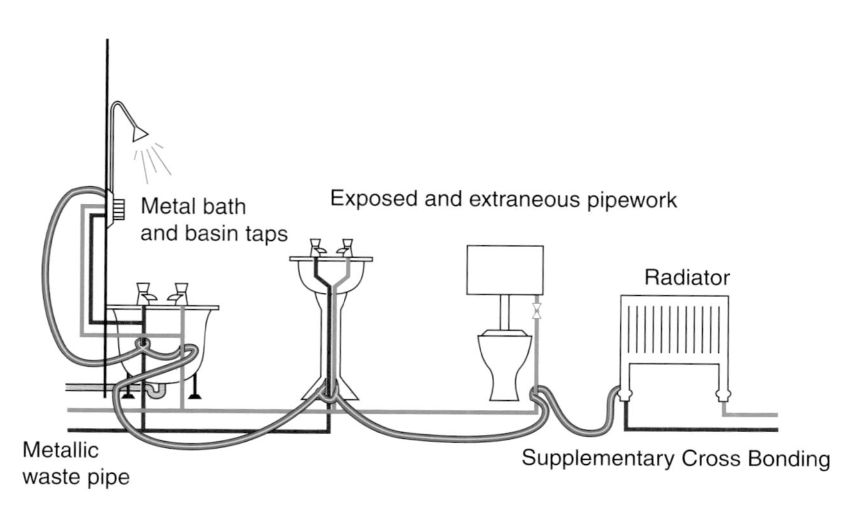

Figure 9.17 Earth continuity bonding

Figure 9.18 Temporary bonding clip

bonding was incorrectly wired, and if you removed a section of pipe or fitting and then touched the pipe, you would get an electric shock.

A typical piece of kit is shown in Figure 9.18. This type, using crocodile clips and 10 mm^2 conductor, 250 V rating minimum, is suitable for metal pipework with a diameter of upto 28 mm.

Activity 9.5

What might happen if you fail to secure the temporary bridging of the gap, when removing a length of pipe to replace it? Write your answer here and then check it out at the end of this book.

Test yourself 9.3

Test yourself 9.3 is in the form of an assignment. Here is what to do. Making reference to the text in Unit 3 and using your college or centre resource facilities carry out the following task.

Imagine you are required to carry out a repair job on an existing heating system. Write a short report on:

- What you looked for in terms of making sure that equipotential and supplementary bonding was correctly and adequate installed?
- What you did to ensure that a section of pipework could be safely removed?

Check your learning Unit 9

Time available to complete answering all questions: 30 minutes

Please note that some of these questions may not be covered in the text, they should be completed once you have completed the electrical content of the Technical Certificate.

1. Specific regulations regarding electrical installations are covered by:
   a. Electricity Supply Regulations 1988 ☐
   b. 16th Edition of the IEE Wiring Regulations ☐
   c. Health and Safety at Work Act 1974 ☐
   d. Electricity at Work Regulations 1989 ☐
2. What Building Regulation makes specific reference to a person's competence to carry out electrical installation work?
   a. Part P ☐
   b. Part H ☐
   c. Part L ☐
   d. Part G ☐
3. The details of the wiring requirements for the controls of a condensing boiler would be found in:
   a. WRAS Guide ☐
   b. Building Regulations ☐
   c. British Standard 6700 ☐
   d. Manufacturers' catalogues ☐
4. When working in a domestic dwelling, the current rating for wiring circuits and appliances could be obtained from:
   a. British Standard 1010 ☐
   b. Manufacturers' catalogues ☐
   c. The consumer unit ☐
   d. Asking the customer ☐
5. Information specifying the minimum conductor size for electrical circuits can be obtained from:
   a. British Standard 7671 ☐
   b. The Electrical Contractors Association ☐
   c. Manufacturers' catalogues ☐
   d. Building Regulations Part L ☐
6. A branch from a ring final circuit is usually described as a:
   a. Spur ☐
   b. Socket outlet ☐
   c. Residual current device ☐
   d. Consumer unit ☐
7. The incoming supply to the consumer unit is known as:
   a. Design current ☐
   b. Three phase ☐
   c. Single phase ☐
   d. Alternating current ☐
8. What is the mcb rating in amps for an immersion heater?
   a. 8 ☐
   b. 16 or 20 ☐
   c. 36 ☐
   d. 45 or 50 ☐
9. The main connection to an immersion heater should be directly from the:
   a. Consumer unit ☐
   b. Fused spur outlet ☐
   c. Ring main ☐
   d. Junction box ☐

(Continued)

Check your learning Unit 9 (Continued)

10. Which of the following items of equipment should be used to test if the electrical circuit is live?
    a. Neon screw driver ☐
    b. Electromagnetic mcb ☐
    c. Multimeter ☐
    d. Home made test lamp ☐

11. Equipotential bonding is a method of:
    a. Testing an electrical circuit for earth leakage problems ☐
    b. Connecting electrical appliances that are not earthed ☐
    c. Isolating the mains before working on the power supply ☐
    d. Earthing all exposed metal work in a domestic dwelling ☐

12. Which one of the following cable types would be used on domestic and general wiring where a circuit protective conductor is required for all circuits?
    a. Single-core PVC insulated and sheathed. ☐
    b. Single-core PVC insulated unsheathed. ☐
    c. PVC insulated and sheathed flat wiring. ☐
    d. Heat resisting PVC insulated and sheathed flexible cords. ☐

13. Which one of the following should be positioned within reach of a person using an instantaneous shower?
    a. Two-way switch ☐
    b. One-way switch ☐
    c. Double pole wall switch ☐
    d. Double pole pull cord switch ☐

14. The colour coding of the wiring in a modern three core flex is made up of blue, yellow, green and:
    a. Brown ☐
    b. White ☐
    c. Black ☐
    d. Red ☐

15. On a 32 amp ring main, the socket outlets installed in a domestic wiring system should be rated at:
    a. 6 amps ☐
    b. 13 amps ☐
    c. 25 amps ☐
    d. 32 amps ☐

16. When installing surface-mounted cables, the best method of ensuring that it is kept straight is by:
    a. Clipping the cable to a pre-drawn level pencil line ☐
    b. Running the thumb over it before clipping ☐
    c. Pre-fixing the clips to a level line before fixing the cable ☐
    d. Offering a spirit level against the cable before clipping ☐

17. When making the connection to an immersion heater, the consumer controls should include:
    a. One-way wall switch ☐
    b. Two-way wall switch ☐
    c. Double pole wall switch ☐
    d. Double pole pull cord switch ☐

18. When working on any existing electrical system, the first test to be carried out should be:
    a. Safe isolation ☐
    b. Earth continuity ☐
    c. Insulation resistance ☐
    d. Polarity ☐

Check your learning Unit 9 (Continued)

19. What type of test would be conducted to ensure that phased conductors are not crossed somewhere?
    a. Polarity ☐
    b. Earth continuity ☐
    c. Safe isolation ☐
    d. Insulation resistance ☐

20. The main reason that the minimum overload protection to an instantaneous shower has to be between 30 and 40 amps is because of the unit's:
    a. Exposure to water ☐
    b. Kilowatt rating ☐
    c. Amperage ☐
    d. Voltage rating ☐

Sources of Information

We have included references to information sources at the relevant points in the text; here are some additional contacts that may be helpful.

In terms of the electrical requirements of specific controls or appliances, these are usually supplied with the manufacturers' instructions, so if you have already received information when you have been carrying out research on the other units, particularly things like pumps, boilers, showers, heating controls, etc have a look at them.

SummitSkills are the sector skills council for the building services sector, which includes the electrical industry. They should be able to offer advice on short courses that are available on electrical training.

They can be contacted on: 0870 3514620. Their website is: www.summitskills.org.uk.

In terms of electrical materials, cables fitments, etc, one of the major wholesalers is City Electrical Factors Ltd. Have a look on their website on: www.cef.co.uk.

This is a useful publication produced by the HSE. It can be downloaded from their website, www.hse.gov.uk; this booklet is free of charge.

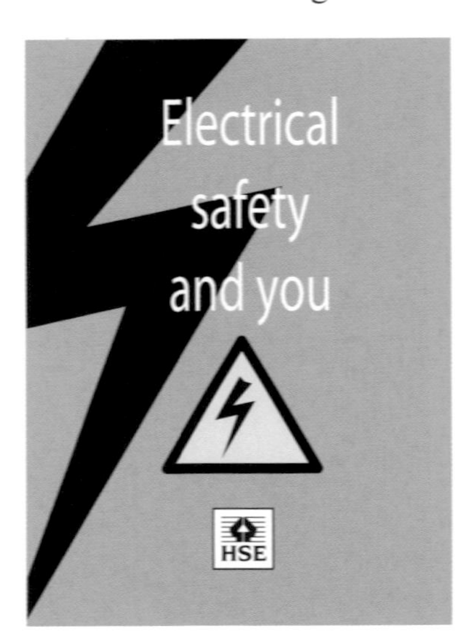

We would also recommend that you obtain an electrical text book either from your local library or purchase one from a book shop. A book related to the 16th edition which covers the requirements of BS 7671 would be particularly useful.

UNIT 10

Sheet Lead Work

Summary

The main area that a plumber will encounter in terms of lead work on site will be on weatherings to chimneys and soil vent pipe penetrations through roofs. This is done either by fabricating or working sheet lead by bossing (working the lead into shape using a range of mallets and dressers), or by welding it together using oxy-acetylene equipment. This subject is also a requirement of the Technical Certificate for both theory and practical. So, in this unit you will study:

- Sheet lead design considerations:
  - Types of sheet lead
  - Codes and sizes
  - Design/installation considerations.
- Working with lead sheet:
  - Tools and equipment for sheet lead installation
  - The main safety factors when working with sheet lead
  - The techniques for lead welding and bossing including the safe use of welding equipment
  - Finish and quality of work.

- Domestic applications of sheet lead work:
  - The principles of sheet weatherings to chimneys
  - Types of abutment flashings
  - Lead slates
  - Fixing techniques
  - Testing for water tightness.

We would like to express our thanks to the Lead Sheet Association for their kind permission in reproducing some of the images from their manual. If you have internet access have a look at their website now; it contains excellent information and a number of free information sheets. The contact details are at the end of this unit.

Sheet lead – Design considerations

Traditionally, plumbers have always been involved in carrying out sheet lead work. At one time, this may have involved working on large jobs such as covering complete roofs in lead. This type of work is now mostly undertaken by specialist lead work contractors, but plumbers are often required to carry out sheet weathering jobs on domestic dwellings. This usually takes the form of:

- Chimney flashings
- Simple abutment flashings
- Lead slates.

Activity 10.1

What do you think is the meaning of the term lead slate? Jot down your thoughts and then check it out at the end of this book.

Key point

Casting was how sheet lead was produced originally. It is still produced in relatively small amounts by specialist lead working firms, by running molten lead over a bed of sand. There are no British Standards for this material, and its sheet sizes and thickness vary.

Hygiene and working with lead

Try this

You may recall that you covered lead safety in the Health and Safety unit. You do not need to go through it all again now, but refresh your memory by looking over the main areas of the Health and Safety unit.

Types of sheet lead

There are two types of sheet lead:

- Cast sheet lead
- Rolled sheet lead.

Rolled sheet lead to BS 12588:1999

Lead manufactured using this process is used by plumbers for carrying out sheet weatherings. It is formed by passing a slab of lead back and forth on a rolling mill between two closing rollers, until it is reduced to the required thickness. The sheet is then cut to a standard width ready for distribution.

Table 10.1

BS 12588 Code No	Thickness (mm)	Weight (kg/m²)	Colour Code
3	1.32	14.97	Green
4	1.80	20.41	Blue
5	2.24	25.45	Red
6	2.65	30.05	Black
7	3.15	35.72	White
8	3.55	40.26	Orange

BS 12588:1999 codes and sizes for rolled sheet lead

Table 10.1 shows the code number, the thickness, weight and colour code of the sheet.

For most flashing applications, Codes 3, 4 and 5 are fine. Code 4 is considered adequate for forming a chimney back gutter if lead welding techniques are used, but if you were to shape it by bossing the lead with tools, Code 5 would be more appropriate. Code 3 would be used for lead soakers.

Do not worry about terminology like back gutter, etc. You will become familiar with these as you work through the unit.

For the purpose of flashing, sheet lead is supplied in rolls (coils) in widths ranging from 150 to 600 mm, going up from 150 mm in steps of 30 mm, i.e. 150, 180 and so on. It is usually supplied in 3 or 6 m lengths. Larger widths and lengths are also available from the manufacturer/supplier.

Design/installation considerations of sheet lead

Here is a list of the properties applicable to sheet lead:

- Durable
- Allows for thermal movement
- Resistant to corrosion
- Fatigue and creep resistance
- Patination

- Resistance to fire
- No problems with contact with other materials
- 'Recycle-ability'.

Durability

You only have to look at the roofs on historic buildings and the original sheet lead weathering details to appreciate the durability of lead. When sheet lead weathering fails it is usually due to poor workmanship or design. Lead sheet is extremely resistant to corrosion, particularly in towns and coastal areas.

Key point

In towns the atmosphere contains or would contain chemicals such as sulphur, which is absorbed in rain water to produce dilute sulphuric acid. In coastal areas the atmosphere contains salt. Both these substances are extremely corrosive to building materials.

In addition to being durable, lead sheet is also malleable. You may remember from your key plumbing principles unit that malleability is 'a material's ability to be worked without fracture'; a quality that is quite helpful in the case of lead sheet!

Thermal movement

Lead has a high coefficient of expansion at 0.0000297 for 1°C. It is important to include regular expansion joints in lead flashings to allow for expansion and contraction due to changes in temperature.

Flashings are often secured into brickwork joints by lead wedges thus restricting its movement so expansion joints would be needed here as well.

Expansion joints for flashings are usually in the form of laps. To minimise the thermal movement at each lap it is important that an individual piece of flashing is not longer than the dimensions given in Table 10.2.

Table 10.2

Code No	Thickness (mm)	Use	Maximum Length (m)
3	1.32	Soakers	1
4	1.80	Flashings	1.5
5	2.25	Flashings	1.5

Most of the weatherings you will work on will be of relatively short dimensions, but you may come across a job where a pitched roof needs weathering against a brick wall. This is called an abutment flashing.

Resistance to corrosion

Sheet lead is resistant to forms of corrosion such as acidic atmospheres. It is also able to resist the effects of lichen growth and moss. The acid run-off from lichen or moss on a roof may cause small holes to appear in the lead sheet under the drip-off point from tiles or slates. The rate of corrosion on the lead is very slow, and is not usually a problem on domestic applications. The solution to the problem is to treat the growth with a chemical fungicide.

Condensation is not really a problem for domestic situations as it is mainly associated with flat roofs.

Fatigue and creep resistance

Activity 10.2

By now you should be getting a 'feel' of lead sheet as a material. What do you think is the meaning of the terms 'fatigue and creep resistance'? Jot down your thoughts and then check them with what we have put at the end of this book.

Patination

Figure 10.1 shows what we mean by the term 'patination', the outcome of which is the light-coloured streaks that appear on the slates.

Over a period of time, lead develops a strongly adhering and insoluble patina (or sheen) which is silver-grey. In rainy or damp

Figure 10.1 Patination (Reproduced with permission of Lead Sheet Association)

conditions, new lead sheet flashings will produce an initial, uneven white carbonate on the surface. Not only does this look unsightly, but the white carbonate can also be washed off by rain, causing further staining on materials (e.g. brickwork) below the flashings. The use of patination oil, applied evenly with a cloth as the job progresses, will avoid any future staining.

Resistance to fire

Lead is incombustible, but melts at 327°C.

Contact with other materials

Lead sheet can be used in situations where it comes into contact with other building materials. This may include:

- Metals
- Sealants
- Masonry and mortars.

Wood is another building material, but would not really be relevant at Level 2.

Metals
In practice, lead sheet can be used in contact with copper, zinc, iron, aluminium and stainless steel without significant risk of bi-metal corrosion.

Key point

If using silicon, be sure to use neutral cure and not acid cure as the latter can cause a white corrosion product to form on the surface of the lead.

Sealants
Sealants have become popular as an alternative for using mortar to seal the joint between lead flashings and the brickwork. This is an acceptable practice using silicone or polysulphide.

Masonry and mortars
Mortars made from Portland cement or lime can initiate a slow corrosive attack on lead in the presence of moisture. However, in the case of flashings, the turn in to a masonry joint should only be 25 mm, so the relatively small surface area will mean the carbonation of the free lime will be rapid and the risk of attack will be non-existent.

'Recycle-ability'

Sheet lead is totally recyclable. In UK, scrap lead is recovered from buildings that are due to be demolished by a national network of reclamation merchants. It is then returned to the manufacturers, where it is carefully refined for reuse in the rolling mills. Owing to this well-established recovery network, the sheet lead industry has a solid reputation for environmental awareness.

Test yourself

1. What is the main type of lead sheet used by plumbers for carrying out roof weatherings?
2. What code of lead is considered adequate for forming a chimney flashing if using lead welding techniques?
3. What colour code is used for the answer to question 2?
4. What code of lead is considered adequate for lead soakers?
5. What colour code is used for the answer to question 4?
6. What is the maximum length of flashing in order to minimise thermal movement?
7. It is not acceptable practice to use neutral cure sealants for making good the joints between brickwork and lead sheet. True or false?

Working with lead

Forming lead sheet into the shapes required to produce weatherings can be achieved by:

- Bossing, or
- Welding.

Generally, lead work on larger sites will be done using welding techniques; however, there will be occasions when you might not have access to welding equipment, and you will need to know how to shape the lead by bossing.

The plumbing Technical Certificate requires candidates to be able to weather a chimney and fabricate a lead slate. Part of the chimney weathering includes a front apron which has to have one corner bossed. You will look at the various components and the method of carrying out this task in this unit.

Tools and equipment used for sheet lead

Bossing tools

The tools used to boss or form sheet lead are shown in Figure 10.2.

Originally produced in boxwood, the tools are now available in durable plastic. In addition to the specific bossing tools you will need:

Spirit-based marker pen and straight edge/steel rule: Used for marking and cutting lines without damaging the surface of the lead.

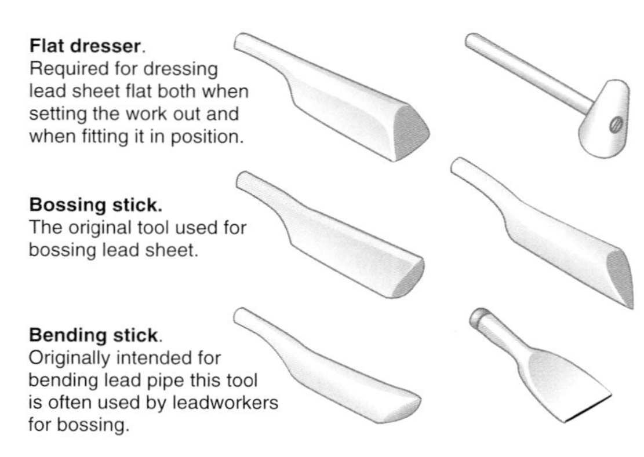

Figure 10.2 Bossing tools (Reproduced with permission of Lead Sheet Association)

Lead knife: Used for cutting sheet lead

Snips: Used for cutting and trimming

Plugging chisel: Used to chisel out mortar from brickwork to let in the lead work

Club hammer: For use with the plugging chisel.

Try this

In the Common Plumbing Processes unit you looked at various tools used by plumbers. You may have obtained tool manufacturers' catalogues. Have a look at these catalogues, or have a walk around your local plumber's merchants or DIY outlets and see what the tools mentioned above look like.

Lead sheet bossing

The term 'bossing' owes its origins to the middle ages; then, the term 'boss' meant 'to beat out metals into a raised ornament'. From that 'bossing' became the term used to describe the shaping of malleable (soft) metals.

Lead is an ideal material for roof weatherings due to its properties, resistance to moisture, etc. Added to this is its malleability. Lead at ordinary ambient (surrounding) temperatures is only 300°C below its melting point, compared to the melting point of copper at ambient temperatures of 1056°C below melting point. Lead behaves in similar ways at lower temperatures to harder metals at higher temperatures.

Lead is a suitable metal for bossing because:

- It is the softest of the common materials
- It is ductile in that it will stretch quite a lot before fracturing or splitting
- It does not harden much when it is been worked on (work hardening), and is self-annealing at ambient temperatures.

Bossing techniques

Lead sheet bossing is not a skill that you can learn by reading about it. You will need to spend some time observing a skilled lead worker, and then practising the techniques yourself. You can, however, gain the underpinning knowledge required to do the job from this unit.

We are going to look here at two techniques for bossing lead sheet: an *internal* corner and an *external* corner and see what a finished job looks like.

(a)

(b)

Figure 10.3 (a) Internal corner (b) tool for use in lead sheet bossing (Reproduced with permission of Lead Sheet Association)

Internal corner

Figure 10.3a shows an internal corner in the process of being 'bossed'. This is for a flat roof detail, but the principle is the same for any internal corner.

Bossing is used to achieve the required shape and it is important to remember not to make the lead too thin during the process.

An internal corner is produced by:

- Marking out the lead to the required dimensions (note: 100 mm minimum for the up-stand)
- Folding lines are set-in with a setting in stick or dresser and the surplus lead is cut off (Figure 10.4(a))
- The sides are partly pulled up and the corner is set-in by directing a few blows downwards with a bossing stick to form the base of the corner (Figure 10.4(b))
- The corner is then worked up, a mallet being held on the inside of the angle and the blows from the bossing stick directed from the base of the corner to work up the lead gradually to the top of the corner (Figure 10.4(c))
- Surplus lead forms as the bossing proceeds and this is trimmed off when the required shape and angle of the corner have been reached (Figure 10.4(d)).

Bossed internal corner

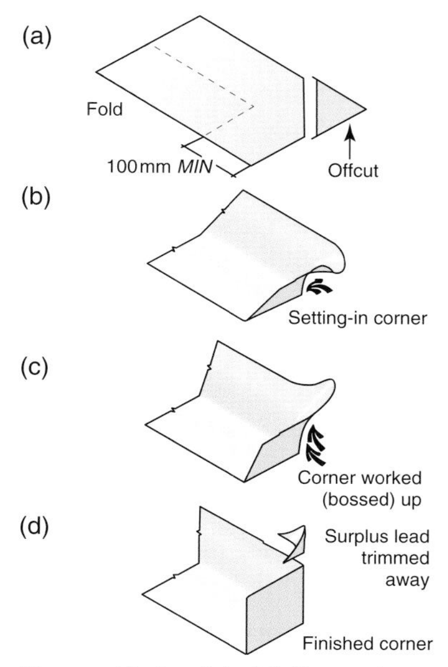

Figure 10.4 (a)–(d) Bossed internal corner details (Reproduced with permission of Lead Sheet Association)

External corner

Referring to Figures 10.5(a)–(c), bossing an external corner is very labour intensive and there can be a risk of undue thinning at the corner. The preferred method of producing this detail would be by lead welding, which is more economical in both material and labour. However, the bossing process is:

- When bossing the external corner, extra lead is required to form the angle and at least 75 mm is added to the up-stand for this purpose. The lead sheet should be set out, as in Figure 10.5(a)
- After setting in the fold lines, the sides are pulled up as far as possible without stretching the lead and then bossed from both sides in the directions, as shown in Figure 10.5(b)
- The bossing allowance is gradually driven towards the corner so that, when finished, the thickness of the lead at the angle is approximately the same as that for the lead being used for the work
- When the corner has been fully bossed to the angle required, the surplus lead is trimmed off, as shown in Figure 10.5(c).

Bossed break (external) corner

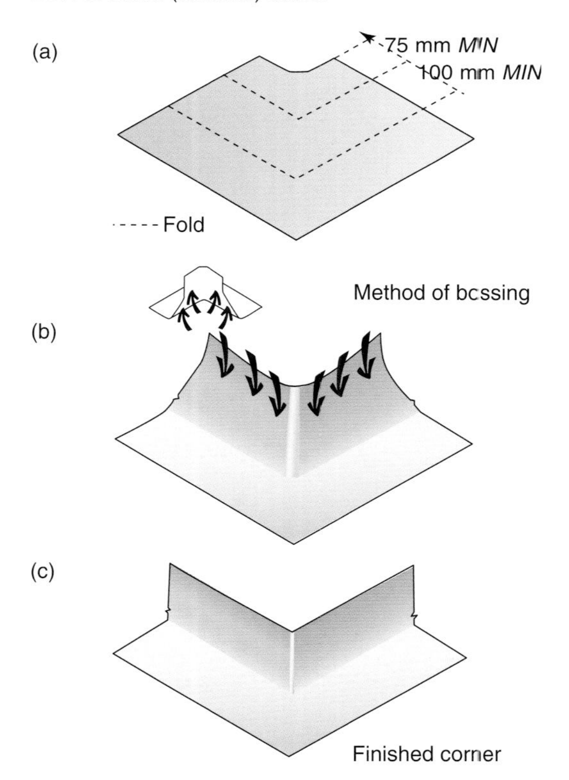

Figure 10.5 (a)–(c) Bossed external corner details (Reproduced with permission of Lead Sheet Association)

Lead welding

Lead welding is a process of joining two pieces of lead by melting the two edges of the lead together (called the parent metal) while a filler rod of lead is added. This is called *fusion welding.* The technique can be used to form an internal or external corner, as an alternative to bossing them into shape.

First, let us take a look at the internal corner.

Referring to Figures 10.6(a)–(d):

- To weld an internal corner the sheet is set out and cut, as shown in Figure 10.6(a)
- If the corner is to be welded in position, the up-stands are folded with the part marked X to the inside, as in Figure 10.6(b), and then welded with angle and inclined seams, as in Figure 10.6(c)
- In most cases however, the corner can be formed on the bench and turned on its side and welded using a flat lapped seam.

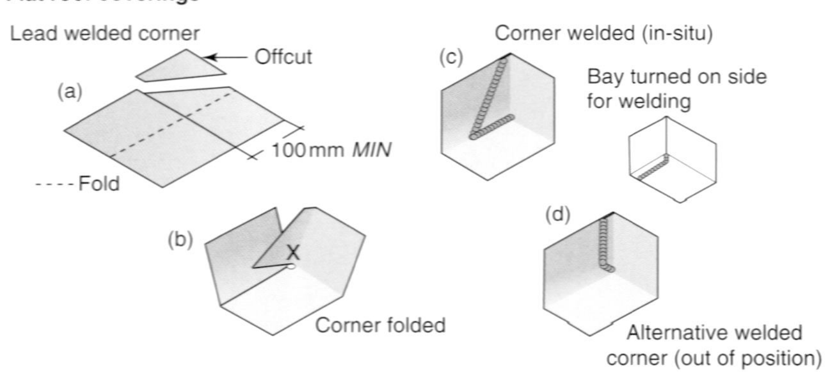

Figures 10.6 (a)–(d) Setting out for an internal corner for welding (Reproduced with permission of Lead Sheet Association)

For an external corner, referring to Figures 10.7(a)–(c):

- The lead sheet is set out, as in Figure 10.7(a). The Figure 10.7(a) shows the setting out for a flat roof; for a pitched roof the shape of the gusset will vary depending on the pitch of the roof.
- The sides are turned, the gusset folded and then welded using an inclined lap seam if welded in position or a flat lapped seam if working on the bench, as shown in Figures 10.7(b) and 10.7 (c).

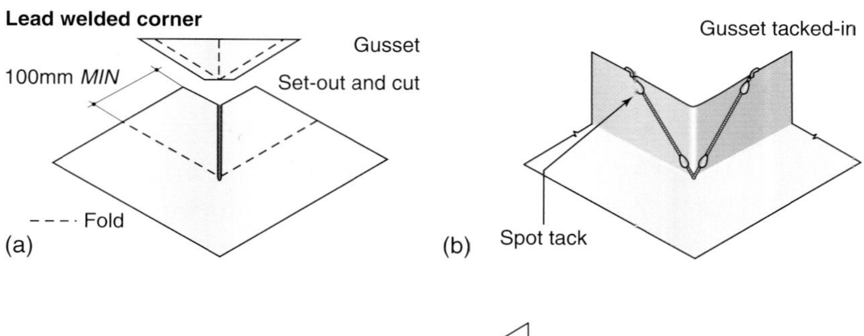

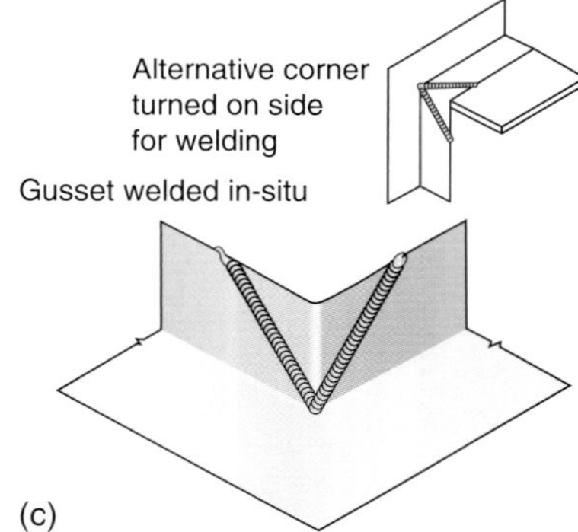

Figure 10.7 (a)–(c) Setting out for an external corner for welding (Reproduced with permission of Lead Sheet Association)

The actual welding process

There are three main types of lead-welded joints:

- Butted seam
- Lapped seam
- Inclined seam.

Figure 10.8 shows an example of a butted and lapped seam.

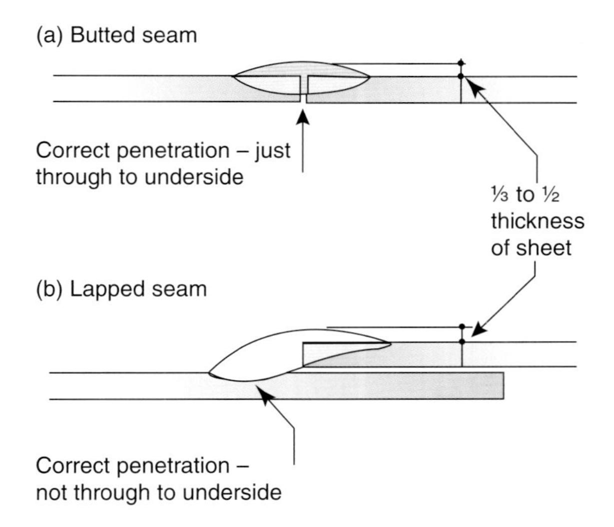

Figures 10.8 (a) and (b) Butted and lapped seam (Reproduced with permission of Lead Sheet Association)

Key point

Some plumbers still refer to welding as lead burning. This term comes from the time when crude welding techniques were used for jointing lead.

Key point

It is essential that the meeting edges and faces of the lead are clean, including the underside meeting faces of lap joints. This is done using a shave hook, and a straight edge such as a steel rule as a guide. You do not need to use flux.

Before looking at the welding process, the basic principles for welding are:

- Clean surfaces
- Correct penetration
- Correct thickness
- Weld width
- Avoid undercutting
- Correct flame.

Clean surfaces

Figure 10.9 shows what a shave hook looks like.

Figure 10.9 Shave hook

Correct penetration

With flat butted seams, the weld should fully penetrate through the thickness of the sheet lead. For lapped seams, the weld should penetrate the surface of the lead but should not penetrate to the underside.

Correct thickness

The correct thickness of a lead seam should be between a third and half thicker than the lead sheet.

Weld width

This will be based on the thickness of the lead sheet, the type of seam and the number of loadings. For example, for Code 4 the minimum width should be 10 mm. There is no set width for a weld, but for aesthetic (appearance) reasons the width of the weld should be the same width throughout the length of the weld.

Avoid undercutting

Undercutting means reducing the thickness of the lead at the side of the weld, and must be avoided because it creates a weakness in the sheet resulting in cracking along the weld line. Figure 10.10 shows examples of undercutting, where the most likely cause is holding the flame too long in the molten pool and an inadequate amount of filler rod.

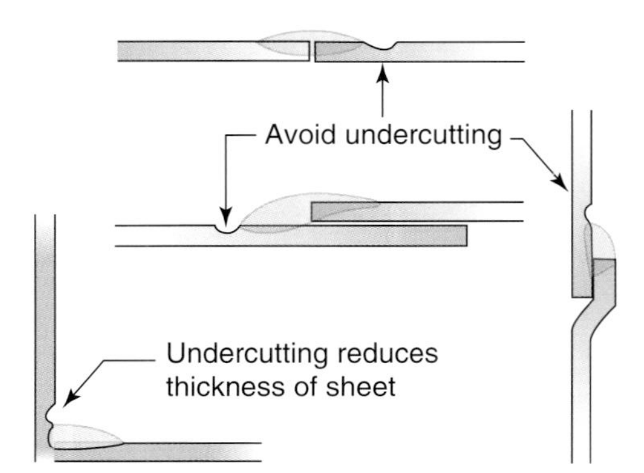

Figure 10.10 Undercutting (Reproduced with permission of Lead Sheet Association)

Correct flame

Referring to Figures 10.11(a)–(c), the gases most commonly used for lead welding are oxygen and acetylene. Portable equipment is readily available making in situ work relatively easier.

Oxy-acetylene flame

A pressure of 0.14 bar (2 lb/sq in.) should be set on both gauges. A small blowpipe is used, which can accept a range of nozzles from 1 to 5. The working part of the flame is immediately in front

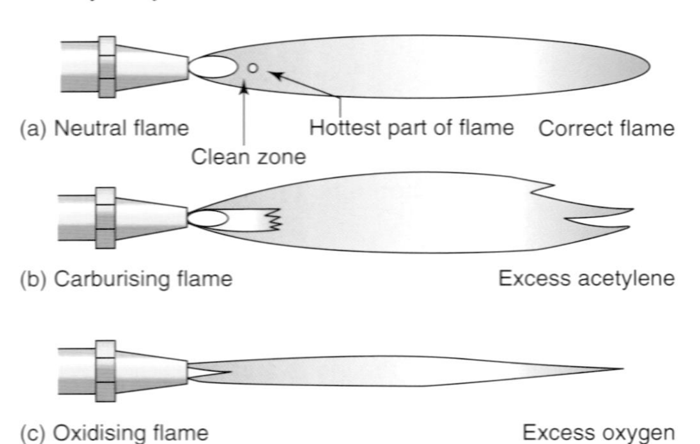

Figure 10.11 Correct flame (Reproduced with permission of Lead Sheet Association)

of the small cone at the tip of the nozzle when the flame is correctly adjusted, as shown in Figure 10.11(a). At this point the temperature is far in excess of the melting point of lead.

When lighting the blowpipe you should:

- Turn on the acetylene first, followed by the oxygen to produce a clean neutral flame, as in Figure 10.11(a)
- Avoid excess acetylene as this will give a carburising flame, as in Figure 10.11(b)
- Avoid excess oxygen as this will produce an oxidising flame, as in Figure 10.11(c).

Key point

Carburising and oxidising flames are both unsuitable for welding sheet lead.

The size of the nozzle used on the blow pipe will depend on the:

- Thickness of the lead sheet
- Type of seam
- Skill of the welder.

Key point

Generally speaking, nozzle sizes 2 and 3 are used for Codes 4 and 5.

Lead welding tools and equipment

To get started you will need oxy-acetylene welding equipment (see Figure 10.12) including:

- Gas bottle set-up
- Blow pipe
- Correct size nozzle.

Figure 10.12 Gas bottle set-up

You will also need:

- Snips, for cutting the lead and lead filler strips
- Flat dresser, for flattening the lead strips and edges of the lead intended for the weld
- Shave hook for cleaning the lead filler strips, and edges of the lead sheet
- Steel rule/straight edge, for measuring and setting out, and as a guide when using the shave hook.

Welding equipment and safety

Safety checklist

- Gas cylinders should be stored in a fireproof room. If possible, store oxygen and acetylene separately. Empty and full bottles should also be stored separately.
- Acetylene gas bottle should be stored upright; this is to prevent leakage of gas
- Oxygen cylinders are filled under high pressure. They should be stored and handled carefully to prevent falling. If the valve is sheared, the bottle will be shot forward with great force
- Keep the oxygen cylinder away from oil or grease as these materials will ignite in contact with oxygen
- Check the condition of the hoses and fittings. If they are punctured or damaged replace them. Do not try to repair or piece them together with approved pipe and fittings.
- Do not allow acetylene to come into contact with copper. This produces an explosive compound.
- Make sure the area where you are welding is well ventilated
- Erect any signs or shields to warn and protect people from the process
- Always have fire fighting equipment on hand
- Wear protective clothing:
  - goggles – clear are fine for lead welding
  - gloves
  - overalls.
- Make sure that hose check valves are fitted to the blow pipe and flashback arrestors to the regulators. This prevents possible flash back on the hoses and the cylinders.
- Allow the acetylene to flow from the nozzle for a few seconds before lighting up
- In the event of a serious flash back or fire, plunge the nozzle into water, leaving the oxygen running to avoid water entering the blowpipe.

Key point

Remember oxygen cylinders are coloured black, and acetylene, maroon.

So far you have covered tools and equipment and looked at how to use the welding equipment safely. Now take a look at the actual process.

- Clean the surfaces that you are going to weld together, and if it is a lap joint, you will have to clean the piece underneath where the faces of the lap joint meet.

- Mark a width of about 10 mm and using your metal straight edge as a guide, shave the surface of the metal with the shave hook.
- Cut strips to use as filler rods. Cut a thin strip of lead about 3–5 mm thick and 300 mm long; it needs to be shaved clean. Alternatively lead rods up to 6 mm can be obtained from a supplier.

Now that everything is prepared, you can set up the welding equipment.

Activity 10.3

What is the required pressure for the oxygen and acetylene, and what is the procedure for lighting the blowpipe? What type of flame should be selected to get the best results? Write down your answers in your portfolio and then check them out at the end of the book.

Lead welding techniques

For Level 2 you will cover:

- Flat butted seams
- Flat lapped seams
- Inclined seams.

Can you remember about weld penetration and weld width for the above welds? You do not have to write anything down here but think about this before moving on.

- Flat butted seam welds should fully penetrate through the thickness of the lead sheet
- Lapped seam welds should penetrate the surface of the lead, but not through to the underside
- The thickness of the seam should be between a ⅓ and ½ thicker than the sheet
- The width of the weld will depend on the thickness of the lead and the seam pattern. When using Code 4, the minimum width of a flat butted seam should be 10 mm.

Key point

Whether jointing lead using a butted or lapped seam you should tack the pieces together. This prevents any movement, and keeps the meeting surfaces in close contact.

Welding flat butted seams

This seam is used for jointing pieces of flat supported lead sheet, as shown in Figure 10.13.

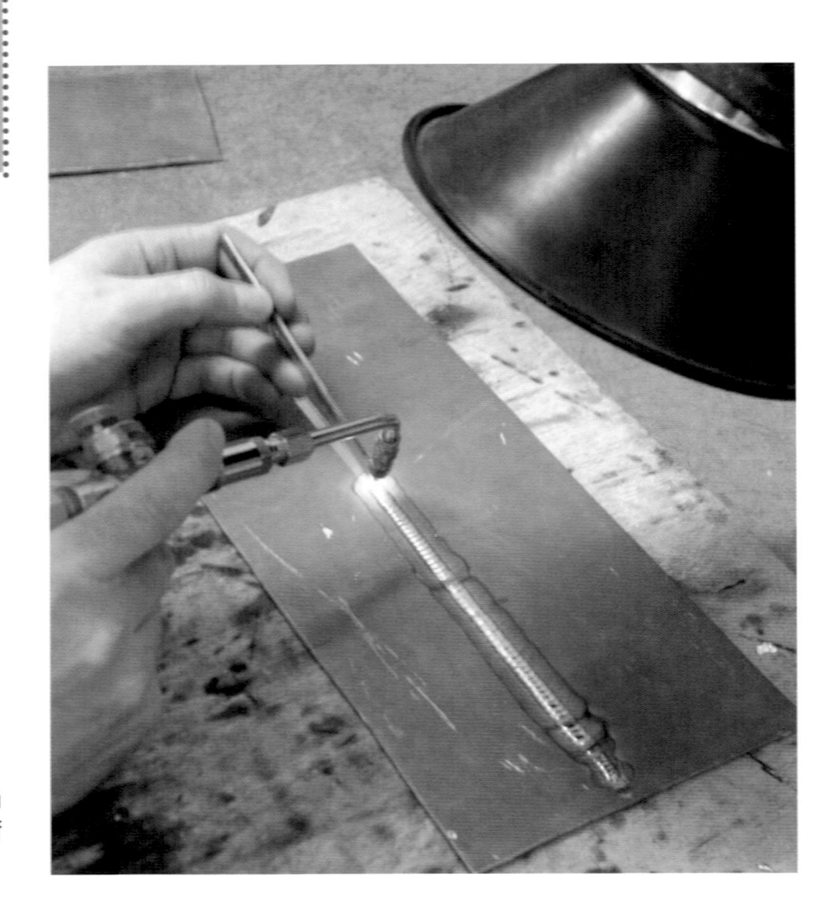

Figure 10.13 Flat butted seam (Reproduced with permission of Lead Sheet Association)

The position of the blowpipe and filler rod should be as shown in Figure 10.14.

- The tip of the cone of the flame should be just clear of the molten lead
- Lead is melted off the welding rod into the weld area
- A seam of a thickness between ⅓ and ½ thickness of the sheet is built up
- The flame is directed into the centre of the seam and is moved forward either in a straight line or slightly from side to side. This will set the pattern of the seam as shown in Figures 10.15(a) and (b).

Key point

You will know if you have got it right if the weld has penetrated just through the underside of the lead for a width of about 10 mm if you are welding Code 4.

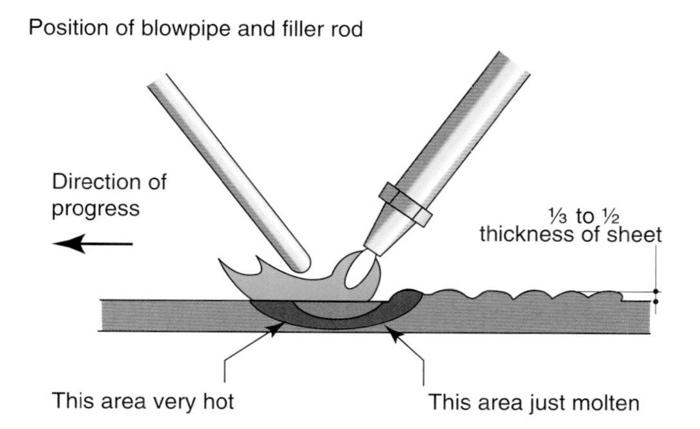

Figure 10.14 (Reproduced with permission of Lead Sheet Association)

Seam pattern

(a) Flat butted seam – straight-line progression

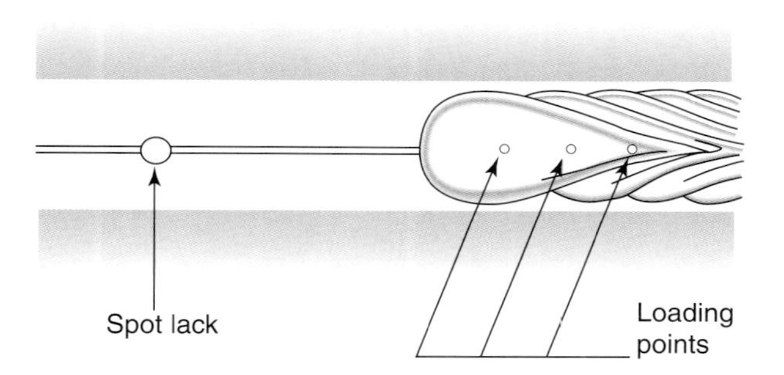

(b) Flat butted seam – side-to-side progression

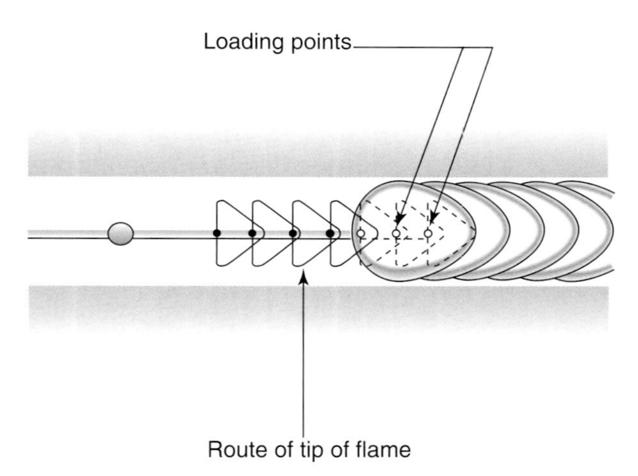

Figure 10.15 (Reproduced with permission of Lead Sheet Association)

Welding a flat lapped seam

This technique is an alternative to butted seams. It is often preferred when working on-site where there is a risk of fire during the welding process, because the flame will not make contact with the material beneath the lead. Most experienced plumbers would also admit that the technique is slightly easier.

Inclined seam on vertical face

You should prefabricate (make prior to installing) sheet lead components before fixing them if you can. There will be occasions, however, when you may have to weld a joint in position. A typical job would be an inclined seam on a vertical face as shown in Figure 10.16.

Key point

This joint can be made without using filler rods by using the overlapping edge to make the seam, but it is recommended that a filler rod should be used.

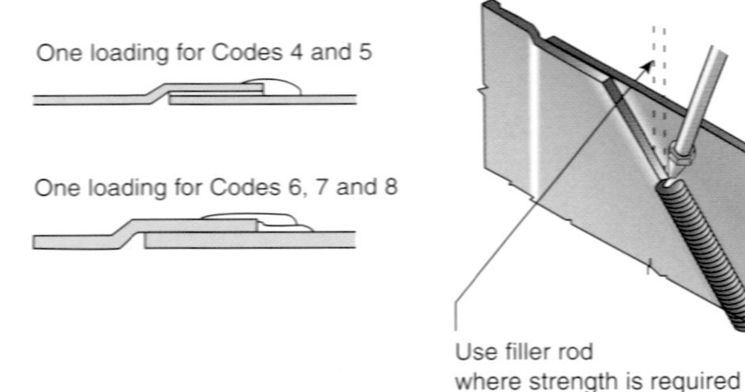

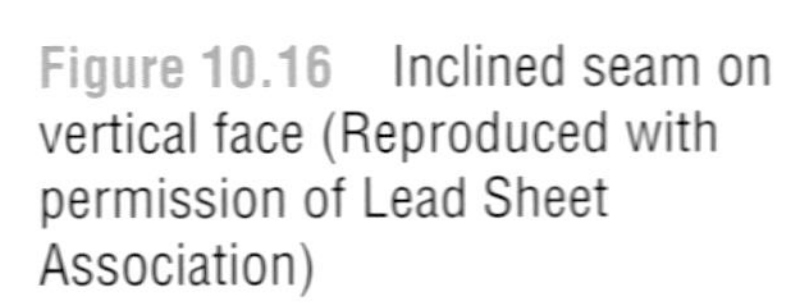

Figure 10.16 Inclined seam on vertical face (Reproduced with permission of Lead Sheet Association)

Fire safety when lead working

Where possible, you should prefabricate lead components before welding. If working on-site, you should try to use lapped joints. The reason for this is because it reduces the risk of fire if using a butt joint against combustible surfaces. However, on occasions this may be unavoidable. Where this is the case you should:

- Wet the timber area beneath the weld
- Alternatively place a non-combustible material beneath the weld.

Finish and quality of work

Bossing

Bossed internal and external corners should have a smooth and even finish without creases or thinning of the lead sheet.

Welding

For internal and external corners, there should be no pinholes in the weld or undercutting of the parent metal.

Test yourself

1. Make a list of the specific tools that are used to boss sheet lead.
2. Which of the following terms would best describe the workability of lead?
   a. Patinated ☐
   b. Annealed ☐
   c. Tempered ☐
   d. Malleable ☐
3. What is the minimum up-stand for a bossed internal corner?
4. What is a shave hook used for?
5. What is meant by the term fusion welding?
6. What are the three main types of lead-welded joint?
7. State three of the basic principles of good practice when carrying out lead welding.
8. The colour code for acetylene bottles is:
   a. Black ☐
   b. Maroon ☐
   c. Red ☐
   d. Blue ☐
9. What size nozzles would be suitable for use in Code 4 or 5 lead?
   a. 1 and 2 ☐
   b. 2 and 3 ☐
   c. 3 and 4 ☐
   d. 4 and 5 ☐
10. What fire safety precautions should be observed when welding lead?

Check your answers at the end of the book.

Assignment
Using bullet points, make a checklist specific to welding equipment and safety. You can check your list against the one contained in this session.

Domestic applications of sheet lead work

So far you have looked at the basic principles of sheet lead, its basic properties, use of tools and fabricating techniques for bossing and welding.

In this session you will look at applying some of this knowledge. The Level 2 Certificate in Plumbing requires that you can competently work on installing chimney weatherings and lead slates, so they will be the main focus of this section.

Installation requirements of a chimney weathering set

This is a mandatory requirement of the practical aspect of the Technical Certificate, and will be carried out in a college or centre workshop. The actual dimension of the various components will vary depending on what facilities are available within a particular college or centre.

A chimney weathering set consists of:

- Front apron
- Soakers
- Side flashing
- Back gutter
- Cover flashing.

Take a look at them one by one.

Front apron

The Technical Certificate will require a bossed internal corner and a lead welded corner. Figure 10.17 shows a front apron with both corners bossed. The lead joint in the brickwork should be at least

Front apron in position

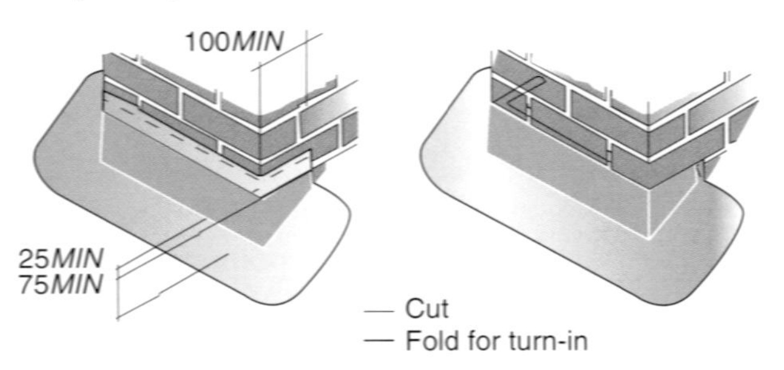

Figure 10.17 Front apron (Reproduced with permission of Lead Sheet Association)

75 mm above the surface of the tiles or slates, and an extra 25 mm added to turn into the mortar joint. The side of the apron needs to be at least 100 mm.

The bossed part of the apron

The lead sheet association recommend that the piece of lead used for the joint apron 'be not less than 300 mm wide – 150 mm for the up-stand against the chimney plus 150 mm for the apron over the tiles'.

The length of the piece will be the width of the chimney plus a minimum of 150 mm for the side. If the roof is covered with deeply contoured tiles, this should be 200 mm.

Shaping a front apron is done mainly on the bench following the techniques as outlined in Unit 2.

The lead welded part of the apron

Obviously, this will be set out based on the same dimensions as the bossed section (see Figure 10.18). You have already covered the marking out and fabrication techniques earlier in Unit 2.

Side flashings

This is the first time that you have encountered side flashings. There are two applications for side flashings:

- Side and cover flashings. These are used where it is not possible to incorporate soakers, such as contoured tiles.
- Side flashings using soakers. These are used on roofs covered with slate or double lap plain tiles.

Side and cover flashings

These follow the same techniques as setting out side flashings using soakers as you will see shortly, but now because the lead will be extended over the roof covering, a width of not less than 150 mm, the overall width of lead will be greater. When installing the step flashing on slates or double-lap plain tiles the allowance for each side of the chimney is 150 mm. If the roof is covered with single lap tiles a measurement of 200 mm is required.

Figure 10.21 shows how the side and cover flashing looks like in position over single lap tiles.

Key point

A key difference to side flashings using soakers is that the piece of lead that is turned around the joint of the chimney will have to be fabricated, either by bossing or lead welding, as shown in Figures 10.19 and 10.20.

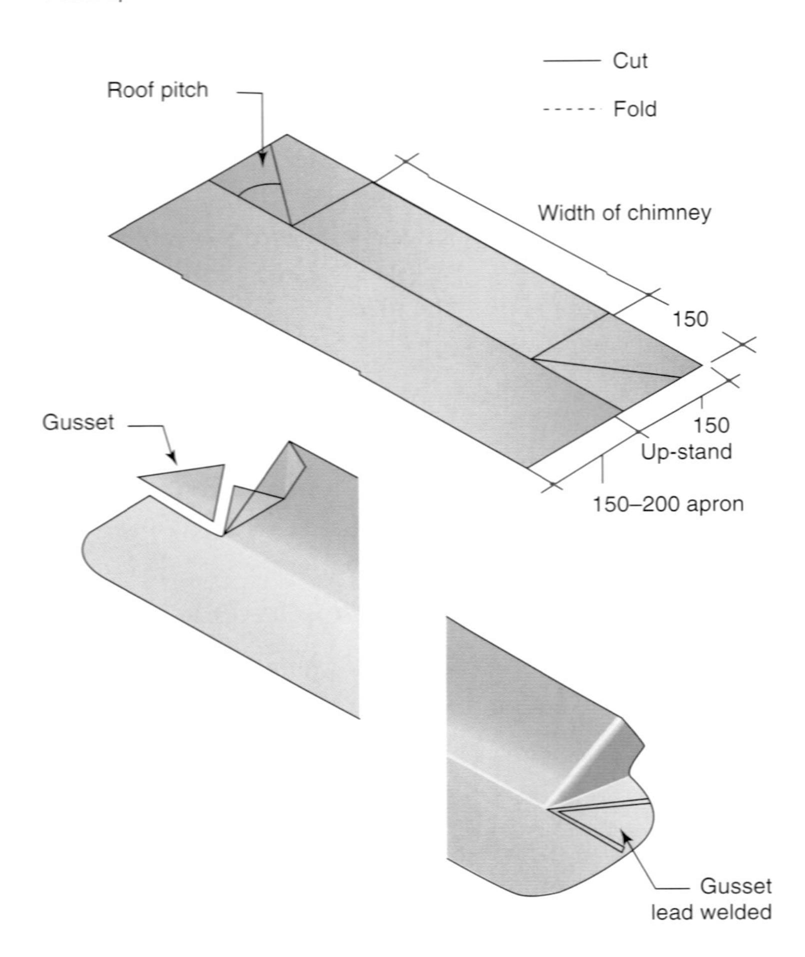

Figure 10.18 Welded part of apron (Reproduced with permission of Lead Sheet Association)

Side flashings with soakers

A soaker is a piece of lead (Code 3), with one side fitted between the slates or tiles, the other turned up the side of the chimney. These are usually fitted by the roofer as the tiles or slates are fixed. They are then covered by the step flashing and in effect provide a 'secret' waterproof gutter between the roof and the chimney. We have included an image of a soaker in Figures 10.22(a)–(c) which shows how they are applied to a plain tiled roof, including how marking out and fitting are done.

The length of a soaker equals the gauge (that is the centre of the battens) plus the lap (the dimension by which the tile or slate overlaps the next but one below it). For example, the length of a soaker for a standard plain tile, laid to a gauge of 100 mm and with a

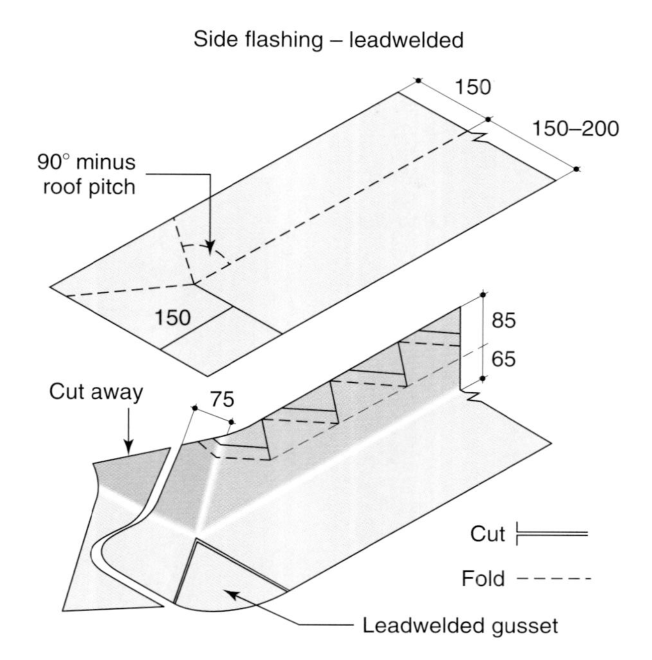

Figure 10.19 Side flashings lead welded (Reproduced with permission of Lead Sheet Association)

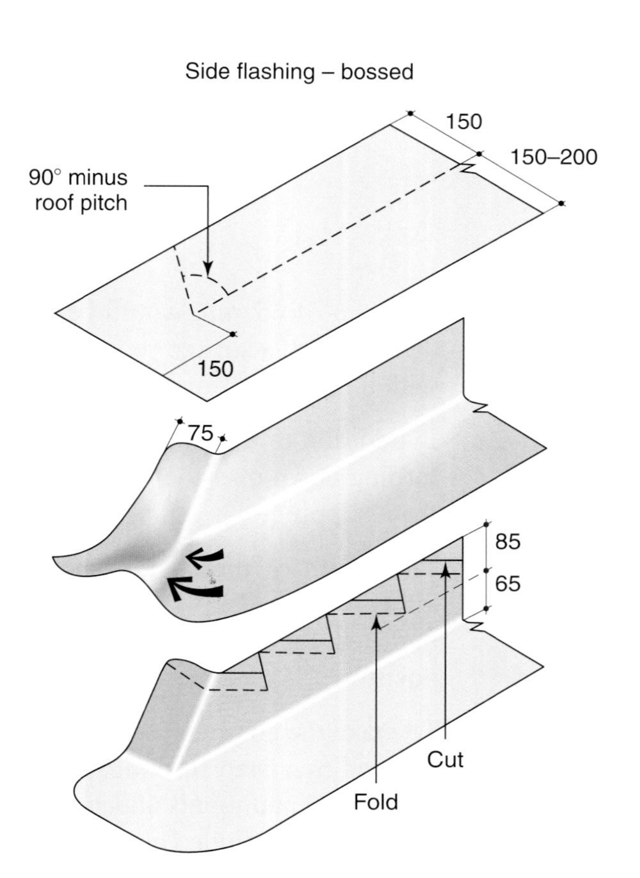

Figure 10.20 Side flashings bossed (Reproduced with permission of Lead Sheet Association)

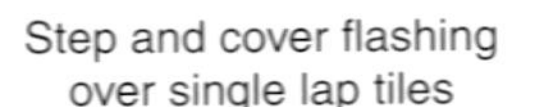

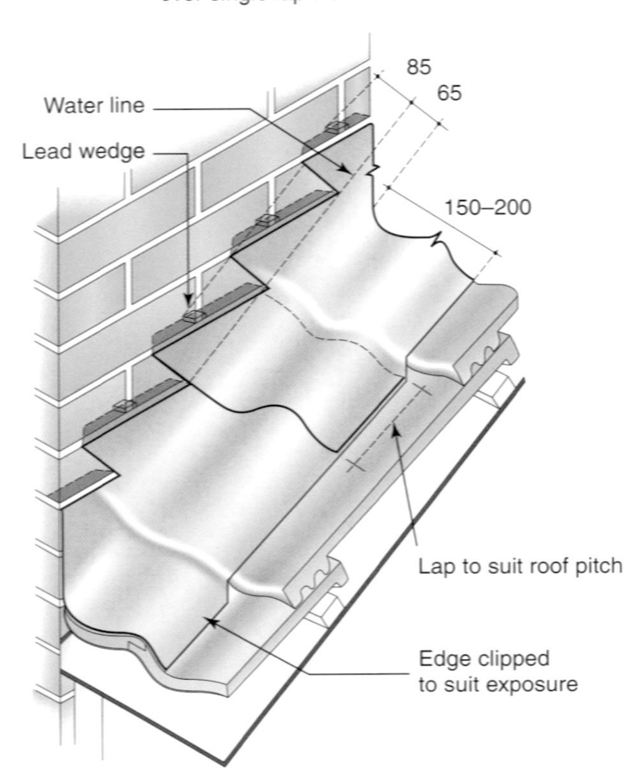

Figure 10.21 Side and cover flashings in position over single lap tiles (Reproduced with permission of Lead Sheet Association)

head lap of 65 mm, would be 100 + 65 = 165 mm. What would the length of a soaker be for a slate laid to a gauge of 200 mm and with a lap of 100 mm? If you said 300 mm, you would be right!

In addition to the calculated length, a further 25 mm is added for turning down over the top of the slate or tile to prevent the soaker from slipping.

The width of the soaker should be a minimum of 175 mm to allow a 75 mm up-stand against the wall, and 100 mm under the tiles. Figure 10.22(a) shows a soaker and several soakers in position.

To make the abutment weather tight, the up-stand of the soakers is covered by a step flashing. A 150-mm-wide piece of lead is held against the abutments and steps are marked, cut, and then the turn in of each step folded and then fitted, as shown in Figure 10.22(b).

Key point

The step flashing must cover the soaker up-stand by not less than 65 mm. Each piece of step flashing should not exceed 1.5 m in length and the laps between the pieces should not be less than 100 mm. The step flashing should be fixed using the lead wedges, as shown in Figure 10.22(c).

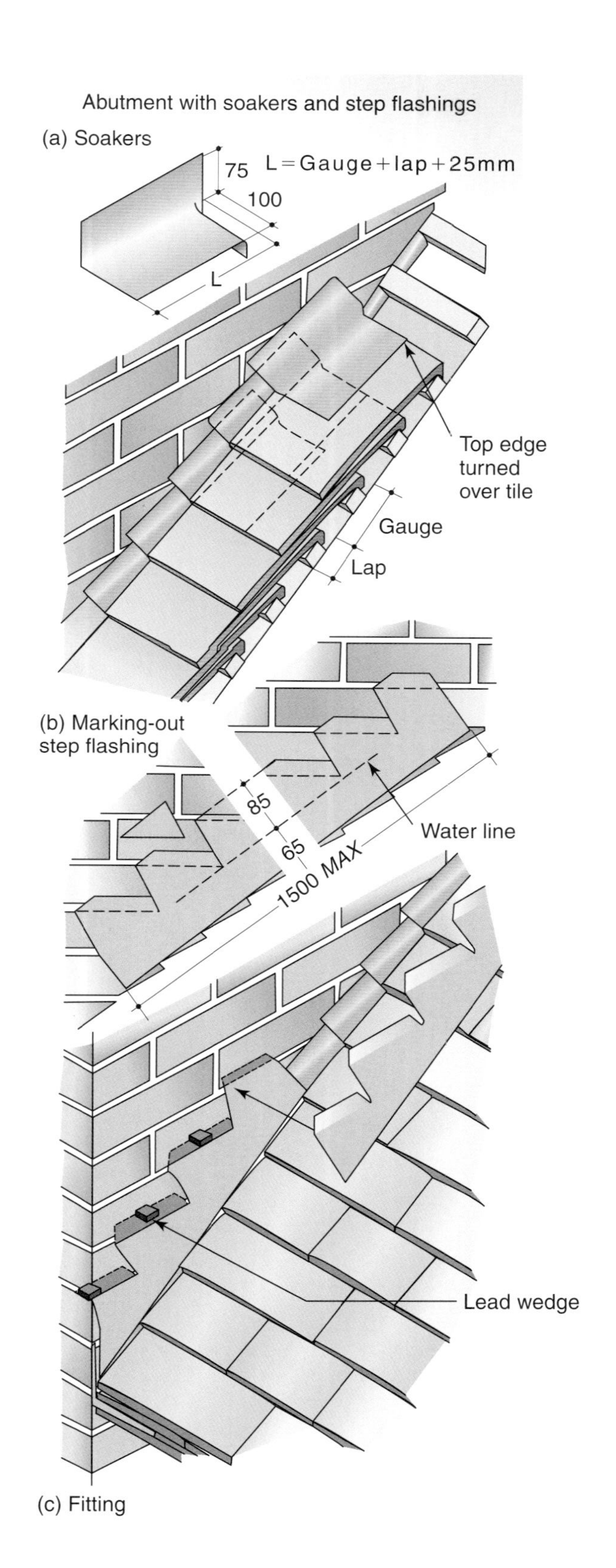

Figure 10.22 (Reproduced with permission of Lead Sheet Association)

Back gutter

The back gutter is the last item to be made and fitted. Figure 10.23 shows what the finished job looks like.

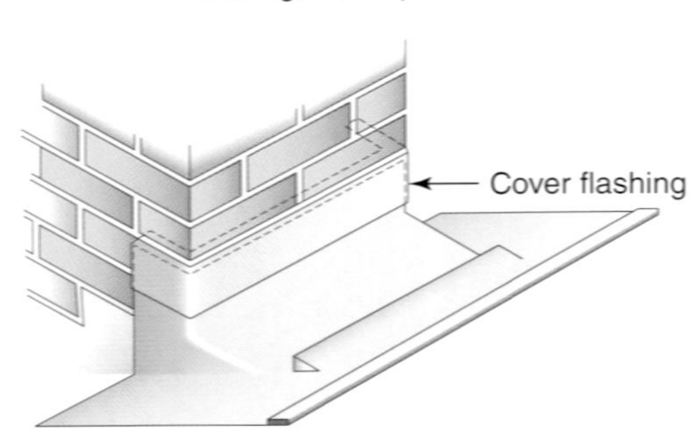

Figure 10.23 Reproduced with permission of Lead Sheet Association

For the Technical Certificate, the back gutter is fabricated using lead welding techniques.

Setting out

Setting out for a lead-welded back gutter is shown in Figure 10.24.

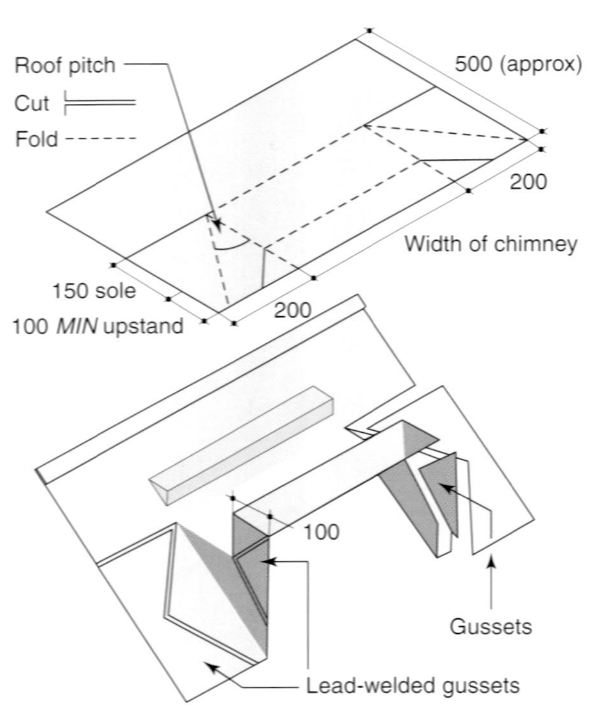

Figure 10.24 Reproduced with permission of Lead Sheet Association

When installing the back gutter on slates or double-lap plain tiles, the allowance for each side of the chimney is 150 mm. If the roof is covered with single lap tiles, a measurement of 200 mm is required.

Once cut, the back gutter is folded into shape and the gussets inserted and welded. The final part of the back gutter is to fit a cover flashing over the up-stand of the back gutter. A minimum of 100 mm over the width of the chimney is allowed on each side for trimming around the corners, and the bottom of the flashing is left about 5 to 10 mm from the gutter base.

Try this

Have a go at producing a back gutter out of card; you can insert the gussets using adhesive tape.

Fixing the components of the chimney weatherings

Throughout this unit, a mention has been made about leaving 25 mm on each measurement for turning into the mortar joints. Hopefully, on new work the joints will have been left 'raked out' to a minimum depth of 25 mm. If not, this is when the club hammer and plugging chisel come in.

If you are doing it, make sure *all* the mortar is removed from the joint, for the type of work you are doing, lead fixing wedges will be adequate. These are made using the off-cuts of lead you have saved while preparing the flashings.

They are cut into strips about 22 mm wide, rolled up and squashed so they are thicker than the gap between the lead turn in and the brick work, about 15 mm. Flatten one edge of the lead to provide a leading edge and place this in the joint. The lead wedge should then be driven in using a wooden chase wedge.

The distance between fixings will depend on the condition of the material you are fixing to. Fix at least one to each side flashing 'step', and for other components, fixings should be placed at between 300 and 450 mm centres.

Once the lead has been fixed, it needs to be pointed with a sand and cement mortar mix of 1 part cement to 4 parts sand. Care must be taken to match the original mortar so that it does not look out of place.

Lead slate

Figure 10.25 shows a lead slate in detail for a plastic soil vent pipe as it penetrates a tiled roof.

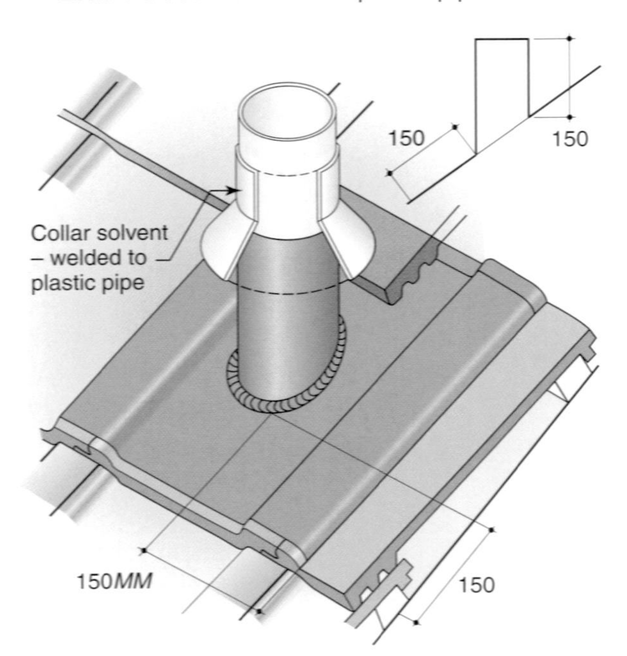

Figure 10.25 Reproduced with permission of Lead Sheet Association

Although this can be formed by bossing, the chances are that they will be made using the lead welding technique as this is much quicker. The Technical Certificate requires you to fabricate a lead slate using lead welding techniques.

The size of the base will vary depending on the roof covering. A typical slate for a 100 mm pipe would be 400 mm wide. The base should extend 150 mm from the front, and be not less than 100 mm under the slate. The height of the up-stand should not be less than 150 mm.

Making the slate

- Cut a piece of lead wide enough to give the height of the up-stand. This will be the circumference of the pipe plus about 5 mm for tolerance. So for a 100 mm pipe this would be calculated as:

Circumference of a circle = πD
3.142 × 100 mm = 314 mm + 5 mm for tolerance = 319 mm (say 320 mm)

- The edges' up-stand are prepared for a butt weld, and then it is turned around a rigid pipe and butt welded
- One end of the up-stand is then cut to the pitch of the roof. This can be done using a bevel and taking the actual angle from the roof. Another method is to 'develop the piece' using a drawing. (Further discussion on that in later sections.)
- When the up-stand is cut to the required angle, the edge is then dressed to form a flange. This is then placed on the base, a hole marked and cut, and the up-stand and base are prepared and welded together.

Testing sheet weatherings

The easiest way to test for water tightness is by applying water to the completed installation by a hosepipe or bucket. Then check the roof internally for any leaks.

This exercise concludes the session on domestic applications. Have a look at the progress check to see how you are doing, and then check your answers at the end of the book.

Test yourself

1. What component is missing from the chimney flashing set below?
   - Soakers
   - Side flashing
   - Back gutter
   - Cover flashing
2. When bossing a front apron, what is the minimum up-stand against the chimney?
3. What are the two main applications for side flashings?
4. The minimum code for lead used for soakers is:
   a. 3 ☐
   b. 4 ☐
   c. 5 ☐
   d. 6 ☐
5. When installing a back gutter on a roof covered with single lap tiles, what is the minimum allowance for the lead on each side of the chimney?

Check your learning Unit 10

Time available to complete answering all questions: 30 minutes.

1. Details relating to the coding, thickness and weights for sheet lead can be found in the:
   a. British Standard 8000 ☐
   b. Building Regulations ☐
   c. British Standard 6700 ☐
   d. British Standard 12588:1999 ☐
2. 'Rolled Sheet Lead: The Complete Manual' is published by the:
   a. Lead Contractor's Association ☐
   b. Lead Sheet Association ☐
   c. Association of Plumbing and Heating Contractors ☐
   d. Institute of Plumbing and Heating Engineers ☐
3. Sheet lead for use in flashings is normally supplied in rolls (coils) in widths of:
   a. 75–300 mm ☐
   b. 100–450 mm ☐
   c. 150–600 mm ☐
   d. 150–750 mm ☐
4. The 25 mm turn-in allowance to a masonry joint for a lead flashing is recommended for:
   a. Back gutters only ☐
   b. Step flashings only ☐
   c. All weatherings ☐
   d. Front aprons only ☐
5. The principle of forming a bossed internal corner can be described as the:
   a. Movement of lead from one place to another ☐
   b. Shaping of lead using homogeneous techniques ☐
   c. Removal of surplus lead as work progresses ☐
   d. Shaping of lead using stretching techniques ☐
6. Overexposure to lead together with insufficient protection is likely to lead to:
   a. Eczema ☐
   b. Chronic illness ☐
   c. Dermatitis ☐
   d. Wiel's disease ☐
7. With regard to oxy-acetylene welding, which of the following materials, if allowed to come into contact with oxygen, will cause it to ignite?
   a. Oil or grease ☐
   b. Cleaning flux ☐
   c. PTFE tape ☐
   d. Copper pipe ☐
8. Which of the following materials can produce an explosive compound if it comes into contact with acetylene gas?
   a. Copper ☐
   b. Aluminium ☐
   c. uPVC ☐
   d. Low carbon steel ☐
9. The correct colour for an oxygen bottle is:
   a. Green ☐
   b. Maroon ☐
   c. Black ☐
   d. Blue ☐
10. What is British Standard colour for Code 4 lead?
    a. Red ☐
    b. Blue ☐
    c. Brown ☐
    d. Green ☐

Check your learning Unit 10 (Continued)

11. On a chimney apron, the minimum recommended distance to the lead joint in the brickwork from above the surface of the tiles or slates at the front of the apron is:
    a. 50 mm
    b. 75 mm
    c. 125 mm
    d. 150 mm

12. Plain roof tiles laid to a gauge of 150 mm, and with a lap of 55 mm would require a length of soaker of:
    a. 95 mm
    b. 205 mm
    c. 215 mm
    d. 230 mm

13. On a chimney weathering set, the cap flashing would be located above the:
    a. Back gutter
    b. Step flashing
    c. Front apron
    d. Soakers

14. The most accurate method of setting out the steps on a step flashing to brickwork is to use a:
    a. Spirit level
    b. Water level
    c. Tape measure
    d. Folding rule

15. The most suitable material for fixing wedges used to secure a chimney weathering set is:
    a. Sheet copper
    b. Hard wood
    c. Plastic wall plugs
    d. Sheet lead

16. One of the most important characteristics which makes sheet lead ideal for lead bossing fabrication techniques is its:
    a. Malleability
    b. Compressive strength
    c. Durability
    d. Tensile strength

17. A higher acetylene than oxygen pressure in a lead welding mixture results in:
    a. Neutral flame
    b. Oxidising flame
    c. Aerating flame
    d. Carburising flame

18. Which of the following nozzles would be used for general purpose lead sheet welding using Code 4 or 5 lead?
    a. 4 or 5
    b. 2 or 3
    c. 5 or 6
    d. 7 or 8

19. The usual cause of undercutting when welding sheet lead is:
    a. Not cleaning the joint correctly, preventing a clean weld
    b. Holding the flame for too long on the vertical surface
    c. Cutting too deep when preparing the joint with a shave hook
    d. Making the flame too hot due to excessive use of oxygen

20. The main cause of fatigue in sheet lead is the effect of:
    a. Overexcessive bossing when forming internal corners
    b. Capillary action allowing trapped water to freeze in winter
    c. Not allowing for sufficient penetration during lead welding
    d. Expansion and contraction due to temperature change

UNIT 10

Sources of Information

We would strongly recommend that you obtain the excellent publication 'Rolled Sheet Lead: the Complete Manual' from the Lead Sheet Association. Here are their details:

- Lead Sheet Association
  Hawkwell Business Centre
  Maidstone Road
  Pembury
  Tunbridge Wells
  Kent
  TN2 4AH
  Tel: 01892 823 003
  Website: www.leadsheetassociation.org.uk

UNIT 11

ENVIRONMENTAL AWARENESS

Summary

Plumbers specify and install boilers and control systems for hot water and central heating systems. This includes making decisions about the energy efficiency of boilers, efficiency of various control system combinations, and installing the system so that it operates to design specification. Ultimately, this relates to the burning of fossil fuels, and subsequently carbon emissions.

Plumbers also generate waste products, such as scrap metals and old bathroom and kitchen appliances that have to be disposed off, so if the materials can be recycled, it will help the environment.

In this unit we will look at the topic of environmental awareness and how this affects the work of a plumber:

- Environmental awareness and plumbing includes:
  - Energy conservation
  - Types of renewable energy including solar power.
- Environmental awareness in practice:
  - Part L1A and LB 2006 and improving energy efficiency
  - Customer advice
  - Reducing waste and waste disposal
  - Environmental hazards.

Key point

A plumber's job will have some effect on the environment – can you think of any such examples?

Activity 11.1

What do you understand by the term 'environmental awareness'? Write down your thoughts here and then check with that given at the end of this book.

Environmental awareness and plumbing

Energy conservation

Energy conservation, and in particular energy efficiency in hot water and central heating, was covered in some detail earlier in the book. We referred to the relevance of Part L 2006 of the Building Regulations to energy conservation and how the Water Regulations and BS specifications ensure that working practices, system design and use of materials are of the highest standards in order to maximise energy conservation, as well as helping to ensure public health and the welfare of the consumer.

Activity 11.2

What do you think is the meaning of the term 'energy conservation'? Jot down your thoughts for your portfolio and then check them at the end of this book.

There are also a number of other general initiatives aimed at improving energy conservation, these include:

- Solar-powered hot water heating systems; these are large panels which consist of a system of pipes located behind a glass panel. The water in the pipe is heated by the sunlight and is piped to the hot water storage cylinder.
- Heat produced as a by-product of the power generation process that would normally be lost to the environment. Combined heat and power (CHP) can increase the overall efficiency of fuel use by as much as 70–90%, compared with 35–52% for normal electricity generation. CHP plants can also use the heat from incinerating refuse, to heat hospitals, factories and blocks of flats.
- Solar photovoltaics; this also uses the Sun's power, but photo-electric cells set in panels are used to produce electricity

Key point

One example of the Government's determination to reduce CO_2 emissions outside the building industry is that vehicles are now taxed on the basis of the amount of CO_2 they produce while burning the fuel.

Activity 11.3

Which of the factors below do you think would affect energy efficiency in domestic buildings? Tick as many as you think are relevant and then check your answer against the one at the end of this book.

- Draught proofing around doors and windows
- Type of boiler
- Age of boiler
- Dripping hot tap
- Insulation in cavity walls
- Boiler interlock
- Insulation level in loft/roof space
- Double glazing
- Thickness of window glass
- Material used for window frames
- Thermostatic radiator valves
- Programmer/time switch
- Long piperuns to hot water outlets
- Insulation on hot water cylinder
- Insulation around hot pipes from boiler.

- Wind farms, which are large windmill-type structures that produce electrical power as the blades rotate
- CHP is not limited to any one fuel. It can use biofuels, such as wood pellets, agricultural crops or even farm refuse.
- Geothermal systems which take heat from below the Earth's surface, and transfer it to buildings using heat exchangers.

'Waste not, want not'

Keeping waste to a minimum is also important as waste can be harmful to the environment, and requires energy to transport it and treat it. Recycling waste is a way of reducing landfill and a means of returning the waste back into the production cycle. The Government has also brought in measures to control waste disposal and encourage recycling.

Here are a few examples of recycling in plumbing:

- Old Belfast sinks or cast iron baths that are being replaced; if in good condition, they can be sold on to people looking to create an original look in their homes

- Similarly sink, bath and basin taps
- Materials, such as brass, copper, aluminium and lead have a 'scrap value' and can be melted down and recycled into new components.

Try this

Can you think of an example of a plumbing material or component that could be recycled?

Solar-powered hot water

Solar water heating systems are recognised as a reliable way to use the Sun's energy in what is referred to as 'renewable solar energy'. At the time of writing there are around 42,000 currently in use throughout the UK. The technology is relatively straight-forward.

If you think about the solar system as supplementing a traditional domestic hot water system, it works on the same principle as indirect hot water heating from the boiler. You can see what we mean by looking at Figures 11.1 and 11.2. As most of the heat energy is generated in summer, it is necessary for the boiler to supplement the solar system in the winter months.

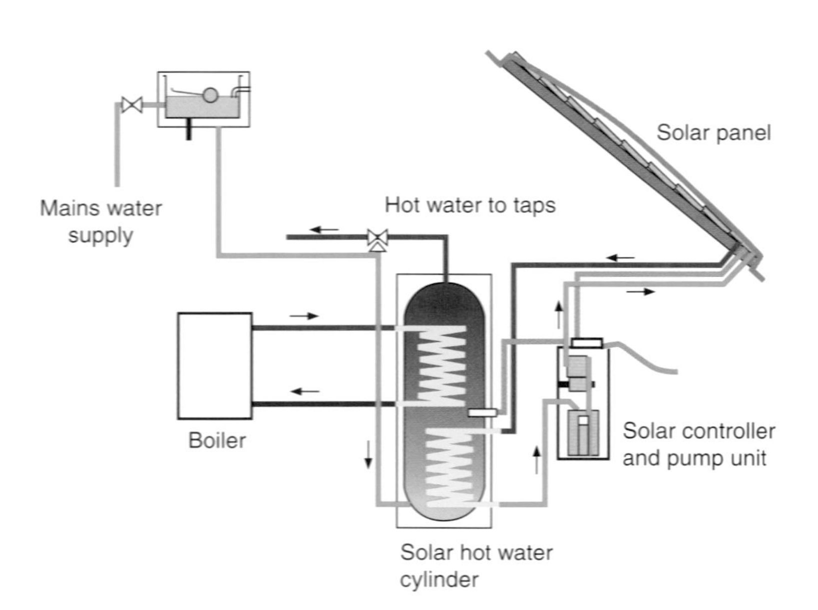

Figure 11.1 Solar-powered hot water system

Figure 11.2 Example of a solar panel installation in a plumbing workshop

A common size solar panel for domestic use is about 3–4 m^2. This will deliver about 1000 kWh per year and will heat just over half the annual water demand of a typical domestic dwelling.

A typical solar water heating system comprises collectors (panels), a hot water storage system and connecting pipework. There are some systems that rely on gravity circulation from the collector to the storage vessel, but in the mains, most systems use a circulating pump.

Activity 11.4

Figure 11.1 shows how a solar-powered hot water system works. What do you think are the advantages and disadvantages of this system? Write down your answer for your portfolio, then check it against the suggested one at the end of this book.

Water Regulations

Water Regulations are covered as a specific topic at Level 3, although by now we have made enough references to them throughout the course for you to be quite familiar with them. In terms of energy conservation, we want to show you here how the Water Regulations make an impact.

Activity 11.5

11.5.1: Which of the following are the Water Regulations designed to prevent?

- Contamination of water supply
- The wastage of water
- The misuse of water supply
- Undue consumption of water
- Erroneous measurement (fiddling with the meter).

11.5.2: From the above which two do you think would have the most effect on energy conservation?
Check your answer against the one at the end of this book.

Waste of water is from dripping taps, leaking joints and overflows, on both hot and cold water supplies. On a hot water system, a dripping tap means (see Figure 11.3) that the hot water is wasted. It is then replaced by cold water which then has to be reheated, thereby wasting boiler energy.

Key point

There are around 18 million homes in this country, so if a leaking hot tap lost an average for each dwelling of just half a litre of water a week, overall you would have to heat 9 million extra litres of water.

On cold taps and cistern overflows, wasted water has to be replaced, which in the wider picture means additional treatment and distribution for water companies, using extra electrical and mechanical energy. As all water is treated to make it wholesome, wasted water also wastes the chemicals and all the overheads used in its treatment.

Figure 11.3 Dripping tap

Waste of water can also be due to burst pipes brought about by freezing through lack of insulation, or piperuns in exposed conditions.

Undue consumption may be caused by bad design, in particular dead legs. This results in high volumes of cold water being run off before hot water arrives at the tap.

Water Regulations also make recommendations about the insulation of hot water pipes and storage vessels which is another measure to reduce heat loss and decrease the load on the boiler.

Test yourself 11.1

1. In terms of alternative methods of improving energy conservation, what is meant by the term 'Solar Photovoltaic'?
2. What area of the Building Regulations covers energy conservation?
3. State two examples of how the water regulations can affect energy conservation.

Check your answers against the suggested ones at the end of this book.

Environmental awareness in practice

In the last unit on environmental awareness and plumbing, we looked briefly at energy conservation and the legislation and at its effects on what you do as a plumber. In this section we will consider how environmental awareness works in practice.

Part L in practice

The relevant sections of Part L1A and 1B 2006 as far as the plumber is concerned are:

- Replacement of heating systems (including boilers and hot water storage cylinders)
- Heating and hot water controls
- Commissioning of heating systems and the provision of instructions for users
- Certification of heating and hot water systems to show that they have been correctly installed and commissioned, and that operating instructions have been left for the user.

Again, the detailed practical implications of Part L1A and 1B 2006 have been covered in the 'Central Heating' unit, but questions related to Part L may occur in the Technical Certificate assessments.

Try this

Rather than repeat the text here, we strongly recommend that you spend some time revisiting the 'Central Heating' unit, and as a revision exercise familiarise yourself with Part L1A and 1B 2006 and how it applies to:

- Boilers
- Central heating systems design
- Central heating controls.

We also suggest that you take time to read the Energy Efficiency Best Practice in Housing publications.

You must complete this activity as there will be check your learning questions later.

Other methods of improving energy efficiency

- All systems pipework should be insulated and installed to comply with Water Regulations
- Systems and components should be serviced and maintained regularly to:
  - ensure that they are working to design specification
  - ensure that they are not wasting water.
- Systems should be designed to:
  - keep dead legs as short as possible
  - avoid overcapacity, i.e. heating high volumes of hot water storage that may not be required.

Customer advice

It is good business practice as well as a requirement of L1 (central heating systems) to provide information on the control and operation of a new or replacement system installation.

Activity 11.6

You have just completed a central heating and hot water installation in an occupied domestic dwelling. Write a check list of what you should do for the customer once the installation has been completed, tested and commissioned. Please check the answer given at the end of this book.

> **Key point**
>
> When you are carrying out day-to-day plumbing work it is important that you get into the habit of keeping waste material to a minimum. Throwing away the excess odd ends of copper tube when carrying out pipework installations might not seem much on a daily basis but just say you discard about 250 mm of 22 mm on an average working day, multiply this by the number of average working days (210), then you have lost 52.5 m of pipe! At today's prices this would be about £60.

Customers should also be given advice on how to use central heating controls like the ones shown in Figure 11.4.

Methods of reducing waste during plumbing

Can you think of ways of keeping waste to a minimum? Here is our check list:

- Take time to carefully measure and set out pipework for bends. This will reduce the amount of wasted pipe when cutting it to a length (remember our 52.5 m of pipe in the key point).
- Look after tools in terms of wear and tear – defective tools have to be replaced, and oh, so do stolen ones!
- If it is economically possible, try to repair an appliance or component rather than replace it. New components have to

(a)

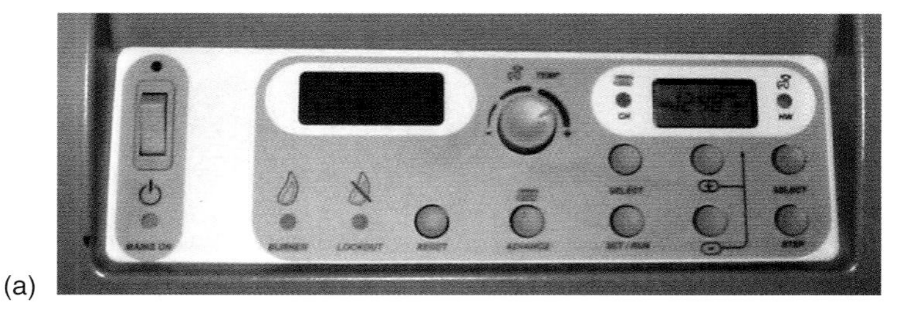

(b)

Figure 11.4 Central heating controls (programmer + thermostat) (a: Reproduced with permission of Ravenheat)

be made, and the manufacturing process uses energy. Old appliances have to be transported and disposed of (although replacing an inefficient boiler would be a good move).
- While using capillary integral solder ring fittings do not use additional wire solder on the joint
- Treat screws and other fixings like money – too often they are left all over a job
- Be extra careful when fitting sanitary ware – a broken one has to be disposed off and replaced
- This also applies to the storage of sanitary ware and other easily damaged goods
- Think about how you use water when you are working on a system. Have a thorough check around all the fittings to make sure they have all been soldered. You do not want to be filling and draining systems unnecessarily.
- Do not overdo the use of jointing compounds, fluxes and other materials
- Take care when taking floorboards up; avoid having to replace damaged boards with new ones
- Similarly, when cutting chases or drilling holes in brickwork, blockwork or other building materials keep the dimensions to the minimum to reduce the amount of refilling.

Waste disposal

There are many special arrangements for the disposal of some types of waste. On larger sites, skips are usually provided for building and plumbing waste (refer to Figure 11.5). These are taken to licensed sites for disposal all at once. Do not overfill the skip, and do not put items in them that are dangerous, e.g. flammable material.

On smaller jobs you might still use a skip, for the disposal of an old bathroom suite, for example. If you try to take 'industrial waste' to your local tip you are likely to get turned away or have to pay for its disposal. It might be tempting to set fire to the contents of a skip to reduce the volume, but the skip hire company would not be too pleased about this.

Plastics such as pipework or tanks should be put in a skip or taken to a licensed Dump-it site. A hot water cylinder can be taken to a scrap metal merchant – even a factory foam-insulated one can be dismantled and the materials reused. Recycling is an energy-efficient and cost-effective way of getting rid of scrap waste.

Activity 11.7

Why is setting fire to the contents of a skip dangerous and environmentally damaging? Write your answers down for your portfolio before comparing them with the suggested answers at the end of this book.

Plumbers work with copper, brass, lead, cast iron and low carbon steel. All these have a value at the scrap merchant's establishment. Remember, some older fittings may even have a high value at a salvage or reclamation yard. Antique cast iron baths fetch hundreds of pounds when renovated! A plastic F&E or cold water tank does not have a resale value, but can be 'recycled' by donating it to a school or community garden, for use as a planting container.

Key point

What do you think is the meaning of the term 'fly tipping'? 'Fly tipping' is the disposal of waste in places other than a registered commercial or private site, and is illegal.

Figure 11.5 Site skip

> **Try this**
>
> If you have not been around an architectural salvage yard before (or for a while) why not visit one, and make a note of what is in demand (and the prices) – you might see 'waste disposal' in a new light.

Environmental hazards

The main environmental hazard as far as plumbing is concerned is asbestos. This was dealt with in detail in the Health and Safety unit. Other topics, such as lead and LPG, are also covered in the relevant units.

A popular pastime on some of the larger sites is burning waste material. All sorts of things get thrown into a site fire – plastics, paint cans and so on. This activity is not environment friendly and is against site rules.

Guide to new technologies

Bio-energy – Biomass is derived from plant material and animal residues/wastes. It can be used to generate electricity and/or heat, and to produce transport fuel. This energy is called bio-energy. There's a load of 'biomass' that can be used for energy purposes, e.g. straw and crop residues, crops specially grown for energy production – willow, oil seed rape and wastes from a range of sources including food production. The nature of the fuel will determine the way that energy can best be recovered from it.

Carbon dioxide (CO_2) – Carbon dioxide contributes around 60% of the potential global warming effect of human-made emissions of greenhouse gases worldwide. Not surprising then, that it's global enemy no. 1. The burning of so-called fossil fuels – oil, gas, coal – releases CO_2 and increases its concentration in the atmosphere.

Carbon Trust – This is an independent, not-for-profit company set up by the British government with support from businesses to encourage and promote the development of low carbon technologies. Its key role is to support British business in reducing carbon emissions through funding, support of technological innovation and by encouraging more efficient working practices.

Combined cycle gas turbines – These use both gas and steam turbine cycles in a single plant to produce electricity with high conversion efficiencies and relatively low emissions.

Combined heat and power (CHP) – CHP is the simultaneous generation of usable heat and power in a single process, thereby discarding less waste than conventional generation.

Decent Homes Standard – This has been set by the Office of the Deputy Prime Minister (ODPM) and the Decent Homes Standard is a minimum standard that all social housing in England should achieve by 2010. A decent home is 'wind and weather tight, warm and has modern facilities'.

Energy Saving Trust (EST) – The EST is a not-for-profit organisation set up and largely funded by Government to manage a number of programmes to improve energy efficiency, particularly in the domestic sector.

Fuel cells – Fuel cells produce electricity from hydrogen and air, with water as the only emission. Potential applications include stationary power generation, transport (replacing the internal combustion engine) and portable power (replacing batteries).

Fuel poverty – The common definition of a 'fuel poor' household is one needing to spend in excess of 10% of household income to achieve a satisfactory heating regime (21°C in the living room and 18°C in other occupied rooms).

Greenhouse gas – A greenhouse gas is one that contributes to global warming. The most significant greenhouse gases are carbon dioxide, methane and nitrous oxide.

Ground source heat pumps (GSHP) – Ground source heat pumps do exactly what they say they do – they extract heat from the ground – by circulating water (or another fluid) through pipes buried either in the ground in trenches, or in vertical boreholes. The pipes extract heat from the ground and a heat exchanger within the pump extracts the heat from this fluid. The compression cycle is employed (also used in refrigerators) to then raise the temperature to supply hot water to the building.

Heat pumps – These work like a refrigerator moving heat from one place to another. Heat pumps can provide space heating, cooling, water-heating and sometimes exhaust air heat recovery.

Heat recovery – A technique for maximising efficiency by making use of heat that would otherwise be wasted, e.g. in hot exhaust gases.

Micro-CHP – CHP at the scale of a single dwelling, used in place of a domestic central heating boiler.

Photovoltaics (PV) – This is the direct conversion of solar radiation – sunlight – Into electricity by the interaction of light with the electrons in a semi-conductor device or cell.

Renewable energy – Energy flows that occur naturally and repeatedly in the environment. This includes solar power, wind, wave and tidal power and hydroelectricity. Solid renewable energy sources include energy crops and other biomass. Gaseous renewables come from landfill and sewage waste.

Renewables Obligation (RO) – This is the obligation placed on electricity suppliers to deliver a stated proportion of their electricity from eligible renewable energy sources.

Solar thermal/solar hot water – A system for using solar radiation to heat water, typically roof-mounted panels connected with pipes to a storage tank. Worcester's Greenskies 240 product is a solar thermal system.

Standard Assessment Procedure (SAP) – The SAP is the Government's recommended system for energy rating of dwellings. It is used for calculating the SAP rating, on a scale of 1 to 100, based on the annual energy costs for space and water heating; and for calculating the Carbon Index, on a scale of 0.0 to 10.0, based on the annual CO_2 emissions associated with space and water heating.

Sustainable Development Commission – The Commission's main role is to advocate sustainable development across all sectors in the UK, review progress towards it and build consensus on the actions needed if further progress is to be achieved.

Units of Energy – Energy is the ability to do work. 1 Watt hour (Wh) is the amount of energy used by a 1 W device operating for an hour. A kilowatt-hour is 1,000 Wh and a megawatt-hour is 1,000,000 Wh.

Test yourself 11.2

1. What are the four main sections of Part L that are relevant to plumbing?
2. Briefly describe the term 'cylinder thermostat'.
3. As well as being a good business practice, what should a plumber do in order to meet the requirements of L1A and L1B 2006 when providing customer advice on completion of a new central heating installation?

Check your learning Unit 11

Time available to complete answering all questions: 20 minutes

Tick the answer that you think is correct. Some of these questions are included as revision.

1. In terms of environmental efficiency, what does SAP stand for?
   a. Seasonal Adjustment Procedures ☐
   b. Standard Architectural Practice ☐
   c. Seasonal Adjustment Protection ☐
   d. Standard Assessment Procedure ☐
2. What Building Regulations cover the conservation of fuel and power?
   a. Part K (Approved document K1) ☐
   b. Part J (Approved document J1) ☐
   c. Part M (Approved document M1) ☐
   d. Part L (Approved document L1A & L1B) ☐
3. One of the main aims of energy conservation is to reduce:
   a. Emissions of CO_2 into the atmosphere ☐
   b. Domestic fuel bills ☐
   c. Emissions of CO into the atmosphere ☐
   d. CFC levels ☐
4. Which of the following would help to increase the energy efficiency of a central heating system?
   a. Lockshield radiator valve ☐
   b. Cold water servicing valve ☐
   c. Two-way gate valve ☐
   d. Thermostatic radiator valve ☐
5. Which feature will reduce the energy efficiency of a domestic hot water system?
   a. TRVs ☐
   b. Boiler interlocks ☐
   c. Dead legs ☐
   d. Three port mid-position valves ☐
6. Which of the following is the most energy-efficient appliance?
   a. Combination boiler ☐
   b. Condensing boiler ☐
   c. Combined primary storage unit ☐
   d. Economy 7 boiler ☐
7. A plumber suspects the presence of asbestos on-site; what first actions should be taken?
   a. Contact the HSE directly and ask them to send out an inspector ☐
   b. Thoroughly soak the material in water and continue with work ☐
   c. Stop work immediately and report to the immediate supervisor ☐
   d. Ring the company office and ask for appropriate PPE ☐
8. In terms of energy efficiency, the term SEDBUK relates to the:
   a. Seasonal efficiency of a boiler ☐
   b. Type of energy used to power a domestic heating system ☐
   c. Maximum energy output of a boiler ☐
   d. Amount of harmful emissions from a domestic property ☐

(Continued)

UNIT 11

Check your learning Unit 11 (Continued)

Time available to complete answering all questions: 20 minutes

9. The term boiler interlock means:
   a. A boiler fault which prevents it from firing up ☐
   b. A situation in which the boiler fires continually ☐
   c. A failure of the boiler feed to the hot water cylinder ☐
   d. An arrangement of system controls ☐

10. On completion of an installation job, what information should always be left with a customer?
   a. The delivery notes for the appliances fitted ☐
   b. Guidance materials from manufacturers ☐
   c. An assortment of spare parts for each new appliance ☐
   d. Receipts for all materials used ☐

Sources of information

There are no additional sources of information.

UNIT 12

Effective Working Relationships

Summary

Plumbing forms part of the building services engineering sector, all of which is encompassed by the construction industry. You will need an idea of what the construction industry is all about, together with an insight into some of the people that you are likely to meet. This unit will give you an overview of the construction industry covering the work undertaken by the other trades you will encounter on site, and what workers from other trades actually do.

Probably, the first priority in your career is to make sure that you become a top class plumber, and that is a good ambition. However, there are other skills that you need to learn in support of your technical ability – 'personal skills'. Personal skills includes how you deal with people, how you communicate, verbally, in writing and visually. This unit covers:

- The construction team:
  - The construction industry, the size and type of businesses
  - The construction team: the various people who control construction work and visit sites; other trades and what they do
  - Information in support of the construction industry.
- Communication skills:
  - Methods of communication (verbal, written, ICT (Information and Communication Technology), visual)
  - How to get the best out of work colleagues?
  - How to deal with site visitors and co-workers?

The construction industry

The nature of a plumber's job is that you are likely to meet a wide range of people, as the job is varied and may include working on a number of different construction sites as well as in people's homes. Hence, it is important that you know how to conduct yourself in all sorts of company and situations. It is also important that you can communicate clearly with other people so that you can do your job as efficiently as possible.

The Construction Industry Training Board (CITB) is the Sector Skills Council (SSC) for the construction industry. According to the CITB, there are around 165,500 companies working in the industry. Of these, 95% employ between 1 and 13 people, 4% between 14 and 299, and 1% 300+.

It is a similar situation in the plumbing industry – around 85% of plumbing businesses are registered as sole traders, with a high proportion of these (80%) employing between 1 and 4 people.

There are other similarities between the construction and the plumbing industry in the type of work that companies do. Small companies will concentrate on one-off house building, extensions, maintenance and repair work. They also sub-contract to larger construction companies, which usually means providing a specific service, such as bricklaying or plastering.

In the same way, small plumbing companies carry out one-off jobs such as doing all the plumbing installations in a new domestic house build, as well as system replacement, repairs and maintenance in people's homes. The larger contractors will develop construction projects from start to finish. These could include anything from private or public housing developments to major construction projects!

Can you think of any current construction projects that fall into the 'major construction project' category? To do this type of work they employ a range of managers, designers, office-based staff, etc as well as people who run the job on site.

Larger contractors are also involved in what is termed 'facilities management'. This is the on-going management of all aspects of running a large building (e.g. a hospital) or an estate of houses (Local Authority, Housing Association or private). You will see what a facilities manager does later.

Many of the larger contractors carry out work on behalf of insurance companies. At one time, it was left to the householder to arrange for a contractor if, say, their ceiling was seriously damaged by a water leak. Now, most insurers place this work directly in the hands of a contractor who handles everything from receiving the emergency call to doing the job.

Company structures

The construction industry then is diverse and encompasses companies of various sizes which take on different ranges of work on varying scales.

Whatever the size of a company, it has to have a 'legal status'. This means it will be one of four types:

- Sole trader
- Partnership
- Limited company (private)
- Public limited companies (PLCs).

You may hear these terms in the course of your career, so you should know a little about each.

Sole trader

A high proportion of both construction and plumbing businesses are sole traders. One person will own the business and look after everything; quoting for work, arranging the jobs, organising materials, doing the job and then sorting out all the paperwork, invoices, taxes and so on.

Quite often, a sole trader may employ one or two people, and maybe an apprentice.

A sole trader business has advantages, the main ones being that the owner is entitled to all the profits made by the business, and generally speaking there is less paperwork to deal with as the business does not have to be registered with Companies House, which sets out the rules for a registered business.

The main disadvantage is that the sole trader is liable for the debts of the business (e.g. bank loans taken out to help set the business up), so if the business fails, the owner runs the risk of losing personal savings or assets (which could include their house) set against the loan.

Partnerships

Partnerships are a popular model for small construction and plumbing businesses. When you see Smith and Jones builders or plumbers on the side of a van, it is likely to be a partnership.

Quite often, this develops from tradespeople working together and deciding to go into business. A joiner and bricklayer partnership is a typical example, but it does not have to be separate trades of course – often two, or even more plumbers set up partnerships.

Partnerships do not have to be registered with Companies House (all limited companies have to be registered with Companies House as a legal requirement), and because there are two or more people involved, the responsibilities of running the business are shared (but so are the profits).

Limited companies (private)

Most businesses, if they begin to expand, tend to go limited. There are two main reasons:

- Tax advantages
- The individual(s) running the company is not personally liable for any debts should the business fail.

These are sometimes referred to as a private limited company, which means they cannot trade their shares on the stock market. Limited companies also have to be registered with Companies House and submit their business accounts on an annual basis.

Public limited companies (PLCs)

These are allowed to trade their shares on the stock market provided they have a share capital of at least £50,000. Some of the major contractors in the construction industry are PLCs.

Structure of businesses

The structure of sole traders and partnerships is pretty straightforward, but large companies are more complicated. Limited companies and PLCs usually have a Board of Directors, and depending on size, will have a Chairman, a Managing Director or Chief Executive Officer.

The Managing Director (MD) or Chief Executive Officer (CEO) is usually responsible for overseeing the management of the business, as well as planning the strategy (tactics) of the business.

A PLC is also required by Companies House to have a professionally qualified company secretary.

Company structures will vary, but Figure 12.1 gives an idea of what a structure might look like for a company employing around 80–100 craft operatives.

What the construction team does?

The structure in Figure 12.1 shows the team members in a typical medium- to large-sized company. *In an industry as large and diverse as construction, job descriptions and responsibilities may vary from company to company.*

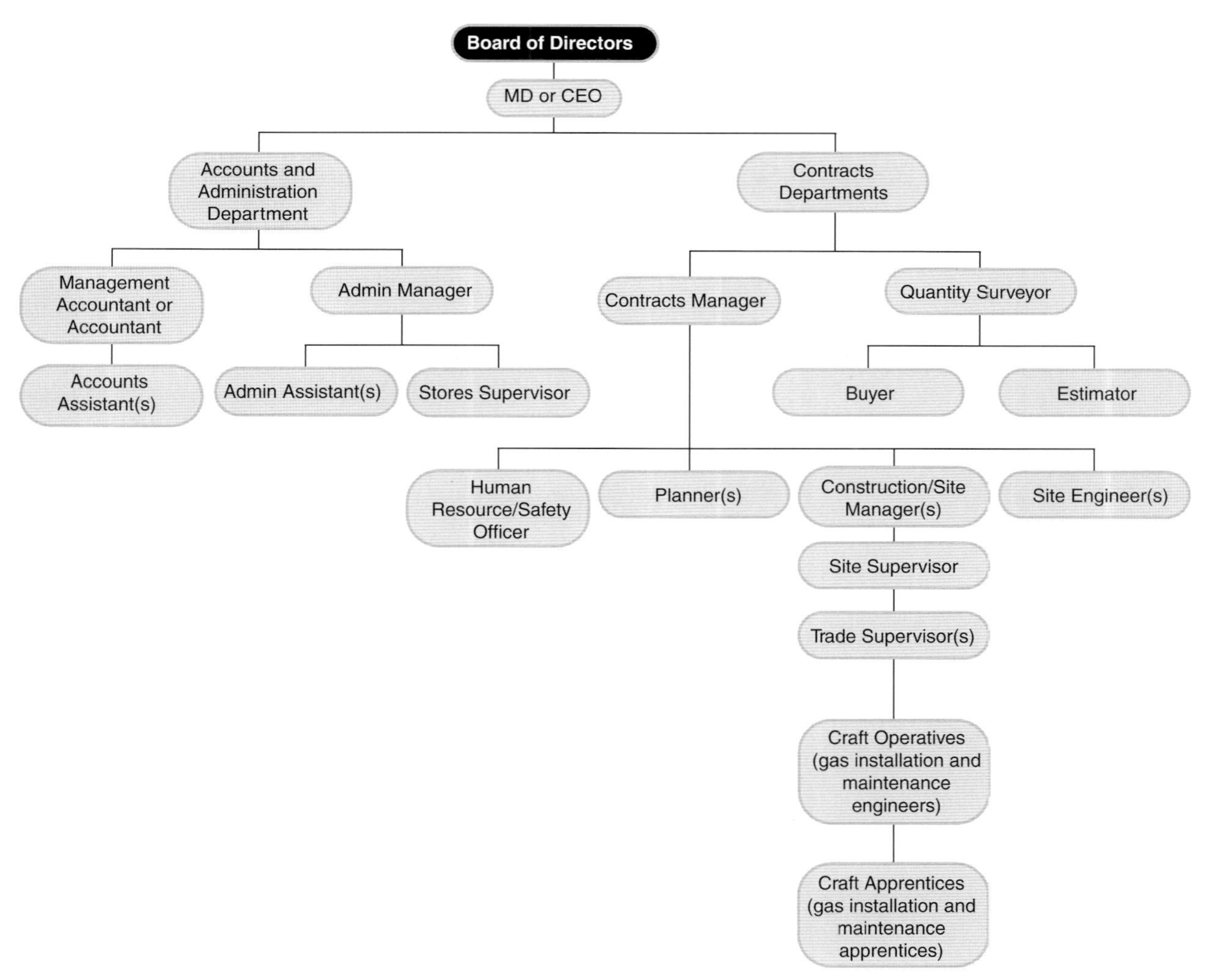

Figure 12.1 Company structure

Contracts Manager

The Contracts Manager is usually mainly office-based, and may have overall responsibility for running several contracts. They work closely with their construction management team, and provide a link back from the other sections of the business and the MD/CEO.

Their job will also involve site visits to make sure the job is running to cost and programme. In some companies, this job might not exist, the duties being shared between the CEO/MD and Construction Managers.

Construction Manager

Construction Managers are responsible for running a construction site or a section of a large project. Other titles used to describe Construction Managers include Site Manager, Site Agent and Building Manager.

A Construction Manager's work includes:

- Developing a strategy for the construction of the project and a plan
- Planning ahead to solve problems before they happen
- Making sure the site and construction processes are carried out safely
- Communicating with the clients' representatives to report progress and seek further information
- Motivating the workforce to get the best out of them.

Supervisory staff

On larger contracts the Construction Manager could have the support of a:

- Site Supervisor (sometimes called Foreman), who would be responsible for the supervision of the Trade Supervisor/Trade Operatives, and day-to-day running of the job
- Trade Supervisor (sometimes called a Charge Hand) who reports to the Site Supervisor, but looks after a number of operatives from a specific trade (e.g. bricklayers). This is usually on really large jobs.

Quantity Surveyor (contractor)

Quantity Surveyors, often called Commercial Managers or Cost Consultants, advise on and monitor the costs of a project. They report to the contractor.

A Quantity Surveyor's work includes:

- Organising the allocation of work to smaller, more specialised sub-contractors according to which offers the best value
- Managing costs to ensure that the initial budget is not exceeded
- Negotiating with the client's private Quantity Surveyor on payments and final account
- Arranging payments to sub-contractors.

Planner

Planners work with Construction Managers and organise the sequence and timing of construction activities to ensure projects are completed on time and within budget. Once construction starts, plans may be updated and modified to ensure the project stays on track. A planner's work includes:

- Working closely with Estimators to establish working methods and costs
- Planning the most effective use of time, people, plant and equipment
- Scheduling events in a logical sequence
- Visiting sites to monitor progress
- Rescheduling projects, if necessary, to bring them back on target.

Site Engineer

Site Engineers ensure the technical aspects of the construction projects are correct. They have a key role in ensuring things are built correctly and to the right quality. A Site Engineer's work includes:

- Setting out the site so that things are in the right place
- Interpreting the original plans, documents and drawings
- Liaising with the workforce and sub-contractors on practical matters
- Checking quality by inspecting and checking measurements
- Referring queries to the relevant people
- Providing 'as built' details
- Supervising parts of the construction.

Estimator

Estimators calculate how much a project will cost, taking into account plant, materials and labour. This will form the basis of a tender the contractor submits to a client. An Estimator's work includes:

- Identifying the most cost-effective construction methods
- Establishing costs for labour, plant, equipment and materials
- Calculating cash flows and margins
- Liaising with other professionals in the contractor's organisation
- Seeking clarification on contract documents, where information is missing or unclear, from the client's representatives.

Buyer

Buyers, sometimes known as a Procurement Officer, purchase all the construction materials needed for a job. A Buyer's work includes:

- Identifying suppliers of materials
- Obtaining quotations from suppliers
- Negotiating on prices and delivery
- Placing orders with suppliers
- Resolving quality or delivery problems with suppliers
- Liaising with other members of the construction team.

Accounts and administration

These departments and staff are responsible for:

Accounts

- Invoicing for work carried out
- Payment of suppliers and sub-contractors
- Payment of staff
- Producing financial reports, budget forecasts
- Dealing with Inland Revenue and VAT.

Administration

- Working closely with accounts to keep financial records
- Dealing with customer, and other enquiries
- Recording and filing work records, time sheets, etc.

Most administration systems are electronic, but companies often keep hard copy records.

Some businesses will also employ a Human Resource Officer, who could look after health and safety matters as well (if not, health and safety would be covered by the Construction or Contracts Manager). Their job is to look after the employees' training and development needs, help with the recruitment and selection of new staff and deal with industrial relations matters.

Hot tips

Construction is a big industry! There is more . . . . There are a number of occupations that play an important role in the overall construction process. Figure 12.2 helps you build up an overview. As you read each paragraph, think about the occupation titles on Figure 12.2.

The other occupations

Building Control Officer (Building Inspectors)

Working for Local Authorities, Building Control Officers ensure that buildings conform to regulations on public health, safety, conservation and access for the disabled. The job involves the inspection of plans and of work-in-progress at various stages relating to Building Regulations. A Building Control Officer's work includes:

- Checking plans and keeping records of how each project is progressing
- Carrying out inspections of foundation, drainage and other major building elements
- Issuing a completion certificate when projects are finished
- Carrying out surveys on potentially dangerous buildings
- Meeting with architects and engineers at the design stage
- Using technical knowledge to talk to people on site.

Architect

Architects plan and design buildings. The range of work varies widely and can include the design and procurement (buying) of

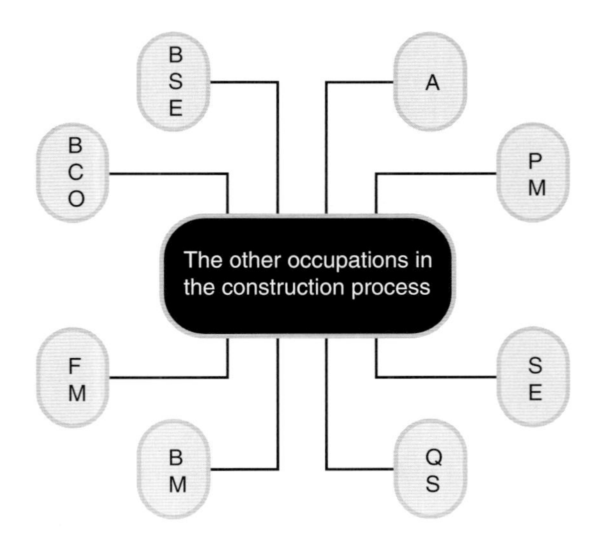

Figure 12.2 Occupations in the construction process. A, architect; PM, project manager; SE, structural engineer; QS, quantity surveyor; BM, building manager; FM, facilities manager; BCO, building control officer; BSE, building services engineer

new buildings, alteration and refurbishment of existing buildings and conservation work. An Architect's work includes:

- Meeting and negotiating with clients
- Creating design solutions
- Preparing detailed drawings and specifications
- Obtaining planning permission and preparing legal documents
- Choosing building materials
- Planning and sometimes managing the building process
- Liaising with the construction team
- Inspecting work on site
- Advising the client about the company they should get to do the work.

Project Manager/Clerk of Works (client)

The Project Manager/Clerk of Works takes overall responsibility for the planning, management, co-ordination and financial control of a construction project. They work for Architects, clients such as Local Authorities, or as a consultant. They ensure the client's requirements are met and that the project is completed on time and within the budget. Depending on the project, responsibilities can start at the design stage and go right through to completion and handover to the client.

A Project Manager's/Clerk of Works' job includes:

- Representing the client's interests
- Providing independent advice on the management of projects
- Organising the various professional people on the project
- Making sure that all the aims of the project are met
- Ensuring quality standards are met
- Keeping track of progress
- Accounting, costing and billing.

Building Surveyor

Building Surveyors are involved in the maintenance, alteration, repair, refurbishment and restoration of existing buildings. A Building Surveyor's work includes:

- Organising and carrying out structural surveys
- Legal work including negotiating with Local Authorities
- Preparing plans and specifications
- Advising people about building matters, such as conservation and insulation.

Structural Engineer

Structural Engineers are involved in the structural design of buildings and structures, such as bridges and viaducts. The primary role of the Structural Engineer is to ensure that structures function safely. They are also involved in the assessment of existing structures. This may be for insurance claims, advising on repair work or analysing the viability of alterations and adaptations. A Structural Engineer's work includes:

- Determining the appropriate structural forms for a project
- Making detailed calculations and drawings for the structure
- Using computer-aided design technology
- Investigating the most suitable materials, such as steel and reinforced concrete for a project
- Working as a team member with other construction specialists involved in the design of a project
- Inspecting, analysing and designing solutions to repair, modify or adapt structures.

Quantity Surveyor (private)

Quantity Surveyors, often called Commercial Managers or Cost Consultants, advise on and monitor the cost of a project. Private Quantity Surveyors (PQS) work for the company having the work done. A private Quantity Surveyor's work includes:

- Advising on the potential of a site and working out what a client can afford to build
- Presenting detailed information on the costs of labour, plant, materials, etc
- Identifying the most suitable contractor based on considerations of cost, quality and service
- Acting as a financial adviser and monitoring progress for the client
- Negotiating changes in price with the contractor's Quantity Surveyor and agreeing payments to the contractor
- Advising the client on the cost of maintaining whatever has been built.

Facilities Manager

Facilities Managers ensure that buildings continue to function once they are occupied. They are responsible for maintaining the building and carrying out any changes needed to ensure that the building continues to fulfil the needs of the organisations using it. A Facilities Manager's work includes:

- Planning how the inside of the building should be organised for the people occupying it
- Managing renovation works
- Managing routine maintenance
- Managing the installation and maintenance of computer and office equipment
- Managing the building's security
- Managing the cleaning and general upkeep of the building
- Negotiating with contractors and service suppliers.

You may be particularly interested in the next occupation, especially if you work for a larger building services company.

Building Services Engineer

Building services include water, heating, lighting, electrical, gas, communications and other mechanical services, such as lifts and escalators. Building services engineering involves designing, installing and maintaining these services in domestic, public, commercial and industrial buildings.

A Building Services Engineer's work includes:

- Designing the services, mostly using computer-aided design packages
- Planning, installing, maintaining and repairing services
- Making detailed calculations and drawings.

Most Building Services Engineers work for manufacturers, large construction companies, engineering consultants, architects' practices or Local Authorities.

Activity 12.1

The job of a Building Services Engineer is the one most closely related to the plumbing industry. Thinking about the other job description as well as what you know about plumbing, write down what you think a Building Services Engineer might do and include the information in your portfolio. Check it with the model answer at the end of this book.

It often involves working with other professionals on the design of buildings including architects, structural engineers and contractors. As a result, it usually means working as a member of a team.

Working for a consultant on designing, the job involves spending most of the time in an office. Once construction starts, it means visiting sites and liaising with the contractors installing the services.

Working for a contractor involves overseeing the job and may well mean managing the workforce. It is likely to be site-based.

Working for a services supplier is likely to mean being involved in design, manufacture and installation. It may involve spending a lot of time travelling between the office and various sites.

How plumbing fits within the construction industry

More about the plumbing industry

It is really difficult to categorise the range of businesses that operate in the plumbing industry. Research carried out by SummitSkills shows that most plumbing businesses, about 80%, employ between 1 and 4 people. Many also fall into the category of 'sole trader' which generally means someone that works on their own, or a 'one man band' as they are referred to in the industry (sole traders can also employ staff).

The type of work carried out by most plumbing companies is reflected in the activities described in the last section. You will find that the smaller firms tend to concentrate on domestic plumbing covering installation, servicing and maintenance. Some firms specialise in installation only, such as new house build.

The larger firms in the industry tend to be more multi-disciplined (or skilled), which means they not only provide a plumbing service to their customer, but will also cover electrical installation, and heating and ventilating in domestic, industrial and commercial premises. Some will also cover industrial and commercial gas work. Having said all that, you might also find a sole trader working on larger industrial commercial installations as a sub-contractor to a larger company.

Large-scale industrial commercial installations are likely to include the welding of large-scale diameter pipe.

Plumbing has always been a mobile industry, unlike a factory or retail job; plumbers move from one site to another. Current trends see more companies/plumbers working away from home, particularly with the larger firms.

The internal structure of the business will vary depending on the size of the firm. In a micro-business (usually classed as employing between 1 and 4 people), the owner/manager or sole trader will be responsible for every aspect of running the business, such as procuring the work, pricing the job, ordering the materials, organising the job, performing the job, customer care in addition to managing the staff and finances of the business.

In a larger firm some of the responsibilities of running the business are shared amongst a management team, site managers and supervisors, etc and technical support staff, such as designers and estimators.

Plumbing as a trade, fits into the building engineering services side of the construction industry, along with trades such as plumbing, heating and ventilation, and electrical installation and maintenance.

Plumbing businesses often work for construction companies on a sub-contract basis. This means the main contractor is the construction company, and they contract directly with the customer or client. The main contractor then sublets a part of the contract (the plumbing work) to the plumbing contractor (sub-contractor). The plumbing sub-contract deals directly with the construction company and not the client. The tender price, contract details and payment of work is between the plumber and the construction company.

It is not just plumbing businesses that sub-contract to construction companies. Others include:

- Electrical
- Heating and ventilating – domestic and industrial commercial
- Refrigeration and air conditioning
- Service and maintenance
- Ductwork.

Finally, you should be familiar with where the plumbing industry fits in the 'grand scheme of things'.

Craft operatives

The term 'craft operatives' refers to your other construction industry co-workers, particularly if you work on a large building site. The list of other trades is quite extensive, but here are some of the main ones:

- Bricklayer
- Carpenter and Joiner
- Plasterer
- Roof Slater and Tiler
- Wall and Floor Tiler
- Built up Felt Roofer
- Construction Operative
- Painter and Decorator
- Floor Layer
- Plant Operator
- Scaffolder.

Activity 12.2

This activity involves you doing some research. Using the contact details for the CITB, as well as your local library, write notes about each of the various trades in the bullet points above. The example of the bricklayer shows what a typical answer looks like.

Once you have finished this activity, check your answers against the model ones at the end of this book. You may wish to do this on separate sheets and add them to your folder.

Bricklayer

Bricklayers use bricks and blocks to build the interior and exterior walls of buildings. They also create other types of walling including tunnel linings, archways and ornamental brickwork. A bricklayer's work includes:

- Working on new buildings or the extension, maintenance and restoration of old buildings
- Building foundations, bringing brickwork and blockwork up to damp proof course level
- Working at height from trestles, hop-ups and scaffolds
- The construction of drainage and concrete work.

Information used in support of the plumbing industry

Whilst working through all the units, you will have noticed references to sources of information used in the plumbing industry. What follows is a summary of these various sources. Probably the most important sources of information to a plumber are:

- BS 6700
- Water Regulations.

BS 6700 provides a specification for the design, installation, testing and maintenance of services supplying water for domestic use within buildings and their curtilages. So, for example, where a plumber was looking at the various options for installing a hot and cold water system, guidance could be found in BS 6700. British Standards are produced by the British Standards Institution (BSI).

There are numerous British Standards relevant to plumbing, too many to mention here, but BS 8000 is a BS that encompasses workmanship across all the construction disciplines.

Water Regulations state the legal requirements for installation of hot and cold water systems.

Building Regulations also have a key influence on the plumbing industry at Level 2, and in particular:

- Part H Drainage
- Part J Combustion
- Part L Energy Conservation
- Part P Electrical Safety.

These are the main plumbing-related ones. Part A is about structures and will be applicable to things like specifications for drilling and notching timber joists to receive services pipework.

Other sources of information include:

- Building plans/drawings. These provide details of the layout of a dwelling/building, including dimensions and positions of sanitary appliances, etc.
- Services layout plans/drawings will show the layout of the services within a dwelling or building, which is helpful for other construction trades and plumbing, as this will provide

an indication to the bricklayer, for example, of where to leave holes or chases for pipework
- Work programmes should provide a clear indication of when plumbing activities should start and finish
- Job specifications typically should provide details of the type and number of fittings and components required for the installation of a full plumbing system on a housing development
- Manufacturers' instructions including installation and users' instructions should be provided with appliances and components. It is essential for the plumber to ensure that the appliance is installed, commissioned, serviced and maintained as designed, and explained to the customer so they know how to use it correctly.
- Manufacturers' catalogues are useful for the customer to make decisions about the choice of appliances, for example, type and colour of a bathroom suite and choice of fitments. Plumbers also use manufacturers' catalogues when selecting and ordering fittings, components and appliances.

Construction Skills Certification Scheme (CSCS)

CSCS aims to register every competent construction operative within the UK not currently on a skills registration scheme. Operatives get an individual registration card (similar to a credit card) which lasts for three or five years. The CSCS card also provides evidence that the holder has undergone health and safety awareness training or testing.

CSCS is owned and managed by CSCS Limited, controlled by a management board whose members are from the Construction Confederation, Federation of Master Builders, GMB Trade Union, National Specialist Contractors Council, Transport and General Workers Union and Union of Construction Allied Trades and Technicians.

Observer members include the Department for Education and Employment; the Department for the Environment, Transport and the Regions; the Health and Safety Executive and the Confederation of Construction Clients.

The Management Board has appointed CITB to administer the scheme.

The scheme has also been extended to the building services sector, with plumbers, gas engineers, electricians and heating and ventilating fitters, all included. For plumbers, the scheme is administered by the JIB for PMES.

The JIB-PMES CSCS Card Scheme is designed to ensure that the Plumbing Industry workforce can integrate fully with the CSCS requirements of the wider Construction and Building Engineering Services Sectors, Local Authorities and other Public Sector organisations. JIB-PMES CSCS Card membership aims to provide the following benefits to individual plumbers:

- access to many construction sites
- industry wide recognition of skills, competence and qualifications
- improved knowledge and awareness of workplace health and safety
- enhanced employment prospects
- Confirmation of your professionalism – a card will help to differentiate you from the industry's 'cowboys'.

Test yourself 12.1

1. Who is the Sector Skills Council for the construction industry?
2. What percentage of companies working in the construction industry employ more than 300 people?
3. There are four main legal classifications for a company; which one is missing from the list given?
   - Limited company (private)
   - Public limited companies (PLCs)
   - Partnership
4. A Project Manager is also known as a Contracts Manager – True or False?
5. List five other trade operatives that you may meet at a construction site.

Communication

So far you have looked at how the industry is structured in terms of businesses as well as some of the various people involved in the process of turning a piece of land into a housing estate or a different type of building structure.

Depending on the type of company you work for (including possibly your own) you are probably going to work with any number of these people at some point in time in your career, as well as customers or clients from the public or private sector.

The need to communicate effectively at all levels within the industry is important, as you are likely to be ordering materials, advising customers, recording complaints about bad service from suppliers, dealing with complaints and discussing matters with other trades on site. These are just a few examples; you can probably think of a few

Activity 12.3

What do you think is the meaning of the term communication? What methods of communication are there? Jot down your thoughts here and then check them with that given at the end of this book.

more. Communicating effectively involves verbal (oral), written and visual methods of communication. Whatever the method of communication, remember:

- Whoever is sending the communication (message) must know what they want
- The message must be delivered to the person receiving it in the appropriate style and in a form that can be understood, i.e. verbally, in writing, etc
- The message must get the desired response from the receiver
- The sender must know that the message has been acted upon.

Activity 12.4

How many ways can you think of to send a message? Write down your answer here and then compare it with the one at the end of this book.

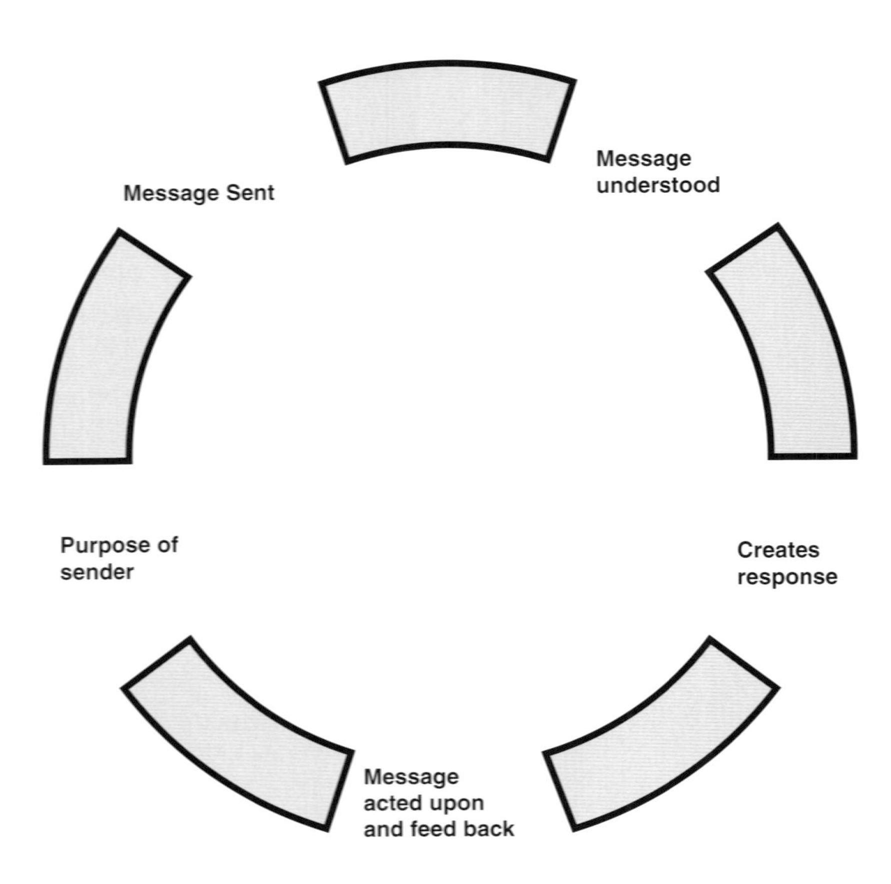

Figure 12.3 Effective communication

This process is shown in Figure 12.3.

Hence, communication is a two-way process between sender and receiver(s). Taking this further, there are five essential 'C's of effective communication:

Clear:	The message should be presented in simple, direct language that the receiver is able to understand
Correct:	The message should contain accurate information
Concise:	Keep the message short without being curt (rudely brief), but
Complete:	make sure it includes all the relevant information, and enough of it, so that the receiver can respond with the information you are after
Constructive:	The message should enable the receiver to respond in a way to achieve their purpose.

As well as the five 'C's there is one 'S' and that is the Style. The message should be in the correct style for the purpose intended and for the person who is receiving it, whether it is formal or informal.

Verbal communication

Verbal communication will probably be the most common form you will use at work, as this method is used almost throughout the day for one reason or another. It has the following advantages:

Advantages

- More personal contact between sender and receiver
- Tone and body language can be used to help convey the message, and interpret the message from the sender
- Feedback is usually instant.

What do you think might be the disadvantages? You do not have to write anything down here, but think about this before moving on.

The disadvantages might include:

- No formal record unless the conversation is recorded
- Not very good if you need to discuss a lot of detailed or technical information
- The receiver can be easily distracted.

Effective verbal communication is based on two factors:

- Speaking
- Listening.

This includes things like speaking face to face(s), and in most cases, over the telephone.

Here are some do's and don'ts:

Do's

- Think before you speak, do not say something for the sake of it
- If you are making a telephone call, write down the points that you want to cover so you do not forget anything
- Speak clearly, at a pace that can be followed and a level that can be heard
- Use different tones of voice to keep the listener interested in what you are saying

- Try to deliver the message in a logical sequence
- Select words that are appropriate for the situation
- Use positive facial expressions (eye contact) and body language, as this helps the listener to understand what you are saying.

Don'ts

- Mumble
- Speak in one tone of voice
- Hog the conversation
- Use jargon (this means language used by a particular profession or industry)
- Use technical terms to non-technical people, e.g. if you are explaining how something works to a customer
- Use body language which distracts the listener, e.g. drumming your fingers on a desk whilst in conversation, twiddling a pen and so on.

Listening skills

Do's

- Concentrate on what the speaker is saying
- Respond to the speaker in small group or one-to-one meetings by confirming you are listening, gently nodding your head, or saying 'OK', 'mmm', at relevant stages
- Ask questions if you do not understand
- Pick up on any changes in voice tone or use of body language where it is used to emphasise a point
- Take notes if appropriate
- People listen faster than the spoken word, so use this time to collect your thoughts before responding.

Don'ts

- Get distracted by other things going on around you
- Interrupt the speaker before they have made a point
- Allow them to go on too long before responding
- Make assumptions about what the speaker is saying, seek clarification
- Day dream, or at the very worst, doze off!

Now you have looked at the do's and don'ts, let's see how we can use verbal communication.

Talking face to face

Your job as a plumber is not likely to thrust you into situations where you are doing presentations to large audiences or running formal business meetings. There will be times though when you are dealing with customers, co-workers and other site visitors, when you will be giving or receiving information.

Think of some occasions when you could be involved in site meetings with people from other trades, meetings at work with your employer, or union meetings, where you could be required to say something, and need to take down information. You might be asked to present information to a small group at work, if, for example, you have been on a short course for a new product, you can update your workmates.

There are a number of skills you will need to develop when dealing with people on a one-to-one basis, as well as handling groups of people. These meetings can be informal or formal.

The telephone

Following this guidance will help your telephone technique.

Plan your call:

- Particularly if it is a business call, write down all the points you need to cover
- Have any reference material at hand
- If you have a number of calls to make, try to finish them all at once so you are not constantly breaking off what you are doing.

Receiving a first-time call:

- Answer the phone promptly
- Say hello or good morning, etc
- Give your name and that of your employer's business if relevant
- Ask how you can help
- Try to speak in an 'up-beat' way.

There are also a few general rules:

- Never eat whilst on the phone
- Do not get too overfriendly on a first-time business call, respect the person on the other end of the phone
- Do not forget to ring back if you have promised to
- Do not talk to other people around you when you are on the phone with someone. Similarly do not interrupt someone who is speaking on the phone.

Meetings

Generally speaking, most of your meetings will be on a one-to-one basis. This might include meeting a customer, or a client's representative on site. If you are meeting someone in relation to your job it is usually to exchange information. Getting the best out of a meeting usually involves:

- Starting meetings
- Getting information
- Providing information
- Ending the meeting.

Starting a meeting

If you are meeting someone for the first time, a customer or new workmate, this usually involves a hello, a handshake and introductions. Try to maintain eye contact, and adopt a pleasant friendly approach, but do not overdo it. Try to remember their name – if you keep forgetting, it gives them the impression you are not really interested in them.

Getting the information

This is similar to the telephone technique really; once you have opened the meeting, you need to know in advance what information you are looking for. You get this by using a series of questions.

Providing information

You will do this a lot in your job, for example, providing operating instructions to customers, explaining how things work, advising co-workers on site, when and where a builder's work is required, chases, access holes, etc. Hence, make sure:

- Information is accurate and clear
- You check their understanding

- You use terminology that the other person understands
- Any written information is provided that supports the meeting, user catalogues for customers, service drawings or details for co-workers.

Ending the meeting

Once all the information has been provided and received, use a phrase such as 'I think we have covered everything'. Then double check that there are no questions, and summarise what has been said or agreed upon. Part company with a 'thanks' and possibly a handshake.

Discussions

You are involved in discussions every day. These are mostly informal; it normally involves one-to-one meetings as above, or small groups, and in some cases will lead to a more in-depth conversation. At this stage in your life, you will have already developed skills that will enable you to:

- Listen and understand
- Question
- Challenge
- Persuade/negotiate/compromise
- Convince
- Manage
- Agree.

Improving these skills will help your performance in discussions (and other meetings). As a plumber, you will be expected to be involved in discussions with your job. *Listening to and understanding* what others are saying is vital if you are going to have a meaningful input to any discussion. You need to have the full facts about the subject, so preparing some notes beforehand may be useful.

Asking *questions* will help you to clarify points you are not sure about, and find out any additional facts which another person has not mentioned (or has intentionally held back!). 'Perhaps you could tell me a bit more about that', or 'Would I be right in thinking …?' are good probing questions.

Be prepared to *challenge* the person speaking if you think they are being untruthful or have got their facts wrong, but do not do it aggressively.

Persuasion is an important skill in any discussion, and in particular negotiation. You might be trying to persuade your boss about a rise in a pay *negotiation*; so, for example, you could use the fact that you have just completed some additional training giving you a wider range of skills that the business can use.

Compromise is common in most discussions. There is no point in forcing a particular point if no one is going to accept it. This situation is usually resolved with a 'middle ground' *agreement* that everyone is happy with.

During the discussion, you may need to *convince* someone that what you are saying is:

- Technically correct
- Current or factual
- Truthful.

You need to *manage* discussions, and by that we do not mean bully, hog or overpower the meeting. Quite often, people may be in disagreement over a point and progress cannot be made. You could say, 'I think we have discussed this point long enough, let us move on and we will come back to this later'. Subsequently, discussion may lead to the point being resolved.

Someone may be going on at length (hogging the discussion). You may politely interrupt by saying, 'That's a good point, but what does Martin think?' Alternatively, asking a question can change direction or you can use a statement like 'We seem to be straying from the point, we were talking about…' or words along these lines.

If it is important to resolve a point or *agree* upon an action, but there is a difference of opinion, the group could put off a decision until a later date to give more time for research, or arrive at a consensus decision or even a compromise!

Written communication

Written communication has advantages and disadvantages.

Advantages

- Permanent record for future reference, particularly:
  - Legal documents
  - Contracts
  - Agreements
  - Key points from meetings.
- Is a more reliable method for reporting complex or technical information

Disadvantages

- Producing written material takes time
- Feedback can be delayed, or not given at all. Chance for misunderstandings.
- Lacks the personal touch
- Can be used for negative reasons, delaying tactics.

Use of ICT/text messages

ICT (Information and Communication Technology) enables written or visual messages to be forwarded quickly using e-mail, which basically is a 'written or graphic' electronic message sent down the phone line.

Documents of all sizes can be forwarded as attachments to the e-mail and then downloaded by the receiver either in hard copy, or read on screen and stored on their system. ICT is constantly improving, and mobile telephones have text facilities and internet connections. This has become even more effective with the use of systems such as Broadband.

Some phones have photo messaging, or software that will enable video messages to be forwarded. Audio files can also be sent using ICT. Text messages are used increasingly as a method of electronic communication.

Activity 12.5

From your experience, what do you find are common faults with written documents such as letters, reports, technical catalogues, brochures and official forms? Write your answers here, and then compare your thoughts with the suggested answers at the end of this book.

Writing skills

Generally people find that it is more difficult to get views, thoughts or ideas on paper than it is to express them verbally. This is because:

- It takes time to think through what you are writing
- What you put down is a permanent record, so it could be used later as evidence for all sorts of reasons. So, the content has to be accurate and unambiguous

- You need to take time to make sure you use the correct:
  - Layout
  - Grammar
  - Punctuation
  - Spelling.

Reading skills

As a plumber, your reading skills are important because you are likely to use or see things like:

- Documents related to your employment, such as:
  - Contract of employment
  - Health and safety policy
  - Equal opportunity policy
  - Employee handbook
  - Memos
  - Letters.
- Regulations
- Technical specifications from clients
- Work programmes
- Manufacturers' catalogues
- Technical publications and journals
- Press releases
- Drawing and design details
- Promotional materials
- Training/learning support material (like this).

Writing style

Any written communication will give the reader a 'feel' of the writing style. This could be formal or informal, depending on its intention.

Use of memos

Depending upon the type of company that you work for, there may be occasions when you might have to send a memo. Any written

Activity 12.6

What do you think is meant by formal and informal writing styles? Jot down your thoughts here and compare them with the suggestions at the end of this book.

Memorandum

To: All Plumbers
CC: Main Office Admin Staff
From: Contracts Manager
Date: 05 November 2004
Re: Completion of Job sheets

Please note that from next month Acme Gas Plumbers will be introducing a new job sheet. All engineers should ensure that they pick up a supply of the job sheets from the Main Office by the end of November at the latest. To ensure the new job sheet is used correctly, all engineers must attend a familiarisation workshop in the Training Room at 9 am on 14th November.

Figure 12.4 Layout of a typical memo

communication should clearly state its purpose. In the case of a memo this should be kept to a single subject but it may have more than one purpose.

The purpose of a memo is to act as a memory jogger, a reminder for action or instruction – it means something to be remembered. Memos are used to send short messages between members of the same company. They are usually sent on pre-printed paper and follow a set format (see Figure 12.4).

A memo should be short and to the point (about four or five lines), and should keep to a single subject. The memo serves as written proof that a message has been passed on. Generally, the style of the memo should be formal, particularly if it is going to your boss, or its intention is to inform on business matters.

Reprimands would also be in a formal style. Some memos can be more informal, for example, if you are trying to persuade some one to do you a favour, you are unlikely to use a tone that would upset them.

E-mail

This is replacing the written memo, and is not just restricted to internal use within a business, but is also used externally to send messages to other companies. The content and style of the e-mail should be the same as written memos.

Letters

If you ever write a letter on behalf of your company, remember that it actually represents the business. It is important that the structure, content, spelling and grammar hit the right note, because a poor letter gives a poor impression.

A business letter from your company will usually go on headed paper, which contains details such as company name, address and contact details.

You may wish to write a personal letter, in which case you would write your details in the top right-hand corner of the letter. Make sure the name and address of the person you are sending it to starts below where your address finishes, on the opposite side of the page.

Activity 12.7

Can you think of any faults commonly found in badly written reports? Think about it and compare your thoughts with the suggested answer at the end of this book.

Reports

Reports are used to present facts, and in most cases provide recommendations on a particular subject. The facts contained in the report should be accurate, relevant, and in a logical format. It should be in a language that the intended reader will understand.

Making notes

This is a useful skill in support of listening and reading. It is used for:

- Telephone message details
- Recording what is said at meetings, training events, seminars/conferences
- Preparing for phone calls, meetings
- Preparing for letters, reports, presentations.

Visual communication

Graphics is a form of visual communication. Can you think of examples of graphical communication you will use in your work as a plumber? Do not write anything down here, but think it over before moving on.

Graphics might include safety signs, warning labels, commercial logos, technical drawings, manufacturers' instructions, trade catalogues and hand signals.

Advantages of visual communication:

- Useful for explaining complex technical details – 'a picture paints a thousand words'
- Useful where it is difficult for people to read
- ICT graphics applications can use animations for improved understanding.

Disadvantages:

- No control over interpretation and feedback unless you are there to explain
- No personal contact if graphics are forwarded onto the receiver.

You can use visual communication to your advantage. Often, drawing a sketch can help to explain what is required on a job for the person in another trade to carry out. It could also help a customer understand what you mean in showing them how something works.

Dealing with other people

This includes:

- Effectively dealing with site visitors, customers and co-workers
- Identify factors that are important when dealing with customers.

Site visitors, customers and co-workers

How you deal with co-workers will be different from site visitors as the relationship will be more informal. Your aim though is to get on well, work as a team and avoid confrontation. Requests for help or information should be made courteously, and should be a request not a demand. When the 'boot is on the other foot', you should be as helpful as possible.

Dealing with site visitors (such as the Building Control Officer, Site Engineer, etc) will require a more formal approach. Remember that you are representing your employer and should act accordingly. Requests for help or information should be dealt with promptly and efficiently. Treat any visitor with respect, and make sure that you present yourself in a way that will promote the image of the company.

Customers and clients

Everything that is applied to site visitors applies equally to customers. With some customers though you could be working in occupied property. Make sure that you arrange to start the job well in advance, and that you keep to time for the appointment.

Respect the person's property, do not move items that could be damaged, and protect floor coverings and furniture with dust sheets. Do not have your radio blurting out, disturbing householders and neighbours, and do not get overfriendly, 'gushy', or disrespectful. Using comments like 'has the kettle burst?', hinting at a cup of tea, may offend some people.

Test yourself 12.2

1. Below is an example of what you need to remember when using any form of communication. Whoever is sending the communication (message), must know what they want. State two other examples.
2. State one advantage and one disadvantage of verbal communication.
3. State three examples of good practice when receiving a first-time call.
4. To a plumber, reading skills are quite important; make a list of as many documents you can think of that a plumber might need to use.
5. You are doing a job in someone's house. Jot down some notes on what is important when dealing with the customer.

Check your learning Unit 12

Time available to complete answering all questions: 20 minutes

1. In the construction industry, the number of people employed by 90% of businesses is:
   a. 1 to 13 ☐
   b. 14 to 24 ☐
   c. 25 to 49 ☐
   d. Over 50 ☐
2. A building control officer works on behalf of the:
   a. Client ☐
   b. Architect ☐
   c. Contractor ☐
   d. Local Authority ☐
3. One of the main roles of a Building Surveyor is to:
   a. Measure all materials used on site, and prepare invoices ☐
   b. Prepare detailed information on the costs of resources ☐
   c. Organise and carry out structural surveys ☐
   d. Organise the labour and material required for the job ☐
4. The building engineering services sector covers plumbing, heating and ventilating, and which of the other industries?
   a. Construction ☐
   b. Electrical ☐
   c. Utilities ☐
   d. Engineering ☐
5. The greatest proportion of the construction industry is made up of businesses classified as:
   a. Sole traders ☐
   b. Partnerships ☐
   c. Public Limited Companies ☐
   d. Private Limited companies ☐
6. When working in an occupied dwelling, which of the following should be observed when dealing with the customer?
   a. Reducing the quality of materials to save costs ☐
   b. Getting the job done quickly at all costs ☐
   c. Establishing a good working relationship ☐
   d. Asking the customer to supply refreshments ☐
7. The preferred method to quickly check the delivery date of plumbing appliances would be by:
   a. Fax ☐
   b. Letter ☐
   c. E-mail ☐
   d. Phone ☐
8. The main purpose of a work programme is to:
   a. Assist a plumber to calculate the amount of materials required for a job ☐
   b. State when payments are due for a particular section of plumbing work ☐
   c. Provide a clear indication of when plumbing activities should start and finish ☐
   d. Assist a contractor to plan how many plumbers might be required for a job ☐

(Continued)

Check your learning Unit 12 (Continued)

9. Which one of the following would state the legal requirements for the installation of coldwater services?
   a. Building Regulations ☐
   b. British Standards ☐
   c. Water Regulations ☐
   d. Construction Regulations ☐

10. One purpose of the CSCS scheme is to:
    a. Certificate and register plumbers with CORGI ☐
    b. Identify and register competent plumbers ☐
    c. Provide a progression route for Technical Certificates ☐
    d. Provide an entrance qualification to NVQs ☐

Sources of information

There are no additional sources of information other than the ones given in the text.

- CITB head office Bircham Newton
  King's Lynn
  Norfolk
  PE31 6RH
  Tel: 01485 577577
  Fax: 01485 577793
  Website: www.citb.org.uk

The Construction Skills Certification Scheme for Plumbing and Mechanical Engineering Services can be contacted at:

- JIB for PMES
  www.jib-pmes.org.uk
  Help desk: 0800 197 1060

Answers to Activities, Test Yourself Questions, Check Your Learning and Case Study

UNIT 1

Answers to activities

Activity 1.1

The construction industry includes a number of trades such as:

- Bricklayers
- Carpenters and Joiners
- Plasterers
- Painters and decorators
- Roofers

Answers to Test yourself questions

Test yourself 1.1

1. - Installation
   - *Service*
   - And maintenance

   of a wide range of *domestic* systems such as:
   - Cold water, including underground services to a dwelling
   - *Hot water*
   - *Heating* systems fuelled by *gas*, oil or solid fuel
   - *Sanitation* (or above ground drainage) including the installation of baths, hand wash basins, WCs and sinks
   - Rainwater systems, *gutters* and fall pipes
   - Associated *electrical* systems
   - *Sheet lead weatherings*. At Level 2, this includes things such as chimney weatherings and soil vent pipe weatherings.
2. a. *SummitSkills*
3. a. *Institute of Plumbing and Heating Engineering*

Test yourself 1.2

1. Complete the missing words in the paragraph from the words given below
   Whilst the *SVQ/NVQ* is about proving competence in the *workplace*, the *Certificate in Plumbing* Level 2; Plumbing Studies, is a qualification based on what you are taught *off-the-job*. It covers *practical* and theory. The *practical* is assessed by a number of workshop assignments, and the theory is the same assessments as used in the NVQ *knowledge* assessments.
2. a. *The main method of assessing the job knowledge aspect of the SVQ/NVQ is by using multi-choice questions.*
3. *National Vocational Qualification*
4. *Assessment is a method of finding out if you can do a job competently, and if you understand the theory that underpins the job. There are various forms of assessment. SVQs/NVQs include assessments that are done at the workplace, either watching a person do the job, or assessing what they have included in their workplace recorder/portfolio. It also includes the knowledge assessment, which in the main is done by multi-choice questioning.*
5. *Higher National Diploma, Certificate or NVQ Level 4 in Building Services Engineering*

Answers to Check your learning Unit 1

1. a
2. a
3. c
4. d
5. b
6. b
7. b
8. c
9. c
10. a

UNIT 2

Answers to activities

Activity 2.1

Here are the guidelines.

- Adjust your safety helmet so that it is a comfortable fit
- All straps should be snug but not too tight
- Do not wear your helmet tilted or the wrong way round
- Never carry anything inside the clearance space of a hard hat, i.e. cigarettes, playing cards
- Never wear an ordinary hat under a safety helmet
- Do not paint your safety helmet as this could interfere with electrical protection or soften the shell
- Handle it with care, do not throw it, try not to drop it, etc
- Regularly inspect and check the helmet for cracks, dents or signs of wear
- Check the strap for looseness or stitching that is worn out and also check that your safety helmet is within the specified life date.

Activity 2.2

Roof ladders are used by plumbers to gain access to a roof while carrying out roofing repairs, flashing repairs and while installing chimney flue liners or twin wall flues for gas- and/or oil-fired appliances. Roof ladders need to be light so they can be easily manoeuvred up the roof and are normally manufactured from aluminium (originally they were constructed from wood). It is positioned by turning it on its wheels, pushing it up the roof, and then turning it through 180 degrees so that you can hook it over the ridge tiles.

Activity 2.3

This Act was a 'milestone' in bringing in laws dealing with the health, safety and welfare of persons at work. Duties are placed on all persons connected with Health and Safety at work, whether as employers, employees, self-employed workers, manufacturers or suppliers of plant and materials. Protection is also given to members of the public affected by the activities of persons at work.

In addition to setting out the basic Health and Safety requirements of employers and employees, the HSWA also gives more detailed guidance to employers on aspects such as Health and Safety policy statements, which are required for companies employing five or more members of staff.

Answers to Test yourself questions

Test yourself 2.1

1. a. Warning
2. c. Figure 2.18
3. c. 110 V
4. a. Making trial lifts on similar loads

Test yourself 2.2

1. a. Assess the levels of risk to safety of operatives that may be present in work operations
2. d. Likelihood × Consequence = Risk
3. d. Your employer

Test yourself 2.3

1. a. No damage to toe boards
2. d. 75°
3. c. 5 rungs
4. b. Stiles

5. d. 450 mm
6. a. 4 m
7. False
8. d. Base plate
9. a. Toe boards
   b. Ledger
   c. Standard
   d. Guard rails/hand rails
   e. Putlog
10. a. Mobile scaffold
11. c. 9 m
12. Roof ladder hook must be fixed firmly over the ridge.
    Wheels of roof ladder must be firmly attached and must run freely.
    Pressure plates must be sound.
    Stiles must be straight and in a sound condition.
    Rungs must be in a good sound condition.
13. c. 1 to 3
14. b. 1.2 m

Test yourself 2.4

1. c. Control of substances hazardous to health.
2. d. Smells
3. b. Inhaled into the lungs
4. d. Sealed in suitable containers and correctly labelled

Test yourself 2.5

1. a. Foam
2. c. An accident in the home
3. a. Fuel, oxygen and heat
4. c. Red
5. a. Name of parent or guardian
6. a. Water type

Test yourself 2.6

1. b. 2501
2. c. LPG is heavier than air and a leak can sink to form high levels of concentration
3. a. Two fire extinguishers must be carried
4. b. ABC
5. Airways, Breathing, Circulation
6. b. Pulse
7. c. Flood the burn with cold water

Test yourself 2.7

1. d. Carry out an assessment of risks associated with all the company's work activities
2. c. Construction (Design and Management) Regulations (1994)
3. d. Safe work at height, including use of scaffold
4. Manual Handling Operations Regulations (1992)
5. d. Yourself as the plumber on site
6. a. A Health and Safety policy statement

Answers to Check your learning Unit 2

1. a	11. d
2. c	12. a
3. a	13. b
4. c	14. d
5. b	15. b
6. a	16. c
7. d	17. a
8. c	18. b
9. b	19. d
10. b	20. c

UNIT 3

Answers to activities

Activity 3.1

Job specifications provide details such as:

- Type and quality of components, materials and fittings to be installed
- Type of clips or brackets to be used
- System test specifications
- Any specific installation requirements
- Whose job it is to carry out any associated building work, e.g. cutting holes for pipework, drilling notching joists, and making good.

Activity 3.2

Here is a checklist that should cover most of the preparation items.

- Have you confirmed with the client or customer the starting time and date?
- Are all the tools, materials and equipment required to do the job on site?
- Are electrical tools safe to use?
- Is the work area clean and safe to work in?
- If using access equipment, ladders, etc, are they safe to use?
- If you are working in a loft, do you have adequate lighting to see what you are doing like torch, inspection lamp, etc?
- If working in an occupied dwelling, have you checked the work area for any pre-work damage?
- If you have noted any damage, have you notified the customer prior to starting the work?
- Again, in an occupied dwelling have you made sure you have protected the customer's property before you start?
- This could include the following:
  - Use of dust sheets (but not in positions where they could prove dangerous, e.g. on stairs)
  - Request that vulnerable furniture is moved
  - Request that carpets are taken off, if necessary
  - Cover up sanitary appliances if working in the bathroom area, and do not leave tools such as hammers or spanners where they could be accidentally knocked into and damage an appliance.

Activity 3.3

Here is our safety maintenance checklist:

General

- Make sure the tools are cleaned regularly
- Always use the right tool for the job. Screwdrivers are not chisels!
- Keep file or rasp teeth clean using a wire brush
- Do not use split handles
- Lubricate the working part of tools
- Once cleaned, lightly coat the tools with an oil spray to prevent rusting
- Do not overdo the oiling.

Hacksaws

- Do not use with defective or worn teeth
- While using a large hacksaw, make sure the blade is tightened correctly
- Make sure the teeth are pointing away from the forward cut.

Pipe cutters

- Make sure that the wheel and the rollers are lubricated and move freely
- Use the pipe cutter to de-burr the inside of pipework
- Replace damaged or blunt cutter wheels and use the correct blade for the material being cut.

Wood chisels

- Keep the chisels sharp using a grinder and whet stone
- Never use with split handles
- Make sure handles are not loose
- Always keep the plastic guard on the chisel.

Cold chisels

- Keep the striking end of the chisel free from the 'mushrooming effect' again using a grinder
- Keep the cutting edge sharp using a grinder
- Do not use a chisel which has been ground down too much.

Hammers

- Do not use a hammer with a defective shaft
- Make sure the head is fitted correctly to the shaft.

Pipe grips and wrenches

- Once the teeth become worn out, the tool should be replaced
- Be careful while loosening a joint or pipe that is difficult to move. It might give in suddenly, and you could damage your hands, or even pull a muscle.
- Keep the teeth free from jointing compounds. If they are made up, it could cause the tool to slip.
- Check for wear on the ratchet mechanism while using pump pliers. These often slip when under pressure.

Screwdrivers

- Never use a screwdriver with a defective handle
- Keep flat-ended screwdrivers for slotted screws to a uniform thickness.

Activity 3.4

This includes things like re-fixing floor boards (although you would use screws where access to pipework is required), pipe boxing, skirtings, and boards for clipping piperuns. If you are working for a small business, you may build your own support platforms for storage and heating cisterns. Below is a sample of what you might carry in your tool box:

- Panel pins
- Oval brad/lost head
- Masonry nails
- Round head, plain and galvanised.

All these are available in a range of lengths and diameters. Check out the hardware catalogues for some examples.

Activity 3.5

Building Regulation Part A – Structure

Answers to Test yourself questions

Test yourself 3.1

1. a. Building Regulations
2. Two from:
   - Type and quality of components, materials and fittings to be installed
   - Type of clips or brackets to be used
   - System test specifications
   - Any specific installation requirements, knowing whose job it is to carry out any associated building work.
3. Advise your immediate superior/senior tradesman or employer. It is then your employer's responsibility to reach a solution with the customer.
4. *Quotations* can be used to give a price for a *job* by a *plumber* or from a *merchant.* An *order* is confirmation that a quotation has been accepted. Once a job has been completed an *invoice* is sent requesting *payment.* The

remittance advice confirms that *payment* has been made.

5. a. Activity against time
6. Any three from the following:
   - Are all tools required available
   - All electrical tools are safe to use
   - Work area is clean and safe
   - Access equipment is in good order
   - If required, artificial lighting is adequate
   - Work area is free from pre-existing damage
   - Customer's property is adequately protected
   - Ensure the customer has a sound knowledge of how to operate any new appliances or systems that may have been installed.

Test yourself 3.2

1. Measuring and marking out
   - Ruler, pencil, tape measure, spirit level

   Cutting and preparation
   - Pipe cutters, hacksaw, files

   Fabrication
   - Pipe bending machines, bending springs

   Jointing
   - Wrenches, pliers, spanners, blow torch

   Fixing and making good
   - Trowel, screwdrivers, drilling accessories
2. State the safety and maintenance requirements for:

Hacksaws

- Make sure the teeth are pointing away from the forward cut
- Do not use with defective or worn out teeth
- While using a large hacksaw, make sure the blade is tightened correctly.

Wood chisels

- Never use with split handles
- Keep the chisels sharp using a grinder and whet stone
- Always keep the plastic guard on the chisel
- Make sure handles are not loose.

Pipe grips and wrenches

- Keep the teeth free from jointing compounds. If they are made up, it could cause the tool to slip
- Once the teeth become worn the tool should be replaced
- Check for wear on the ratchet mechanism when using pump pliers. These often slip when under pressure.
- Be careful when loosening a joint or pipe that is difficult to move. It might give in suddenly, and you could hurt yourself.

3. The cutting of sheet metal
   - Straight tin snips

   De-burring a pipe
   - Abrasive pads, wire wool

   Removing a mild steel fitting from a pipe
   - Wrenches, locking pliers

   Notching a floor joist
   - Wood or floor board saw

   Removing an immersion heater
   - Immersion heat spanner
4. a. 110 V
5. PAT (Portable Appliance Test) are maintenance records used to ensure that all portable electrical appliances are in good, safe working order.
6. a. 18
7. - Take care to avoid nails and screws
   - Make sure guard is in place
   - Ensure cutting depth is same depth as the floor.
8. - Always use 110 V supply
   - Check all tools are double insulated
   - Check for test labels to show equipment is safe to use.

Test yourself 3.3

1. c. Low carbon steel tube
2. a. BS EN 1057 – R250 (formerly BS 2871; Part 1 Table x)

3. Bending springs are used to support the tube walls while the bend is made. Internal and external springs are available and can be used to protect the integrity of tubes up to 22 mm.
4. a. 22 mm
5. d. 600 mm folding rule

Test yourself 3.4

1. • Copper
   • Plastic
   • Low carbon steel
2. From the list below, tick two answers which you think are correct.
   The two main types of material used for fittings to joint LCS are:
   • Steel
   • Malleable iron.
3. You have been asked to joint a length of LCS tube to an elbow.
   • Ensure the fitting and the diameter of tube is compatible
   • Cut the length of LCS tube to the correct length using a large frame hacksaw
   • De-burr the pipe and chamfer the end that is going to be threaded
   • Securely fix the LCS pipe in a vice and ensure the die is correctly set up
   • Apply cutting fluid to the chamfered end of the tube prior to cutting
   • Cut the threads into the tube (ensure that approximately two threads will extend beyond the end of the fitting)
   • Wipe away excess cutting fluid
   • Apply appropriate jointing material (paste, pipe sealant, PTFE tape, etc) and screw the fitting into place
   • Tighten pipe and fitting using appropriate tools (e.g. pipe grips, stilsons) to complete the joint.

Test yourself 3.5

This is a self-assessment exercise.

Test yourself 3.6

1. • Floor boards
   • Chipboard.
2. $200 \div 8 = 25$ mm
3. Minimum length $= 7 \times \text{length} \div 100$
   $= 7 \times 4000 \div 100$
   $= 280$ mm
   Maximum length $= \text{length} \div 4$
   $= 4000 \div 4$
   $= 1000$ mm

Answers to Check your learning Unit 3

1. a	11. b
2. c	12. c
3. a	13. b
4. c	14. a
5. a	15. c
6. d	16. c
7. c	17. d
8. b	18. b
9. d	19. b
10. c	20. c

UNIT 4

Answers to activities

Activity 4.1

Area of Circle = 19.64 cm^2 (πr^2)
Area of Triangle = 3.75 cm^2 (½ Height × Base)
Area of Rectangle = 12.5 m^2 (Length × Width)
Volume of Cylinder = 0.2583 m^3 (πr^2 × Length)
Volume of Cuboid = 12 m^3 (Length × Width × Height)

Activity 4.2

Thermal expansion is quite an important aspect of a plumber's job. When water expands in a hot water system as it is heated, where do you think the water expands to, you will find out in Unit 6.

Allowance has to be made on plastic gutter and soil pipe installation. Plastic has a high coefficient of linear expansion, and if space is not allowed in joints for it to expand, then the pipe lengths would buckle as they expanded.

One of the most common plumbing problems is noise from hot water pipework laid under timber floors. If it is laid too tightly against the underside of floor boards, or against notches in timber joists, as it expands it will cause creaking and cracking noises.

Activity 4.3

Examples of materials and where they might be used

- Ceramic sanitary appliances, baths, WCs, etc
- Steel pipework, fittings, radiators
- Lead sheet roof weatherings
- Brass-fittings
- Copper pipework, fittings, hot water storage vessels

Activity 4.4

The British Safety Standard Kitemark:

Figure A1 British Standards Kitemark

Answers to Test yourself questions

Test yourself 4.1

1. a. Mass
2. c. Mass → Grams
3. d. 29.4 N
4. b. 4°C
5. a. 11.4
6. Answer: c

Test yourself 4.2

1. b. Soft
2. c. Acidic
3. c. Acidity and alkalinity
4. a. Sulphur dioxide
5. b. Dissimilar metals are placed in direct contact with each other
6. d. Anode
7. b. Temperature
8. Re-arrange the chart below so that the process matches the correct description.

Process	Description
Conduction	The transfer of heat through direct contact with a heated up substance
Specific heat capacity	The amount of heat required to raise the temperature of 1 kg of a material by 1°C
Convection	Movement of a fluid substance as the result of heating
Radiation	The transfer of heat through waves from a hot body to a cooler one
Thermal expansion	Molecules in a material are heated and move further apart

9. d. Capillary attraction
10. a. The weathering materials are lapped closely together
11. c. Atmospheric pressure
12. c. 3 m 20 mm low carbon steel one elbow
13. d. Methane

Test yourself 4.3

1. b. British Standard Institution
2. d. Newtons
3. c. Iron
4. d. Table y
5. a. PVC
6. d. Lead

Answers to Check your learning Unit 4

1. d
2. c
3. a
4. b
5. c
6. d
7. b
8. d
9. a
10. c
11. d
12. c

UNIT 5

Answers to activities

Activity 5.1

The objectives of the regulations are to:

1. Prevent contamination of the water supply
2. Prevent the wastage of water (i.e. faulty appliances, poor maintenance, leaking fitting, e.g. dripping tap)
3. Prevent misuse of water supply (i.e. the use of energy in the mains supply to provide motive power, or for the generation of electricity)
4. Prevent undue consumption of water (i.e. fittings and appliances that use more water for the purpose for which they were designed)
5. Prevent erroneous measurement (i.e. tampering and interference with the measurement of water passing through the meter, e.g. fiddling the meter).

Activity 5.2

The physical forms of water are:

- Solid – as ice
- Liquid – as water
- Gas – as steam or vapour.

At atmospheric pressure, and at a temperature of between 0 and 100°C water is found as a liquid.

At a temperature of 0°C or below, water changes to ice, expanding immediately in volume by 10% (hence the reason water pipes burst when frozen).

When subjected to a temperature above 100°C, water changes to steam, expanding in volume by approximately 1600 times.

Activity 5.3

Direct systems offer the following advantages:

- Cheaper installation costs, due to less pipework being required
- Drinking water is available from all draw-off points
- Water storage will only be required to feed the hot water storage vessel, so the storage cistern will be smaller in size. The minimum requirement being 115 litres.

The disadvantages are:

- The higher water pressure may make the system suffer from transmission noise, e.g. water hammer
- No reserve of water if the mains or service supply is shut off for a period of time
- Precautions must be taken so as to prevent back siphonage from occurring (water flowing back) or foul water from appliances, contaminating the mains supply. Remember this is a requirement of the Water Regulations (you may have missed this one).

The indirect system offers the following advantages:

- Lower delivery pressure reduces the risk of system noise
- Lower demand on the water main at peak periods
- There is a reserve supply of water should the mains supply be turned off for a period of time
- The wear and tear element on taps and valves is reduced, due to the lower delivery pressure.

The disadvantages are:

- Higher installation cost for larger storage cistern and additional fittings and pipework
- Larger diameter of pipework required for appliances
- Additional structural support will be required for the increased weight larger storage cistern
- The storage cistern will require an increased space or area (minimum capacity is 230 litres)
- Increased risk of damage from frost.

Activity 5.4

The main advantage is that it enables maintenance work to be carried out on the cistern without turning off the whole water supply.

Activity 5.5

The installation is fitted with a drain tap, which allows the pipework to be drained once the stop tap has been turned off. The minimum depth of 750 mm below the ground still applies to installations of this type.

The double check valve fitted to protect against backflow contamination to the water supply meets the requirements of Schedule 2 Paragraph 15(1) of the Water Regulations.

Answers to Test yourself questions

Test yourself 5.1

1. Water Supply (Water Fittings) Regulations 1999
2. One from:

- Prevent contamination of the water supply
- Prevent the wastage of water
- Prevent misuse of the water supply
- Prevent undue consumption of water
- Erroneous measurement.

3. 10%
4. Hard water
5. Two from:
   - Wells
   - Artesian wells
   - Springs.
6. Water evaporates from rivers, lakes, the sea and the ground and forms clouds. Clouds contain water vapour which, when the climatic conditions are right, condense and fall as rain. When rain falls to the ground, some rain water runs into streams, rivers and lakes, the remainder soaks into the ground, where it collects temporarily and eventually evaporates.

Test yourself 5.2

1. 750 mm minimum and 1350 mm maximum
2. - Base exchange water softeners
   - Electrolytic scale inhibitors or Electromagnetic conditioners.
3. - Isolate the water supply
   - Permit maintenance and servicing to be carried out on systems.
4. Two examples from:
   - Supply to hose taps
   - Supply to standpipes
   - Pipe connection to cisterns using Part 1 float valves
   - Supply to shower fitting.
5. The paragraph below describes the difference between a direct and an indirect cold water system, fill in the missing words.
   All the pipes to the draw-off points (kitchen sink, bath, wash hand basin, WC, etc.) in a *direct* system are taken *directly* from the *rising main* and operate under *mains* pressure. With the *indirect* system, one outlet, usually the kitchen sink, is fed directly from the rising main, before it continues to supply the *CWSC*. The remaining draw-off points are fed from this source.
6. - Pipe, components and cistern insulation
   - Trace heating.

Answers to Check your learning Unit 5

1. d
2. c
3. a
4. c
5. a
6. c
7. b
8. b
9. d
10. b
11. c
12. c
13. a
14. a
15. b
16. d
17. d
18. b
19. c

UNIT 6

Answers to activities

Activity 6.1

The reason direct hot water systems are not recommended in hard water regions is because hard water contains dissolved calcium and as the water is heated, it causes calcium to come out of the water and produces scale (calcium carbonate).

This is commonly referred to as limescale, which is deposited on the walls of pipes, boilers and storage vessels. In a direct system, all the system water is constantly being replenished when water is drawn off from a tap, so when fresh water enters the system and is heated limescale will be given off each time. The result is a constant build up of limescale, which reduces system performance and can eventually result in component failure.

Activity 6.2

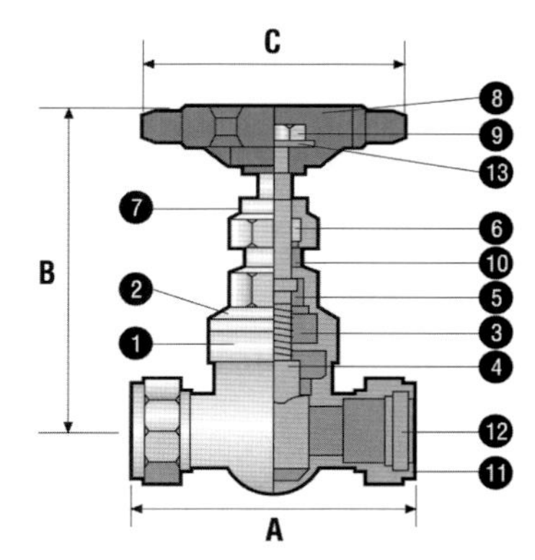

Figure 6.27 Gate valve parts labelled

Specification for the valve

No.	Component
1	Body
2	Bonnet
3	Stem
4	Wedge
5	Stem ring
6	Gland
7	Gland nut
8	Handwheel
9	Handwheel nut
10	Spindle
11	Compression nut
12	Compression cone
13	Rating disc

Activity 6.3

How does the system work?
Water from the secondary system (this is the water within the storage vessel, fed from the cold water storage cistern) feeds the primary system via a special heat exchanger in which an air pocket separates the water from mixing. Water fills the primary circuit through a series of holes in a standpipe located in the heat exchanger.

On filling the primary circuit, the heat exchanger self vents by means of the vent pipe; once filled, the filling of the secondary supply continues. When the secondary supply is full, two air seals are formed within the heat exchanger, providing a seal and preventing the water from mixing.

Once the water is heated, the expansion of water in the primary circuit is taken up by forcing the air from the upper dome.

Activity 6.4

Combination units are commonly found in dwellings that have restricted or no roof space to house a CWSC, such as ground floor flats, maisonettes and dwellings with low-pitched roofs.

The disadvantages with hot water systems of this type are:

- Low water pressure at hot taps
- Not suitable for supplying a low-pressure thermostatic shower.

Activity 6.5

Indirect thermal stores are suitable for both open-vented primary systems and sealed systems, and as the static water store is open vented, they are not required to be installed at the highest point of a system, so this gives greater flexibility when siting the units.

The storage capacity for an average three bedroom domestic dwelling with one bathroom and thermostatic shower would be 120 or 140 l.

Activity 6.6

Typical situations are:

- Where an old system is going to be completely stripped out of a domestic property and replaced completely by a new system, or an alternative system, for example direct hot water system replaced with an indirect hot water system
- Where a system is going to be stripped out permanently, for example prior to the demolition of a building.

Activity 6.7

The procedure for testing rigid pipes (e.g. copper)

- Make sure any open-ended pipes are sealed, e.g. vent pipe
- Once the system has been filled it should be allowed to stand for 30 min to allow the water temperature to stabilise
- The system should be pressurised using the hydraulic testing equipment to a pressure of 1½ to 2 times the system working pressure
- Allow to stand for 1 h
- Check for visible leakage and for loss of pressure. If sound, the test has been satisfactory.
- If no sound, repeat the test after locating and repairing any leaks.

The procedure for testing plastic pipes
BS 6700 shows two test procedures for plastic pipes, Procedure A and B.
Test A procedure:

- Apply test pressure (1.5 times maximum working pressure) by pumping for a period of 30 min and inspect visually for leakage
- Reduce pressure by bleeding water from the system 0.5 times maximum working pressure. Close the bleed valve.
- Visually check and monitor for 90 min. If the pressure remains at or above 0.5 times working pressure the system can be regarded as satisfactory.

Test B procedure:

- Apply test pressure (1.5 times maximum working pressure) by pumping for a period of 30 min. Note the pressure and inspect visually for leakage.
- Note the pressure after a further 30 min. If the pressure drop is less than 60 KPa (0.6 bar) the system can be considered to have no obvious leakage.
- Visually check and monitor for 120 min. If the pressure drop is less than 20 KPa (0.2 bar), the system can be regarded as satisfactory.

Answers to Test yourself questions

Test yourself 6.1

1. Intensity of heat refers to a substance's temperature; the quantity of heat also includes a substance's temperature and takes into account a substance's volume
2. A servicing valve
3. 60°C
4. 1/24
5. Centralised system:
   With this type of system, the heated water can be stored, usually centrally within the building, supplying a system of pipework to various draw-off points. Stored means water held in a vessel until required, with the water temperature usually controlled by a thermostat.

   Localised system:
   With this type of system, the water is heated locally to meet the requirements of the consumer, a typical example being a single-point instantaneous water heater sited over a sink. These are often used in situations where a long distribution piperun from a stored system would lead to an unnecessary wait for hot

water to be drawn off. Not only does this save energy and reduce the wastage of water, it also helps prevent the risk of microbiological growth in the system such as Legionella.

Test yourself 6.2

1. Indirect
2. 22
3. Liquid corrosion inhibitors must not be introduced into the system, as the water in the boiler and storage vessel is not separated, and directly supplies the draw-off outlets
4. Combination storage unit
5. Cylinder thermostat
6. High performance storage vessel
7. Part L1 of the Building Regulations 2002 requires that all pipes that are connected to a storage vessel should be insulated for up to a metre in length, or up to the point where they become concealed. This requirement includes the primary flow and return and vent pipe connections, reducing the heat loss from the storage vessel. The thermal insulation of hot pipework can be achieved by using spilt sectional closed cell type pipe insulation.

Test yourself 6.3

1. List the main types of instantaneous water heaters
    - Single gas-fired instantaneous water heater
    - Multi-point instantaneous water heater
    - Electric instantaneous showers.

Test yourself 6.6

1. Any three from:
    - Suitable for their purpose
    - Have sufficient strength to resist normal and surge pressure
    - Easily accessible to renew seals and washers
    - Made of corrosion-resistant materials
    - Capable of working at appropriate temperatures.
2. They are usually located in low-pressure pipelines such as the cold feed from the CWSC to the hot water storage cylinder, and the cold water supply on an indirect system. You might also find them used on supplies to thermostatic shower valves.
3. Any pipework running horizontally should be level or to an appropriate fall allowing air to escape from the system. It should never be allowed to rise upwards from the CWSC.
4. This term dead leg is used to describe long lengths of pipe from the hot water storage vessel to the appliance. The Water Regulation definition of a dead leg is 'a length of distribution pipe without secondary circulation'. A dead end usually refers to a length of pipework that has been capped off when an appliance has been taken out. These should be cut back to the point where the branch is taken off, or preferably be cut out altogether.
5.
    - Expansion noise
    - Water hammer
    - Flow noise.
6.
    - Visual inspections
    - Testing for leaks
    - Pressure testing
    - Final checks.
7. BS 6700

Answers to Check your learning Unit 6

1. a
2. a
3. b
4. d
5. b
6. c
7. c
8. a
9. b
10. d
11. a
12. b
13. c
14. d
15. a
16. b
17. b
18. a
19. c

UNIT 7

Answers to activities

Activity 7.1

The use of the valve reduces the amount of ventilating pipework on a ventilated discharge stack and means that on internal stacks the need for roof weatherings is avoided.

Activity 7.2

This system is likely to be specified in situations where it is not possible to have sanitary appliances grouped closely together, resulting in the use of long discharge piperuns.

Activity 7.3

A trap is used to retain a 'plug' of water to prevent foul air from the AGDS and underground drains from entering a room.

Activity 7.4

1 Swivel trap, 2 Shower trap, 3 Low-level bath trap, 4 Bottle trap, 5 Straight through trap, 6 'S' trap, 7 Hepworth valve, 8 Running trap, 9 'P' trap, 10 Re-sealing trap

Activity 7.5

Push fit

- Can be dismantled for adjustment
- Can accommodate thermal movement
- Can be assembled quickly
- Can be installed in wet conditions
- A wider range of specialised fittings is available.

Solvent welded

- Needs no protection from ultraviolet light when used externally
- MuPVC exhibits superior fire performance when compared to other thermoplastics
- Shows smoother cleaner lines

Activity 7.6

- Sinks
- WC cisterns and pans
- Wash basins
- Baths
- Showers
- Bidets
- Urinals.

Activity 7.7

The maximum flow rate for three urinals is 7.5 litre per flow per bowl which is 22.5 l/h.

Activity 7.8

No more than four WC pans should be stacked one above the other. Make sure the storage surfaces (battens/pallets) are clean, battens are used for all sorts of things on site, and if they are covered in grit, then storing materials on them will defeat the purpose.

Answers to Test yourself questions

Test yourself 7.1

1. Building Regulation H1
2. Any three from:
   - Conveys the flow of water to a foul water outfall (a foul or combined sewer, a cesspool, septic tank or settlement tank)

- Minimises the risk of blockage or leakage
- Prevents foul air from the drainage system from entering a building under working condition
- Is ventilated
- Is accessible for cleaning blockages.

3. a. BS 8000
4.
   - Primary ventilated stack systems
   - Stub stack or air-admittance valve system.
5.
   - The gradient of the discharge pipes
   - The location of branch connections
   - The connection of the soil pipe to the drain.
6. Cross flow describes what happens when two branches are located opposite to each other on the soil stack.
7. Open
8. Any five from:
   - Self-siphonage
   - Induced siphonage
   - Compression
   - Evaporation
   - Wavering out
   - Foaming
   - Momentum
   - Capillary action
   - Leakage.

Test yourself 7.2

1.
   - Push fit
   - Solvent-welded
   - Compression.
2.
   - Polypropylene
   - PVC
   - uPVC (MuPVC).
3. 0.8 m
4. Access to enable the pipework to be cleaned and clear any blockages is a requirement of the regulations. This can be done by inserting access plugs into waste and soil tee junctions or by installing purpose made fittings. All traps are also provided with a cleaning eye, or in the case of a bottle trap, the bottom portion of the trap can be removed.
5.
   - The combined system
   - The separate system
   - The partially separate system.

Test yourself 7.3

1. Any three from:
   - Materials handling
   - Site storage of components
   - The preparation of work, materials and components
   - The installation of sanitary appliances
   - Inspection and testing.
2. Any three from:
   - Vitreous china
   - Fireclay
   - Cast and formed acrylic
   - Enamelled steel
   - Stainless steel
   - (Cast iron can also be included).
3. Any two from:
   - Close-coupled wash down or siphonic units
   - Wall-mounted
   - Back to the wall
   - Low level
   - High level.
4. 6 litres
5. 10 l/h
6. The principle of the siphonic pan is to create a negative pressure below the trap seal. With the single trap pan, this is done by restricting the flow from the cistern and is achieved by the design of the pan.

Test yourself 7.4

1. a. 4 pans
2.
   - Check the delivery items against what was ordered

- Make sure everything is in good condition and is not damaged.

3. This is a term that is often used by plumbers which describes installing the taps, wastes, and in the case of the bath, the cradle frame or feet. It will also include installing float valves, overflows, siphons or flushing valves and the handle assembly to the WC cisterns.
4. Horizontal pipework must be laid to a fall of 6 mm/m minimum

Test yourself 7.5

1. BS EN 12056
2. b. Air
3. d. 38 mm
4. a. 25 mm
5. Inform the customer that soundness testing is being carried out

Test yourself 7.6

1. Traps can be a potential source of problems as they often accumulate hair, soap residue, tooth paste and other objects that are small enough to fall through the grid of the waste hole and create blockages
2. Force cup or sink plunger, and it used to clear blocked sinks and basins
3. Hand spinner
4. Removing cast iron soil and vent pipework can be dangerous due to its weight, so it needs careful handling, and it is definitely a two-handed job. It is best to try to take it down in short sections by partially cutting it with an angle grinder and then tapping the pipework with a hammer, which will cause it to shear; a rope should be tied to the sections and then the section lowered to the floor.

Make sure no one is in the area where you are working. Fixing lugs can be broken from the joint, and the nails prised out using a wrecking bar. Remember your PPE, including a hard hat!

Once the stack is removed, make sure the joint to the drain is covered to avoid anyone falling over it, or debris entering the drain.

Test yourself 7.7

1. b. Plastic
2. Half round, square, ogee
3. c. 1.00 m
4. d. 1.80 m

Case study 7.1

Here are a few pointers.

- Check out the job to determine exactly what pipe and fittings you are going to need
- Arrange to collect the new wash basin and pedestal a couple of days before starting the job, check that it is not damaged and all the 'bits' are there
- Store it carefully
- Make sure the customer knows what time you will be arriving, and to collect some drinking water as the water supply will be turned off
- On arrival, put down dust sheets in the area in which you will be working, explain to the customer exactly what you will be doing
- Dress the wash basin in preparation for the installation, leave any protective tape on the basin until after the installation is completed. Fit the trap to the basin before fixing it to the pedestal as this will be easier than trying to tighten the trap when it is behind the pedestal.

- Turn off the hot and cold water supply at the point which will result in the least draining down and hence wastage of water. This will depend on the age of the system, or how the system has been installed originally. If servicing valves have been fitted to the basin, great, if not, the next best bet is gate valves to the hot and cold supply. If the cold water supply is direct, then it is a case of turning off the stop valve on the incoming supply.
- Take out the old basin by disconnecting the taps and waste, be careful if it is cracked, as it could break unexpectedly, and the edges of vitreous china are razor sharp. Use sturdy gloves to handle the old basin. If the old fittings are difficult to disconnect, you may be quicker to cut the pipework, as it usually has to be altered to fit the new appliance anyway.
- Once removed, we would cut the pipework back to about 300 mm above the floor level and fit servicing valves to the hot and cold water supply. This would mean you could turn the supply back on for the customer, and that in future the basin could be isolated for any routine maintenance.
- Make sure the work area is really clean before installing the new basin and that the waste pipe is prepared (i.e. a new fitting is installed if you have to extend or alter the waste pipework)
- Fit the basin on the pedestal (using any manufacturers' fitments provided) and place it in position to enable the fixing holes to be marked up
- Drill and plug the wall ready for fixing
- Place the basin back in position, but do not fix it yet
- Pipe up the hot and cold water supplies. You may wish to do this in situ, or alternatively remove the pipework and solder it up separately (if using capillary fittings)
- Complete the waste water connections first before fixing the supply pipework and the basin as this will give you some 'manoeuvrability' with the waste water pipework
- Complete the piping system by tightening the tap connectors, and connecting the other ends to the servicing valves. Tighten the basin fixing screws slightly.
- Make sure the taps are closed. Slowly open the servicing valves one at a time and check all the joints for leaks. Run a small amount of water through the taps and test the waste. Assuming everything is OK, run the taps fully to flush out any flux or debris from the pipework
- Fully fix the basin to the wall
- Clean up and advise the customer that the work is complete.

Answers to Check your learning Unit 7

1. d
2. b
3. b
4. d
5. b
6. b
7. d
8. b
9. d
10. b
11. a
12. b
13. a
14. a
15. c
16. b
17. c
18. d
19. a
20. c

UNIT 8

Answers to activities

Activity 8.1

It could lead to a reduced heating temperature within the dwelling dropping below comfort level, due to the high domestic hot water demand. This could happen in a large family dwelling or in poorly insulated properties.

Activity 8.2

We would recommend a multi-zone system. This type of installation is recommended for use in dwellings that have a usable floor area greater than 150 m^2. The heating circuits should be divided into a minimum of two zones, with a separate zone for the hot water circuit, having its own independent time and temperature control. The use of the two individual two-port valves offers greater flexibility to the design of the heating system, and allows the building to have separate time- and temperature-controlled heating zones.

Activity 8.3

We have explained that the vent will allow hot water to expand into the cistern should the system controls fail. If a valve was fitted and inadvertently turned off, then it acts as a blockage in the pipe. This will prevent the water from expanding and will eventually cause a build up of pressure and a potential explosion. A closed valve would also mean that the system was not operating at atmospheric pressure as it is designed to do.

Activity 8.4

Another checklist

General: you must be competent to carry out any associated electrical work

- First of all check whether it is the circulator that is actually faulty, and not an electrical control fault, or something simple, such as a blown fuse at the fused spur, supplying the circulator. We will assume here that it is a faulty circulator.
- Check first to see if the replacement circulator will fit without any alterations to the existing pipework. Secondly, check that there are isolation valves to the circulator and that they are in good working order. If yes to all, then you will not have to drain down the system. If no to either, then the system must be drained.
- Proceed and isolate the water and electrical supply.
  – Electrical
  This will depend on the type and the age of the system. There may be a fused spur to the circulator, or a fused isolating switch. The circulator could also be part of a system control package. In this case you will have to isolate the fused spur to the wiring centre or junction box.
  In all the cases remove the fuse to isolate the electrical supply. In the cases where wiring centres or junction boxes have not been used (i.e. on older systems), as a safeguard turn off the boiler as well.
- If any pipework alterations are necessary you will have to drain the system partially or fully; this will depend on the location of the circulator. Some circulator manufacturers produce pipe extension make up pieces, to avoid draining the system, so check these out first.

- Disconnect the electrical wiring from the circulator by removing the plastic cover plate. This will reveal wiring that is the same as that in an ordinary electrical plug – Blue to neutral, Yellow and Green to earth and Brown to live. On older installations the wiring could be Black to neutral, Green to earth and Red to live.
- Remove the plastic cable cover making sure there is sufficient bare wire to connect to the circulators' terminal fittings, and that they are in good condition.
- If not in good condition and provided that the cable is long enough, cut back and re-strip the wire. Wiring the circulator prior to fixing in position makes the task easier. If the cable is not of sufficient length it will require re-wiring back to the point of isolation.
- Assuming that you did not have to drain down, turn off the isolating valves and remove the defective circulator. If it has been in a long time, you are going to have fun, sometimes a lot of fun! You may have to apply heat to the circulator valve nuts together with some persuasive action, by using a hammer and cold chisel to loosen the nuts. You will also need some large stillsons or grips on both the valve nuts, to stop the circulator from rotating, whilst undoing. Be careful when doing this to not disturb the joints between the pipe and the circulator valves, as this can result in water leaks.
- Once the circulator is removed, make sure that the isolation valve seatings are free from any sealing washer material
- Fit the new circulator, making sure that new sealing washers are fitted
- If you had to alter the pipework, fit the new valves, sealing washers and new circulator.
- Reinstate the supply of water. If you have drained the system, you will need to re-fill the system, as you did for the radiator valve job. If you did not drain, then all you need do is turn the isolation valves back on.
- The circulator must be vented via the slotted screw head, sited in the body of the circulator.
- If drained and on completion of filling, remember to add a liquid corrosion inhibitor; if an in-line type is not fitted

  [*Note: This does not apply to systems with a self-venting cylinder, e.g. primatic.*]

- Reinstate the electrical supply and test the circulator, setting it on the recommended speed, normally setting Number 2 for a typical fully pumped domestic installation and setting Number 3 for sealed system boilers such as combies
- Advise the customer that the system is back in service.

Most modern circulators are now designed so that when defective, the head unit is available for replacement; this reduces the need to alter and adapt pipework, making the task quicker and reducing labour costs.

Activity 8.11

There is no answer for this activity, but discuss your report with a tutor.

Activity 8.12

There is no answer for this activity, but discuss your report with a tutor.

Answers to Test yourself questions

Test yourself 8.1

This is a self-assessment exercise; check through the text and discuss your answers with a tutor.

Test yourself 8.2

1. a. 1.0
2. a. Convection
3. c. 43
4. Condensing boiler
5. Condensing boiler

Test yourself 8.3

1. Time clock:
   Commonly used on semi-gravity systems, this is a basic switch that is electrically operated, by means of a clock. When set, it offers timed or constant control to both the heating and hot water circuits. If required they can provide separate hot water, but not independent heating.

   Time control programmer:
   Ideal for fully pumped systems, these put the heating on standby, ready to respond to the requirements of the cylinder and room thermostats. They are available as a 7 day full programmer with independent heating and hot water timings, offering up to three on/off periods in a day.
2. Bypass valve
3. • Three-port diverter valve
   • Three-port mid-position valve
   • Two-port valve
4. This is a term used in the Building Regulations Approved Document L1. It is not an actual physical control device but an interconnection of the controls, such as zone valve, room thermostat, cylinder thermostat, programmers, etc; connected in an arrangement in order to ensure that the boiler does not fire when there is no demand for heat.
   In a system with a traditional boiler this would mean the correct wiring of the room stat, cylinder stat and motorised valve(s). It can also be achieved by more advanced controls such as a boiler energy manager. The installation of thermostatic radiator valves alone is not sufficient for a boiler interlock.
5. ¼ to ⅓ from the bottom

Test yourself 8.4

1. c. 1½ times the working pressure and 1 h
2. Any two from:
   - Types of metals used in the system
   - Nature of the water, hard or soft (acidic)
   - The degree to which air is drawn into the system; this is generally associated with poor system design, e.g. pumping over or air being pulled down the open safety vent pipe.
3. Any three from:
   - Prevents the build up of 'black sludge' (magnetite (Fe_3O_4) and haematite ($6Fe_2O_3$)) the major cause of heating problems, which can result in the pin holing of radiators
   - Reduces fuel costs
   - Prevents frequent venting of the radiators
   - Reduces system noise and reduces scale build up
   - Non-acidic neutral formation, so safe to use
   - Harmless to the environment.

Test yourself 8.5

1. Flushing should be done on a yearly basis in conjunction with the servicing of the appliance, or while carrying out system repairs.
2. Use pipe freezing equipment.

3. False, if the cylinder seal fails, the water with the additive would mix with the domestic hot water.

Test yourself 8.6

1. Benchmark
2. Three from:
   - The customer and equipment
   - Who installed and commissioned it
   - Information on the commissioning and checks carried out
   - Handing over the equipment, with a clear explanation and demonstration of system and appliance operation. Leaving all manufacturer's literature with the occupier on completion
   - Regular services and checks carried out.
3. b. 10°C

Answers to Check your learning Unit 8

1. c	8. d	15. c
2. d	9. a	16. d
3. a	10. b	17. d
4. b	11. b	18. a
5. d	12. c	19. c
6. c	13. a	20. a
7. a	14. a	

UNIT 9

Answers to activities

Activity 9.1

The correct fuse to use with the dishwasher is 5 amps (4.1875 amps)

Activity 9.2

The most common causes are:

- Overload
- Short circuits
- Earth faults

Activity 9.3

You may have listed: Accidental severing of a supply cable while an appliance is in use; nailing through a ring main circuit; water ingress into electric plug, short circuiting the 'live' to 'neutral' wires.

Activity 9.4

The correct sequence is:

- Identify sources of supply
- Isolate
- Secure isolation
- Test if the equipment/system is dead
- Begin work

Activity 9.5

You stand a good chance of getting an electrical shock and joining the electrical casualty statistics

Answers to Test yourself questions

Test yourself 9.1

1. a. Ohm
2. c. Potential difference
3. d. Voltage, current, and resistance

4. c. Miniature circuit breaker
5. b. Electromagnetism
6. a. 230 V
7. d. 30 amp

Test yourself 9.2

Test yourself answers are to be self-assessed.

Test yourself 9.3

Is in the form of an assignment

Answers to Check your learning Unit 9

1. b
2. a
3. d
4. c
5. a
6. a
7. c
8. b
9. a
10. c
11. d
12. a
13. d
14. a
15. b
16. b
17. c
18. a
19. a
20. b

UNIT 10

Answers to activities

Activity 10.1

A lead slate is a type of flashing used where a pipe or structure (a vent pipe from a soil stack for example) penetrates a roof. It got its name because it was originally designed to fit in with roofs covered in slates.

Activity 10.2

Fatigue basically means the loss of strength of the lead due to thermal movement, which eventually leads to its cracking. The strength of the lead relies on the grain structure of the metal. The chemical composition of lead is governed by BS 12588 which effectively controls the grain structure to make the lead sheet more resistant to thermal fatigue without effecting malleability.

Creep describes the tendency of metals to stretch slowly in the course of time. It does not mean that a piece of lead has slipped down a pitched roof when the fixings have failed! Making sure you correctly size and fix individual pieces will reduce the risk of creep.

Activity 10.3

A pressure of 0.14 bar (2 lb/sq.m) for both the oxygen and acetylene is required

The blow pipe should be lit by:

– turning on and lighting the acetylene first
– then feeding in the oxygen.

You need to achieve a neutral flame to get the best results

Answers to Test yourself questions

Test yourself 10.1

1. Rolled sheet lead
2. Code 4
3. Blue
4. Code 3
5. Green
6. 1.5 m
7. False

Test yourself 10.2

1. • Flat dresser
   • Bossing stick
   • Bending stick
   • Bossing mallet
   • Setting-in stick
   • Chase wedge
2. d. Malleable
3. 100 mm
4. Cleaning the filler rods and surfaces of sheet lead prior to welding.
5. Fusion welding is a process of joining two pieces of lead by melting the two edges of the lead together (called the parent metal) while a filler rod of lead is added.
6. • Butted seam
   • Lapped seam
   • Inclined seam
7. Any three from:
   • Clean surfaces
   • Correct penetration
   • Correct thickness
   • Weld width
   • Avoid undercutting
   • Correct flame
8. b. maroon
9. b. 2 and 3
10. Where possible, you should prefabricate lead components before welding. If working on-site you should try to use lapped joints. The reason is because this reduces the risk of fire if using a butt joint against combustible surfaces. However, on occasions this may be unavoidable. Where this is the case you should:
    • Wet the timber area beneath the weld
    • Alternatively, place a non-combustible material beneath the weld

Test yourself 10.3

1. Front apron
2. 150 mm
3. • Side and cover flashings. These are used where it is not possible to incorporate soakers, such as contoured tiles.
   • Side flashings using soakers. These are used on roofs covered with slate or double lap plan tiles.
4. a. 3
5. 200 mm

Answers to Check your learning Unit 10

1. d	11. b
2. b	12. d
3. c	13. a
4. c	14. d
5. a	15. d
6. b	16. a
7. a	17. d
8. a	18. b
9. c	19. b
10. b	20. d

UNIT 11

Answers to activities

Activity 11.1

Environmental awareness means understanding why it is important to conserve energy, dispose of waste properly and prevent the wastage of materials. Environmental awareness is very topical; the Government is keen to promote initiatives which help to conserve energy and reduce waste.

It has also improved Building Regulations to make buildings more energy efficient, as well as making the use of building materials more effective by not over-specifying building details.

Activity 11.2

Energy conservation is aimed at reducing the CO_2 emissions into the atmosphere, which has a detrimental effect on the ozone layer, leading to global warming. This needs to be reduced, and the Government has tackled this by bringing in legislation making buildings better insulated and heating appliances more efficient.

Activity 11.3

They all do! Give yourself a pat on the back if you recognised all these factors!

Activity 11.4

The image shows how a solar-powered hot water system works. What are the advantages and disadvantages of this system?

Advantages	**Disadvantages**
Needs no maintenance – panels clean themselves	In winter, there will be little solar gain, so auxiliary heating is needed
The sun's power is free	Does not work in dark, or when covered by snow
Can be fitted onto flat or pitched roof (but should be south-facing and angled towards the sun to get maximum benefit)	Very long payback period, so financially not particularly attractive

Activity 11.5

11.5.1 The regulations are designed to cover all the points listed
11.5.2 b and d

Activity 11.6

Here is a list of our ideas of what should be done for the customer:

- Provide a company folder containing:
  - Emergency contact number for burst pipes, etc
  - General advice line number, you might be able to sort a problem out over the phone
  - Most manufacturers produce customer guidance leaflets. Put these in the folder, but go through them with the customer.
- Label the various fittings or components that the customer may need to use in an emergency, e.g. stop valve, gate valve, etc
- Show them or tell them where the service valves are, and what they do
- 'Walk them around the system'. Show them the various components. Tell them what they can touch and what they must not touch.
- Explain what the components do in simple terms, e.g. 'This is the cylinder thermostat and when the water in this cylinder gets to that pre-set temperature it shuts off the heat to the cylinder, you do not have to alter that setting.'
- When you have finished, do not go before you have given them chance to ask any questions.

Activity 11.7

High temperatures may melt containers, allowing their contents to leak out. Materials which are safe to dispose off in their normal state may combine chemically with other materials in the skip, creating poisonous or hazardous compounds.

Some materials give off toxic or noxious fumes when burnt. Pressurised containers should never be burnt, as they explode like a rocket. Glass melts then welds itself to other materials when it cools. So, a skip full of miscellaneous waste materials could be transformed into a lethal disaster if it is set on fire!

Answers to Test yourself questions

Test yourself 11.1

1. Solar photovoltaics use the sun's power to energise photoelectric cells set in panels which are used to produce electricity.
2. The regulations relevant to energy efficiency are 'The Building Regulations 2000, Part L1' which came into force in England and Wales from 1 April 2002.
3. • Prevent the wastage of water
   • Prevent undue consumption of water

Test yourself 11.2

1. • Replacement heating systems (including boilers and hot water storage cylinders)
   • Heating and hot water controls
   • Commissioning of heating systems and the provision of instructions for users
   • Certification of heating and hot water systems to show that they have been correctly installed and commissioned, and that operating instructions have been given for the user
2. A cylinder thermostat is used to control the temperature of stored hot water. It is usually located on the side of the cylinder about a third of the way up. It is commonly used with a motorised valve to provide close control of the water temperature.
3. • Provide a company folder containing:
     – Emergency contact number for burst pipes, etc
     – General advice line number, you might be able to sort out a problem over the phone
     – Most manufacturers produce customer guidance leaflets. Put these in the folder, but go through them with the customer.
   • Label the various fittings or components that the customer may need to use in an emergency, e.g. stop valve, gate valve, etc
   • Show them or tell them where the service valves are, and what they do
   • 'Walk them around the system'. Show them the various components. Tell them what they can touch and what they must not touch.
   • Explain what the components do in simple terms, e.g. 'This is the cylinder thermostat and when the water in this cylinder gets to that pre-set temperature it shuts off the heat to the cylinder, you do not have to alter that setting.'
   • When you are finished, do not go before you have given them chance to ask any questions

Answers to Check your learning Unit 11

1. d
2. d
3. a
4. d
5. c
6. b
7. c
8. a
9. d
10. b

UNIT 12

Answers to activities

Activity 12.1

The job of a Building Services Engineer is the one most closely related to the plumbing industry. Building services include water, heating, lighting, electrical, gas, communications and other mechanical services, such as lifts and escalators. Building services engineering involves designing, installing and maintaining these services in domestic, public, commercial and industrial buildings.

A Building Services Engineer's work includes:

- Designing the services, mostly using computer-aided design packages
- Planning, installing, maintaining and repairing services
- Making detailed calculations and drawings.

Most Building Services Engineers work for manufacturers, large construction companies, engineering consultants, architects' practices or Local Authorities. It often involves working with other professionals on the design of buildings including architects, structural engineers and contractors. As a result, it usually means working as a member of a team. When working for a consultant on designing, the job involves spending most of the time in an office. Once construction starts, it means visiting sites and liaising with the contractors installing the services.

Working for a contractor involves overseeing the job and may well mean managing the workforce. It is likely to be site-based.

Working for a services supplier is likely to mean being involved in design, manufacture and installation. It may involve spending a lot of time travelling between the office and various sites.

Activity 12.2

Carpenter and Joiner
A Carpenter and joiner positions and fixes the timber materials and components that go into a building – from roofs and floors to doors, kitchens and stairs.

Work on site is usually divided into two phases: the first and second fixes. The first fix is usually completed before the building has been made watertight. The work includes fixing floor joists, boards and sheets, stud partitions, roof trusses and timbers. The second fix takes place when the building is watertight. Staircases, doors, kitchen units, architraves and skirtings are common second fix items.

Plasterer
Plasterers work on site. Their job is to apply plaster to interior walls so that decorations can then be applied. They apply renders on external walls and sometimes lay floor screeds. Some plasterers specialise in fibrous plasterwork. Others also work on the maintenance, conservation and restoration of existing buildings.

Roof Slater and Tiler
Roof slaters and tilers create a waterproof covering for a building by applying individual slates and tiles to a basic framework. Some roof slaters and tilers specialise in the maintenance, conservation and restoration of existing buildings, some specialise in new ones.

Roofing felt is laid over the roof timbers and tacked down. Timber battens are then fixed horizontally at centres to suit the roof dimensions and type of slate or tile.

The roof is loaded with slates or tiles and then they are laid. Tiles have to be cut to fit at valleys, hips and gable ends. Ridge tiles are bedded on mortar.

On re-roofing or maintenance work tiles and slates have to be removed and timber needs to be checked and replaced wherever necessary. If the tiles or slates are to be reused they are checked and sorted before being used again.

Roofers use a range of specialist tools especially to cut slates and tiles.

Wall and Floor Tiler
Ceramic tiles come in a variety of shapes, sizes, textures and colours and may be laid in intricate, decorative patterns. They can be fixed on both exterior and interior walls, as well as other surfaces, such as floors and swimming pools.

On maintenance and restoration work, most surfaces need to be repaired. This may involve removing existing floor or wall coverings and levelling floors.

Setting out the tiles can be critical, especially when intricate and complex patterns are being produced. The tiles are normally laid on adhesive, cut along the edges and around obstacles, and finally grouted up.

Tilers also fix tiles made from ceramics and other materials, such as stone, terracotta or marble.

Built up Felt Roofer
Built up felt roofing is usually used on flat roofs but is also sometimes used on sloping roofs and occasionally on vertical surfaces. It is called 'built up' because it involves putting layers of felt on top of each other using bitumen to form the waterproof surface.

Construction operative
Construction operatives have many skills. They may include concreting, form working, steelfixing, kerblaying and drainage work. Some specialise in such areas such as spraying or repairing concrete.

Painter and decorator
Painters and decorators apply paint, wall coverings and other materials to the inside and outside of buildings.

Most work on the maintenance, conservation and restoration of existing buildings, while others work on commercial and industrial sites.

Painters and decorators must first prepare the area. This is done by removing or protecting items, such as furniture, carpets, floor coverings and ironmongery.

Old paint may need stripping off or at least rubbing down, while cracks and holes may need to be filled and surfaces cleaned.

Paints are applied using brushes, rollers or spray equipment. There is a range of different paints including acrylic, emulsion and gloss, each requiring different techniques. There are also a variety of specialist finishes, such as ragging, graining and marbling.

Decorators also use wallpapers and other wall coverings such as fabrics. These need to be accurately measured, cut and hung using the correct adhesive.

Sometimes painters and decorators help clients choose colour schemes and the type of paint or coverings to be used.

Floor Layer
Floor layers prepare and level the floors for both new and old buildings.

The first job is to carefully prepare the floor that is going to be covered. This may involve scraping, smoothing and cleaning. Sometimes fillers are used to patch areas. The area is then carefully measured and the laying planned. This avoids wasting any materials. The most common materials are carpet, cork, plastic and timber. A variety of different techniques are then used depending on which materials are being laid. These include gripper fixings, glues, adhesives and secret nailing.

There are approximately 8,000 floor layers. They usually either work in pairs for specialist firms or are self-employed. There is little direct interaction with people in other occupations.

Plant Operator

The two main types of plant operator are crane drivers and those who operate the plant for moving and transporting construction materials. Cranes can be mobile, track mounted or tower cranes. Transporting plant includes excavators, specialised earthmoving equipment, forklift trucks and power access equipment.

Scaffolder

Scaffolders put up scaffolding or working platforms for construction workers to use. They also fix edge protection to stop people and materials falling from high working platforms.

The job may involve working on new buildings or on the extension, maintenance, conservation and restoration of old buildings.

Scaffold erection has to be carefully planned. The scaffold needs a firm foundation and needs to be erected to fit in with the requirements of the construction sequence and programme.

The scaffold may be made from traditional tubes and fittings or may be a proprietary (purpose made) system. Working platforms are formed using scaffold boards. Access is usually provided using ladders. Hoists are often incorporated within the scaffold, as are special loading platforms.

Safety nets and guard rails need special attention to protect everyone on, near and underneath the scaffold. Most scaffolds need to be lifted and modified as construction work progresses.

Activity 12.3

Communication is the passing and receiving of information between one or more people, which achieves understanding and gets the required result. This can be done:

- Verbally
- In written form
- Visually (using graphics, body language).

Activity 12.4

This is a list of ways to send a message.

Activity 12.5

Some common faults in written documents, such as letters, reports, technical catalogues, brochures and official forms may include:

- The information does not follow a logical sequence
- The layout and display look untidy
- Use of jargon (this means language used by a particular profession or industry)
- Use of words or phrases that create the wrong image, which could be outdated or slang
- Over-complicated construction of sentences or taking a long time to get to the point, which means the document has to be read a few times before it is understood
- Bad spelling or grammar.

Activity 12.6

Formal writing is the style most commonly used in official business letters, memos, reports, text, legal documents, reference books, business e-mails and so on.

Informal style is most commonly used in conversation, and when written, such things as personal notes, fiction books, personal text messages, personal e-mails, etc are included.

The use of the informal style is not an excuse for using substandard or vulgar English.

Activity 12.7

Here is a list of faults commonly found in poor reports:

- Long winded and difficult to understand
- Poorly structured
- Not enough headings or subheadings so it is difficult to find your way around it
- Heavy going, large chunks of text and long paragraphs
- Pitched at the wrong level

Answers to Test yourself questions

Test yourself 12.1

1. The Construction Industry Training Board (CITB)
2. 1%
3. Sole trader
4. False – A Contracts Manager works for a construction company, is office-based, and provides a link from other sections of the business to the Managing Director, whereas a Project Manager works for or on behalf of the client to ensure the client's requirements are met, and may often be on site.
5. You may meet any five from of the following:
    - Carpenter and joiner
    - Roof slater and tiler
    - Built up felt roofer
    - Bricklayer
    - Plasterer
    - Painter and decorator
    - Wall and floor tiler
    - Floor layer
    - Plant operator
    - Scaffolder
    - Electrician

Test yourself 12.2

1. Any two from:
    - The message must be delivered to the person receiving it in the appropriate style and in a form that can be understood, i.e. verbally, in writing, etc
    - The message must get the desired response from the receiver
    - The sender must know that the message has been acted upon.
2. Any one from the following advantages:
    - More personal contact between the sender and the receiver
    - Tone and body language can be used to help convey the message, and interpret the message from the sender
    - Feedback is usually instant.

   Any one from these disadvantages:
    - No formal record unless the conversation is recorded
    - Not much good if you need to send a lot of detailed or technical information
    - The receiver can be easily distracted.
3. Any three from:
    - Answer the phone promptly
    - Say hello or good morning, etc
    - Give your name and the name of your employer's business if relevant
    - Ask how you can help
    - Try to speak in an 'up-beat' way.
4. List from:
    - Documents related to your employment, such as:
        - Contract of employment
        - Health and safety policy
        - Equal opportunity policy
        - Employee handbook
        - Memos
        - Letters.

- Regulations
- Technical specifications from clients
- Work programmes
- Manufacturers' catalogues
- Technical publications and journals
- Press releases
- Drawing and design details
- Promotional materials
- Training/learning support material (like this).

5. Make sure that you arrange to start the job well in advance, and that you keep to time for the appointment.

Respect the person's property, move items that could become damaged, and protect floor coverings and furniture with dust sheets. Do not have your radio blurting out, disturbing householders and neighbours, and do not get over-friendly, 'gushy', or disrespectful. Using comments like 'has the kettle burst?' hinting at a cup of tea, may offend some people.

Answers to Check your learning Unit 12

1. a
2. d
3. c
4. b
5. a
6. c
7. d
8. c
9. c
10. b

INDEX